NUCLEAR REACTORS

Edited by **Amir Zacarias Mesquita**

Nuclear Reactors
Edited by Amir Zacarias Mesquita

Published by InTech
Janeza Trdine 9, 51000 Rijeka, Croatia

Publishing Process Manager Iva Simcic

Technical Editor Teodora Smiljanic

Cover Designer InTech Design Team

Additional hard copies can be obtained from orders@intechopen.com

Nuclear Reactors, Edited by Amir Zacarias Mesquita
p. cm.
ISBN 978-953-51-0018-8

For EU product safety concerns:
IN TECH d.o.o., Prolaz Marije Krucifikse Kozulić 3, 51000 Rijeka, Croatia,
info@intechopen.com or visit our website at intechopen.com.

Meet the editor

Amir Zacarias Mesquita is Doctor of Science in Chemical Engineering - State University of Campinas/Brazil (2005), Master of Science and Nuclear Technologies - Federal University of Minas Gerais/Brazil (1981) and graduated in Electrical Engineering - Federal University of Minas Gerais/Brazil (1978). He is Titular Researcher at Nuclear Technology Development Center (CDTN), an institute of the Brazilian Nuclear Energy Commission (CNEN). He is scholarship in technological productivity at Brazilian Council for Scientific and Technological Development (CNPq). He is expert of the International Atomic Energy Agency (IAEA), member of the Brazilian Association for Nuclear Engineering (ABEN), professor of training course for Brazilian nuclear reactors operators, and reactor Senior Operator. He is author of more than two hundred of technical papers published in international journals and conference proceedings.

Contents

Preface

Rising concerns about global warming, supply security, and depleting fossil fuel reserves have spurred a revival of interest in nuclear power generation, giving birth to a "nuclear power renaissance" in countries the world over. As humankind seeks abundant and environmentally responsible energy in the coming decades, the renaissance of nuclear power will undoubtedly become reality as it is a proven technology and has the potential to generate virtually limitless energy with no greenhouse gas emissions during operations. According to the International Atomic Energy Agency the number of nuclear power reactors in operation worldwide in 2011 is 433 units, and 65 under construction. A large-scale period of nuclear power plants construction would allow nuclear energy to contribute substantially to the decarbonisation of electricity generation. In addition, basic research and nuclear technology applications in chemistry, physics, biology, agriculture, health and engineering have been showing their importance in the innovation of nuclear technology applications with sustainability.

The renaissance of nuclear power has been threatened by the catastrophe in Japan and the atomic industry faces the challenge of assuring a skeptical public that new reactors are safer than the old ones and nearly disaster-proof. The disaster at the Fukushima Daiichi nuclear plant in Japan demonstrates that older nuclear reactor technology requires strict adherence to quality assurance practices and procedures. Newer nuclear reactor designs promote the use of passive cooling systems that would not fail after power outages, as well as other innovative approaches to managing reactor heat. New reactors use the same principle of power generation as in older water reactors such as the ones at Fukushima: nuclear reactors heat water to create steam that turns turbines to generate electricity. However, technological advances have improved efficiency and stricter safety precautions have made the third-generation reactors more secure. The new generation of pressurized water reactor plants has diesel-powered backup systems that are housed in separate buildings to protect them from any accident that might occur in the main reactor building. The plant must also have access to other sources of electricity if the diesel generators fail.

This book is targeted at nuclear regulatory authorities, environmental and energy scientists, students, researchers, engineers, seismologists and consultants. It presents a comprehensive review of studies in nuclear reactors technology from authors across the globe. Topics discussed in this compilation include: thermal hydraulic investigation of TRIGA type research reactor, materials testing reactor and high temperature gas-cooled reactor; the use of radiogenic lead recovered from ores as a coolant for fast reactors; decay heat in reactors and spent-fuel pools; present status of two-phase flow studies in reactor components; thermal aspects of conventional and alternative fuels in supercritical water–cooled reactor; two-phase flow coolant behavior in boiling water reactors under earthquake condition; simulation of nuclear reactors core; fuel life control in light-water reactors; methods for monitoring and controlling power in nuclear reactors; structural materials modeling for the next generation of nuclear reactors; application of the results of finite group theory in reactor physics; and the usability of vermiculite as a shield for nuclear reactor. Concluding the book is presented a review of the use of neutron flux in the radioisotopes production for medicine.

Amir Zacarias Mesquita, ScD.
Researcher of
Nuclear Technology Development Center (CDTN)
Brazilian Nuclear Energy Commission (CNEN)
Belo Horizonte - Brazil

1

Experimental Investigation of Thermal Hydraulics in the IPR-R1 TRIGA Nuclear Reactor

Amir Zacarias Mesquita[1], Daniel Artur P. Palma[2], Antonella Lombardi Costa[3], Cláubia Pereira[3], Maria Auxiliadora F. Veloso[3] and Patrícia Amélia L. Reis[3]

[1]*Centro de Desenvolvimento da Tecnologia Nuclear/Comissão Nacional de Energia Nuclear*
[2]*Comissão Nacional de Energia Nuclear*
[3]*Departamento de Engenharia Nuclear –Universidade Federal de Minas Gerais*
Brazil

1. Introduction

Rising concerns about global warming and energy security have spurred a revival of interest in nuclear energy, leading to a "nuclear power renaissance" in countries the world over. In Brazil, the nuclear renaissance can be seen in the completion of construction of its third nuclear power plant and in the government's decision to design and build the Brazilian Multipurpose research Reactor (RMB). The role of nuclear energy in Brazil is complementary to others sources. Presently two Nuclear Power Plants are in operation (Angra 1 and 2) with a total of 2000 MW_e that accounts for the generation of approximately 3% of electric power consumed in Brazil. A third unity (Angra 3) is under construction. Even though with such relatively small nuclear park, Brazil has one of the biggest world nuclear resources, being the sixth natural uranium resource in the world and has a fuel cycle industry capable to provide fuel elements. Brazil has four research reactors in operation: the MB-01, a 0.1 kW critical facility; the IEA-R1, a 5 MW pool type reactor; the Argonauta, a 500 W Argonaut type reactor and the IPR-R1, a 100 kW TRIGA Mark I type reactor. They were constructed mainly for using in education, radioisotope production and nuclear research.

Understanding the behavior of the operational parameters of nuclear reactors allow the development of improved analytical models to predict the fuel temperature, and contributing to their safety. The recent natural disaster that caused damage in four reactors at the Fukushima nuclear power plant shows the importance of studies and experiments on natural convection to remove heat from the residual remaining after the shutdown. Experiments, developments and innovations used for research reactors can be later applied to larger power reactors. Their relatively low cost allows research reactors to provide an excellent testing ground for the reactors of tomorrow.

The IPR-R1 TRIGA Mark-I research reactor is located at the Nuclear Technology Development Centre - CDTN (Belo Horizonte/Brazil), a research institute of the Brazilian Nuclear Energy Commission - CNEN. The IPR-R1 reached its first criticality on November

1960 with a core configuration containing 56 aluminum clad standard TRIGA fuel elements, and a maximum thermal power of 30 kW. In order to upgrade the IPR-R1 reactor power, nine stainless steel clad fuel elements were purchased in 1971. One of these fuel elements was instrumented in the centreline with three type K thermocouples. On December 2000, four of these stainless steel clad fuel elements were placed into the core allowing to upgrading the nominal power to 250 kW. In 2004 the instrumented fuel element (IF) was inserted into all core rings and monitored the fuel temperature, allowing heat transfer investigations at several operating powers, including the maximum power of 250 kW (Mesquita, 2005). The basic safety limit for the TRIGA reactor system is the fuel temperature, both in steady-state and pulse mode operation. The time-dependence of temperature was not considered here, hence only the steady-state temperature profile was studied.

This chapter presents the experiments performed in the IPR-R1 reactor for monitoring some thermal hydraulic parameters in the fuel, pool and core coolant channels. The fuel temperature as a function of reactor power was monitored in all core rings. The radial and axial temperature profile, coolant velocity, mass flow rate and Reynolds's number in coolant channels were monitored in all core channels. It also presents a prediction for the critical heat flux (CHF) in the fuel surface at hot channel. Data from the instrumented fuel element, pool, and bulk coolant temperature distribution were compared with the theoretical model and results from other TRIGA reactors. A data acquisition system was developed to provide a friendly interface for monitoring all operational parameters. The system performs the temperature compensation for the thermocouples. Information displayed in real-time was recorded on hard disk in a historical database (Mesquita & Souza, 2008). The data obtained during the experiments provide an excellent picture of the IPR-R1 reactor's thermal performance. The experiments confirm the efficiency of natural circulation in removing the heat produced in the reactor core by nuclear fission (Mesquita & Rezende, 2010).

2. The IPR-R1 reactor

The IPR-R1 TRIGA (*Instituto de Pesquisas Radiativas* - Reactor 1, Training Research Isotope production, General Atomic) is a typical TRIGA Mark I light-water and open pool type reactor. The fuel elements in the reactor core are cooled by water natural circulation. The basic parameter which allows TRIGA reactors to operate safely during either steady-state or transient conditions is the prompt negative temperature coefficient associated with the TRIGA fuel and core design. This temperature coefficient allows great freedom in steady state and transient operations. TRIGA reactors are the most widely used research reactor in the world. There is an installed base of over sixty-five facilities in twenty-four countries on five continents. General Atomics (GA), the supplier of TRIGA research reactors, since late 50's continues to design and install TRIGA reactors around the world, and has built TRIGA reactors in a variety of configurations and capabilities, with steady state thermal power levels ranging from 100 kW to 16 MW. TRIGA reactors are used in many diverse applications, including production of radioisotopes for medicine and industry, treatment of tumors, nondestructive testing, basic research on the properties of matter, and for education and training. The TRIGA reactor is the only nuclear reactor in this category that offers true "inherent safety", rather than relying on "engineered safety". It is possible due to the unique properties of GA's uranium-zirconium hydride fuel, which provides incomparable safety characteristics, which also permit flexibility in sitting, with minimal environmental effects

(General Atomics, 2011). Figure 1 shows two photographs of the pool and the core with the IPR-R1 TRIGA reactor in operation.

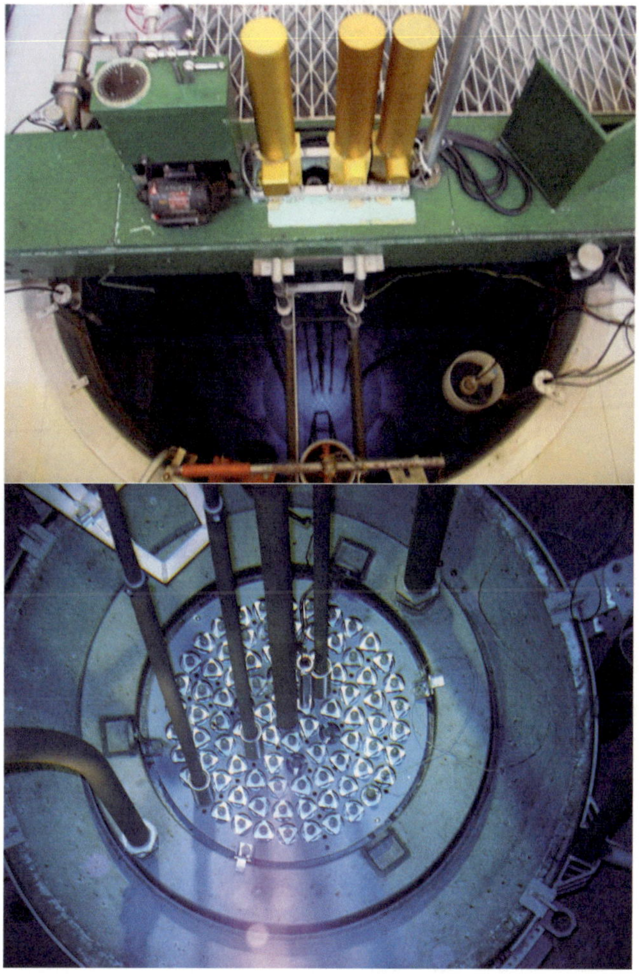

Fig. 1. IPR-R1 TRIGA reactor pool and core

The IPR-R1 TRIGA reactor core is placed at the bottom of an open tank of about 6m height and 2m diameter. The tank is filled with approximately 18 m^2 of water able to assure an adequate radioactive shielding, as shown in Fig. 2. The reactor is licensed to operate at a maximum steady-state thermal power level of 100 kW, but the core and the instrumentation are configured to 250 kW, and waiting the definitive license to operate in this new power. Some of the experiments reported here were performed at power operation of 250 kW. For these experiments was obtained a provisional license for operation to this new power.

The reactor core is cooled by water natural circulation. Cooling water passage through the top plate is provided by the differential area between a triangular spacer block on top of fuel element and the round hole in the grid. A heat removal system is provided for removing heat from the reactor pool water. The water is pumped through a heat exchanger, where the heat is transferred from the primary to the secondary loop. The secondary loop water is cooled in an external cooling tower. Figure 3 shows the forced cooling system, which transfers the heat generated in the reactor core to a water-to-water heat exchanger. The secondary cooling system transfers the reactor core heat from the heat exchanger to a cooling tower. In the diagram is shown also the instrumentation distribution and the forced and natural circulation paths in the pool.

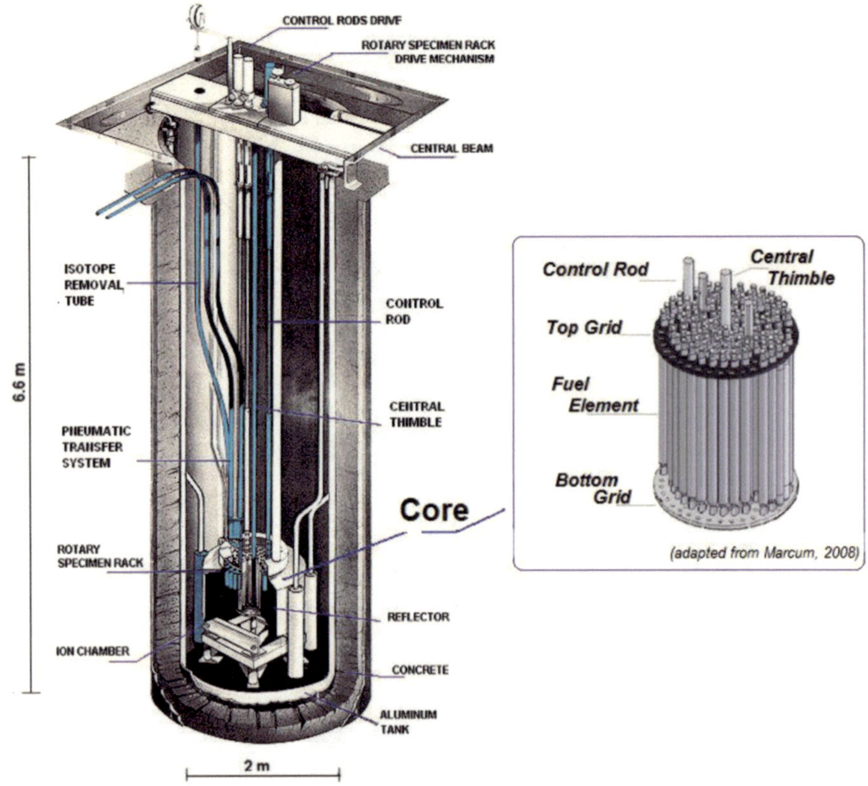

Fig. 2. IPR-R1 TRIGA reactor pool and core

A simplified view of the IPR R1 TRIGA core configuration is shown in the Fig. 4. As shown in the diagram there are small holes in the core upper grid plate. These holes were used to insert thermocouples to monitor the coolant channel temperatures. The core has a cylindrical configuration of six rings (A, B, C, D, E and F) having 1, 6, 12, 18, 24 and 30 locations respectively. These 91 positions are able to host either fuel rods or other components like control rods, a neutron source, graphite dummies (mobile reflector), irradiating and measurement channels (e.g. central thimble or A ring). Each location corresponds to a role in the aluminum upper grid plate of the reactor core. The core is surrounded by an annular graphite reflector and water. Inside the reflector there is a rotary specimen rack with 40 positions for placement of samples to be activated by neutron flux. The top view of the reactor core and the rotary specimen rack are presented in Fig. 5. There is a very high number of reactor loading configurations, so that it is possible to obtain the sub-critical level required simply loading/unloading fuel rods from the core.

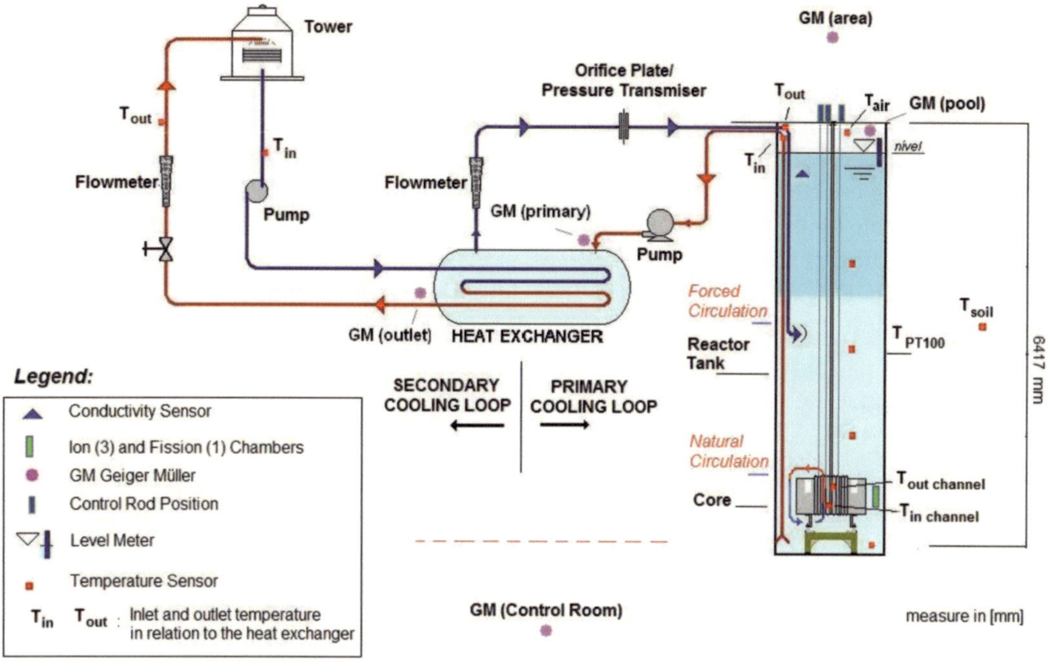

Fig. 3. IPR-R1 TRIGA reactor cooling system and instrumentation distribution

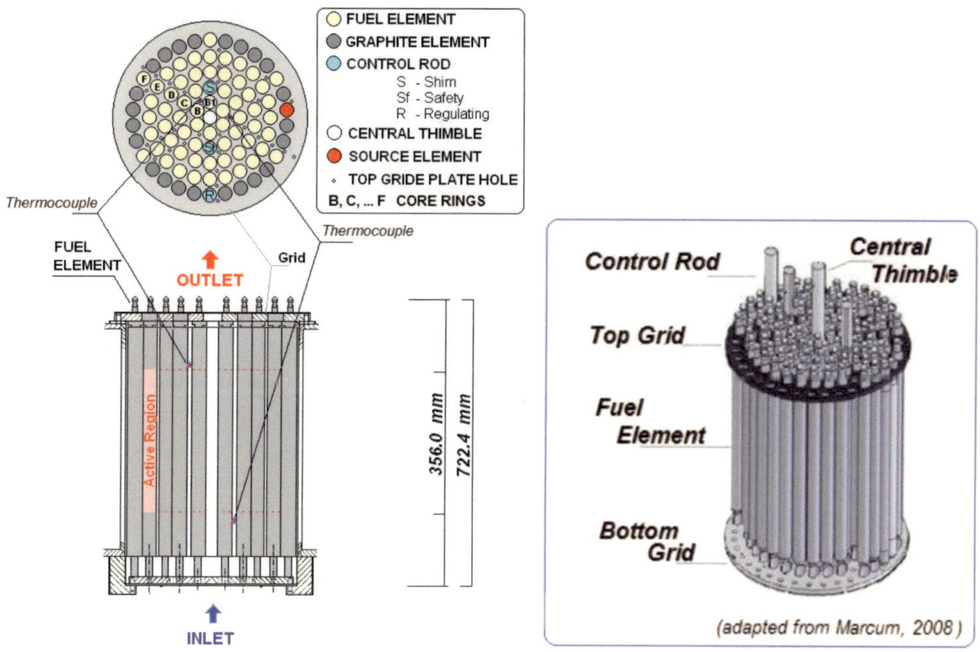

Fig. 4. Simplified core diagram

The prototypical cylindrical fuel elements are a homogeneous alloy of zirconium hydride (neutron moderator) and uranium enriched at 20% in ^{235}U. The reactor core has 58 aluminum-clad fuel elements and 5 stainless steel-clad fuel elements. One of these steel-clad fuel elements is instrumented with three thermocouples along its centreline, and was inserted in the reactor core in order to evaluate the thermal hydraulic performance of the IPR-R1 reactor (Mesquita, 2005). The fuel rod has about 3.5 cm diameter, the active length is about 37 cm closed by graphite slugs at the top and bottom ends which act as axial reflector. The moderating effects are carried out mainly by the zirconium hydride in the mixture, and on a smaller scale by light water coolant. The characteristic of the fuel elements gives a very high negative prompt temperature coefficient, is the main reason of the high inherent safety behavior of the TRIGA reactors. The power level of the reactor is controlled with three independent control rods: a Regulating rod, a Shim rod, and a Safety rod.

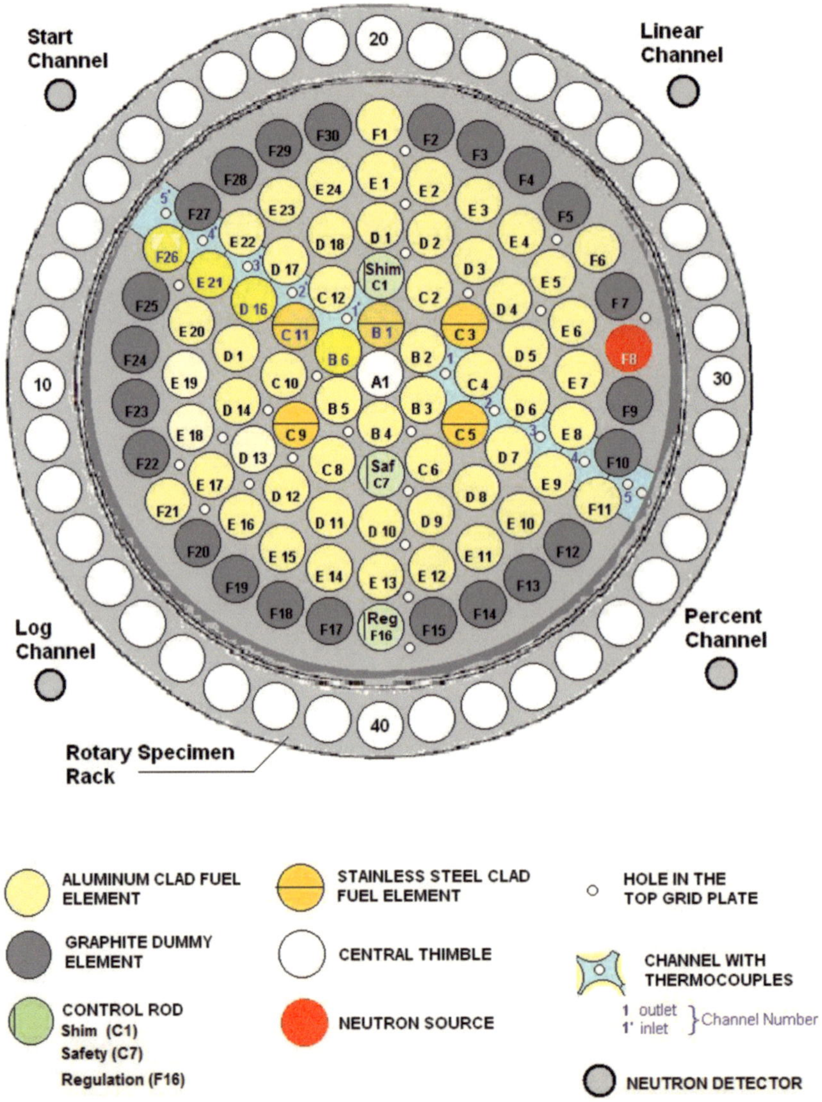

Fig. 5. Core configuration with the rotary specimen rack

3. Methodology

3.1 Fuel and core coolant channel temperatures

Before starting the experiments the thermal power released by the core was calibrated, according with the methodology developed by Mesquita et al. (2007). The calibration method used consisted of the steady-state energy balance of the primary cooling loop. For this balance, the inlet and outlet temperatures and the water flow in this primary cooling loop were measured. The heat transferred through the primary loop was added to the heat leakage from the reactor pool. The temperature measurements lines were calibrated as a whole, including sensors, cables, data acquisition cards and computer. The uncertainties for the temperature measurement circuit were ±0.4 °C for resistance temperature detectors, and ±1.0 °C for thermocouples circuits. The adjusted equations were added to the program of the data acquisition system (DAS). The sensor signs were sent to an amplifier and multiplexing board of the DAS, which also makes the temperature compensation for the thermocouples. The temperatures were monitored in real time on the DAS computer screen. All data were obtained as the average of 120 readings and were recorded together with their standard deviations. The system was developed to monitor and to register the operational parameters once a second in a historical database (Mesquita & Souza, 2010).

The original fuel element at the reactor core position B1 was removed and replaced by an instrumented fuel element. Position B1 is the hottest location in the core (largest thermal power production), according to the neutronic calculation (Dalle et al., 2002). The instrumented fuel element is in all aspects identical to standard fuel elements, except that it is equipped with three chromel-alumel thermocouples (K type), embedded in the fuel meat. The sensitive tips of the thermocouples are located along the fuel centreline. Their axial position is one at the half-height of the fuel meat and the other two 2.54 mm above and 2.54 mm below. Figure 6 shows the diagram of the instrumented fuel element and the Table I presents its main characteristics (Gulf General Atomic, 1972). Figure 7 shows the instrumented fuel element and one thermocouple inside a core channel.

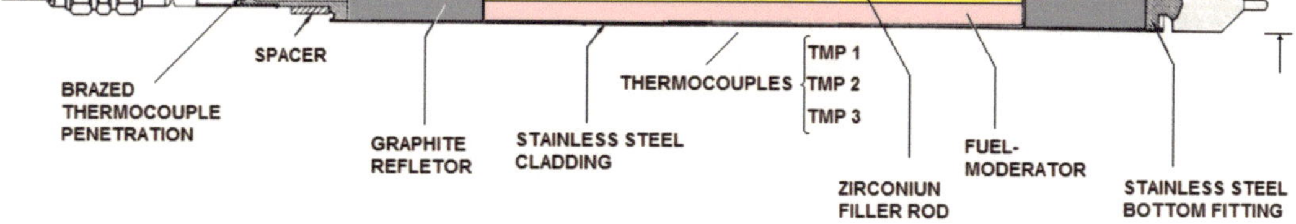

Fig. 6. Diagram of the instrumented fuel element

Parameter	Value
Heated length	38.1 cm
Outside diameter	3.76 cm
Active outside area	450.05 cm^2
Fuel outside area (U-$ZrH_{1.6}$)	434.49 cm^2
Fuel element active volume	423.05 cm^3
Fuel volume (U-$ZrH_{1.6}$)	394.30 cm^3
Power (total of the core = 265 kW)	4.518 kW

Table 1. Instrumented fuel element features

Fig. 7. IPR-R1 core top view with the instrumented fuel element in ring B and one probe with thermocouple inside the core.

The instrumented fuel element was replaced to new positions and measures the fuel temperature in each one of the core fuel rings (from B to F). At the same way, two thermocouples were replaced to channels close to the instrumented fuel element to measure the coolant channel temperature. Experiments were carried out with the power changing from about 50 kW to 250 kW in 50 kW steps for each position of the instrumented fuel element. The fuel and coolant temperatures were monitored as function of the thermal power and position in the core.

In the TRIGA type reactors the buoyancy force induced by the density differential across the core maintains the water circulation through the core. Countering this buoyancy force are the pressure losses due to the contraction and expansion at the entrance and exit of the core as well as the acceleration and friction pressure losses in the flow channels. Direct measurement of the flow rate in a coolant channel is difficult because of the bulky size and low accuracy of flow meters. The flow rate through the channel may be determined indirectly from the heat balance across the channel using measurements of the water inlet and outlet temperatures. Two type K (chromel–alumel) thermocouples fixed in two rigid aluminum probes (7.9 mm of diameter), were inserted into the core in two channels close to position B1 (Channel 1 and 1′ in Fig. 4 and Fig. 5) and measured the inlet and outlet coolant channel temperatures. The probes penetrated axially the channels through small holes in the core upper grid plate. The probes were positioned in diametrically opposite channels, so that when a probe measured the channel entrance temperature, the other one registered the channel exit temperature. In a subsequent run, the probe positions were inverted. This procedure was used also for the Channels 1′, 2′, 3′, 4′ and 5′ (Fig. 5). There is no hole in the top grid plate in the direction of the Channel 0; so it was not possible to measure its temperature. The inlet and outlet temperatures in Channel 0 were considered as being the same of Channel 1. For the other channels there are holes in the top grid plate where it was possible to insert the temperature probes. To found the bulk coolant temperature axial profile at hot channel, with the reactor operating at 250 kW, the probe that measures the channel inlet temperature was raised in steps of 10 cm and the temperature was monitored. The same procedure was done with the reactor operating at 100 kW, but the probe was raised in steps of 5 cm.

3.2 Hydraulic parameters of the coolant

The mass flow rate through the core coolant channels was determined indirectly from the heat balance across each channel using measurements of the water entrance and exit temperatures. Although the channels are laterally open, in this work cross flow or mass transfer between adjacent channels was ignored. Inlet and outlet coolant temperatures in channels were measured with two rigid aluminum probes with thermocouples. They were inserted in the upper grid plate holes (Fig. 5). Figure 8 illustrates schematically the general natural convection process established by the fuel elements bounding one flow channel in the core. The core coolant channels extend from the bottom grid plate to the top grid plate. The cooling water flows through the holes in the bottom grid plate, passes through the lower unheated region of the element, flows upwards through the active region, passes through the upper unheated region, and finally leaving the channel through the differential area between a triangular spacer block on the top of the fuel element and a round hole in the grid. As mentioned, in natural convection the driving force is supplied by the buoyancy of the heated water in the core channels.

In a typical TRIGA flow channel entire fuel element is cooled by single phase convection as long as the maximum wall temperature is kept below that required to initiate boiling. However, at higher power levels the inlet and outlet regions of the core, where the heat fluxes are the lowest, the channels are cooled by single phase convection. In the central region, where the axial heat flux is highest, the mode of heat transfer is predominantly subcooled boiling (Rao et al., 1988 and Mesquita et al. 2011).

The channel heating process is the result of the thermal fraction contributions of the perimeter of each fuel around the channel. So there was an average power of 4.518 kW dissipated in each stainless steel cladding fuel element and 4.176 kW dissipated in each aluminum cladding fuel element at 265 kW core total power. The values are multiplied by the fuel element axial power distribution and core radial power distribution factors as shown in profiles of Fig. 9.

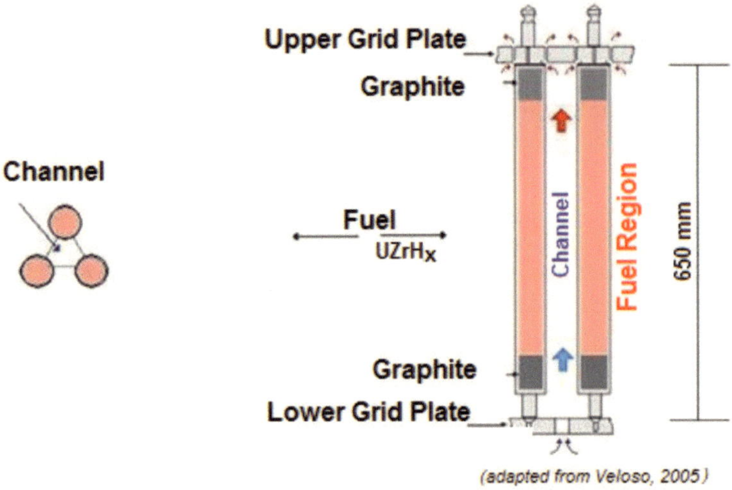

Fig. 8. A scheme of one flow channel in the TRIGA core

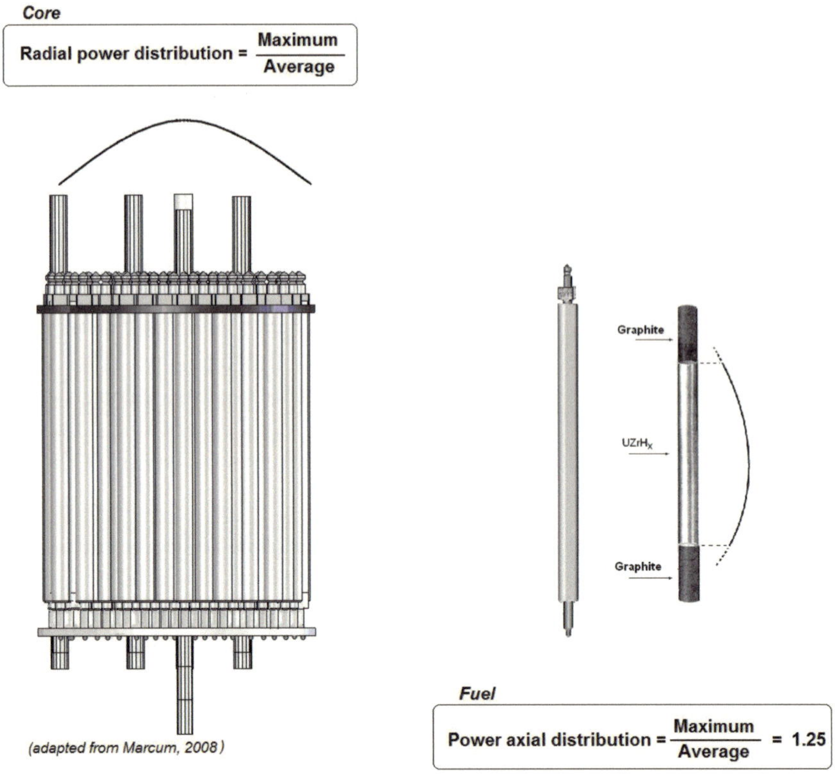

Fig. 9. Core radial and fuel element axial power profiles

The power axial distribution factor in the fuel is 1.25, according with Marcum (2008). Figure 10 shows in detail the coolant channels geometry. The core radial power distribution factors, shown in Fig. 10, were calculated by Dalle et al. (2002) using WIMS-D4 and CITATION codes. The products are multiplied by the fractions of the perimeters of each fuel in contact with the coolant in each channel. The two hottest channels in the core are Channel 0 and Channel 1′. Channel 0 is located closer to the core centre, where density of neutron flux is larger, but there is no hole in the top grid plate in the direction of this channel. Table 1 gives the geometric data of the coolant channels and the percentage of contribution relative to each fuel element to the channels power (Veloso, 2005 and Mesquita, 2005).

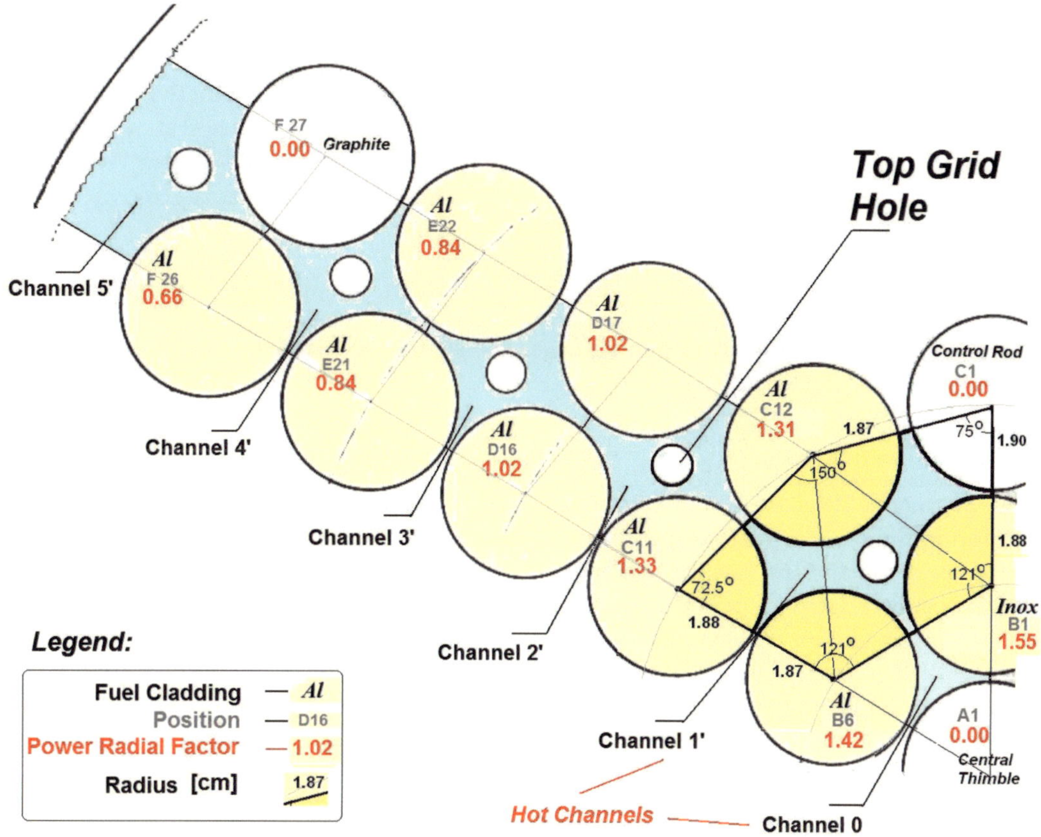

Fig. 10. Core coolant channels geometry and radial power distribution

Channel Number	Area [cm²]	Wetted Perimeter [cm]	Heated Perimeter [cm]	Hydraulic Diameter [cm]	Channel Power [%]
0	1.5740	5.9010	3.9060	1.0669	1.00
1′	8.2139	17.6427	15.1556	1.8623	3.70
2′	5.7786	11.7456	11.7456	1.9679	2.15
3′	5.7354	11.7181	11.7181	1.9578	1.83
4′	5.6938	11.7181	8.6005	1.9436	1.13
5′	3.9693	10.8678	3.1248	1.4609	0.35

Table 2. Channel geometry and hydraulic parameters (Veloso, 2005; Mesquita, 2005)

The mass flow rate in the hydraulic channel ($\dot{m}$) in [kg/s] is given indirectly from the thermal balance along the channel using measurements of the water inlet and outlet temperatures:

$$\dot{m} = \frac{q_c}{c_p \Delta T} \tag{1}$$

Where q_c is the power supplied to the channel [kW], c_p is the isobaric specific heat of the water [J/kgK] and ΔT is the temperature difference along the channel [°C]. The mass flux G is given by: $G = \dot{m}$ / *channel area*. The velocity u is given by $u = G/\rho$, where ρ is the water density (995 kg/m³). The values of the water thermodynamic properties were obtained as function of the bulk water temperature at the channel for the pressure 1.5 bar (Wagner & Kruse, 1988) Reynolds number (Re), used to characterize the flow regime, is given by:

$$\mathrm{Re} = \frac{G D_w}{\mu} \tag{2}$$

Where G is the mass flux in [kg/m²s], D_w is the hydraulic diameter in [m] and μ is the dynamic viscosity [kg/ms].

3.3 Pool temperatures

Nine thermocouples and one platinum resistance thermometer (PT-100) were used to monitoring the reactor pool temperature. The thermocouples were positioned in a vertical aluminum probe and the first thermocouple was 143 mm above the core top grid plate. The reactor operated during a period of about eight hours at a thermal power of 265 kW before the steady state was obtained. The forced cooling system was turned on during the operation. This experiment is important to understand the behavior of the water temperature in the pool and evaluate the height of the chimney effect.

3.4 Temperatures with the forced cooling system turned off

The power of the IPR-R1 TRIGA was raised in steps of about 25 kW until to reach 265 kW. The forced cooling system of the reactor pool was turned off during the tests. The increase of the power was allowed only when all the desired quantities had been measured and the given limits were not exceeded. After the reactor power level was reached, the reactor was maintained at that power for about 15 min, so the entire steady-state conditions were not reached in the core and coolant. The fuel temperature data was obtained by using the instrumented fuel element. The fuel temperature measurements were taken at location B1 of the core (hottest position). The outlet temperature in the channel was measured with thermocouple inserted near the B1 position. One platinum resistance thermometer measured the water temperature in the upper part of the reactor tank. Two thermocouples measured the ambient temperatures around the reactor pool. The IPR-R1 reactor has a rotary specimen rack outside the reactor core for sample irradiation. It is composed by forty irradiation channels in a cylindrical geometry. One type K thermocouple was put during the experiment in Position 40 of the rotary specimen rack (Fig. 5).

3.5 Critical heat flux and DNBR

As the power in the IPR-R1 TRIGA core is increased, nucleation begins to occur on the fuel rod surfaces. The typical pool boiling curve (Fig. 11) is represented on a log-log plot of heat flux versus wall superheat ($T_{sur} - T_{sat}$). At low values of ΔT_{sat} the curve is fairly linear, hence the convective heat transfer coefficient (h) is relatively constant. There is no bubble formation and the heat transfer occurs by liquid natural convection. At about ten to twenty degrees above saturation the heat flux increases rapidly with the increasing of the wall temperature. The increase in heat transfer is due to nucleate boiling. The formation of vapor bubbles increases the turbulence near the heated surface and allows mixing of the coolant fluid in the film region, thus enhancing the heat transfer rate (Haag, 1971). From the shape of the curve, it can be seen that the heat transfer coefficient increases dramatically in the boiling regime.

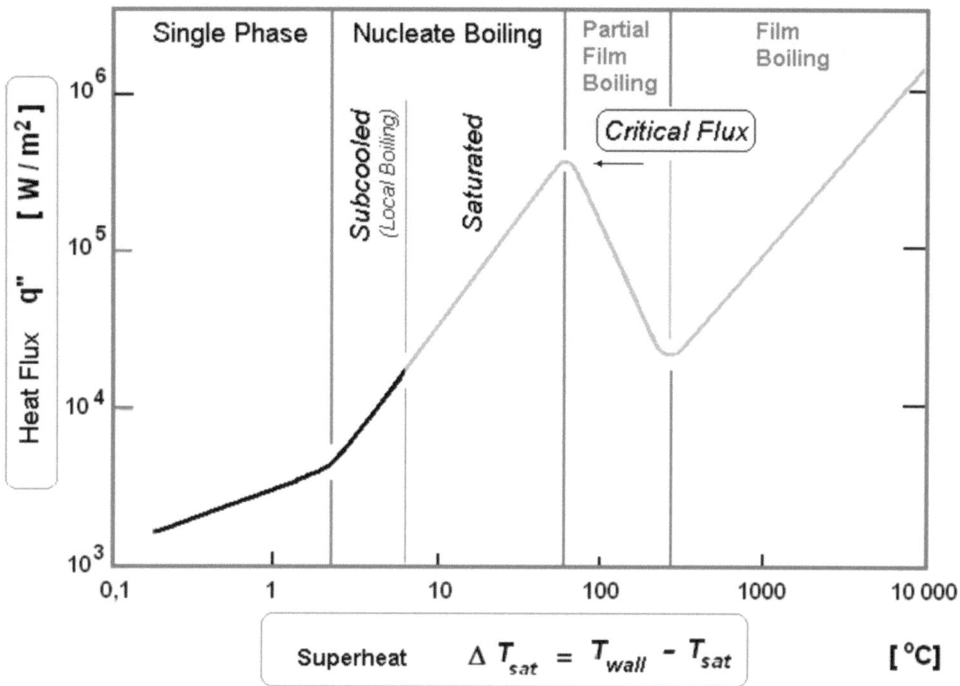

Fig. 11. Typical pool boiling curve for water under atmospheric pressure

Whenever the surface temperature of a solid exceeds the saturation temperature, local boiling may occur even if the bulk water temperature is below the saturation temperature. The water temperature in the boundary layer on the heated surface can become sufficiently high so that subcooled pool boiling takes place. The bubbles will be condensed upon leaving this boundary layer region because the bulk water is below the saturation temperature. By increasing the surface temperature, the heat flux can reach the critical heat flux where the film boiling occurs. At this point the bubbles become so numerous that they form an insulating layer of steam around the fuel element and the heat flux is reduced significantly.

The critical heat flux is the maximum heat flux that a saturated fluid can absorb before acquiring more enthalpy than can be dissipated into its surroundings.

In the fully developed nucleate boiling regime, it is possible to increase the heat flux without an appreciable change in the surface temperature until the point of Departure from Nucleate Boiling (DNB). At this point, the bubble motion on the surface becomes so violent that a hydrodynamic crisis occurs with the formation of a continuous vapor film in the surface and the Critical Heat Flux (CHF) is reached. In subcooled boiling the CHF is a function of the coolant velocity, the degree of subcooling, and the pressure. There are a lot of correlations to predict the CHF. The correlation done by Bernath found in Lamarsh and Baratta (2001) was used to predicts CHF in the subcooled boiling region and is based on the critical wall superheat condition at burnout and turbulent mixing convective heat transfer. Bernath's equation gives the minimum results so it is the most conservative. It is given by:

$$q''_{crit} = h_{crit}(T_{crit} - T_f), \tag{3}$$

where,

$$h_{crit} = 61.84\frac{D_w}{D_w + D_i} + 0.01863\frac{23.53}{D_w^{0.6}}u \tag{4}$$

and,

$$T_{crit} = 57\ln(p-54)\frac{p}{p+0.1034} + 283.7 - \frac{u}{1.219} \tag{5}$$

q''_{crit} is the critical heat flux [W/m^2], h_{crit} is the critical coefficient of heat transfer [W/m^{2}K], T_{crit} is the critical surface temperature [°C], T_f is the bulk fluid temperature [°C], p is the pressure [MPa], u is the fluid velocity [m/s] ($u = \dot{m}$ */channel area/water density*), D_w is the wet hydraulic diameter [m], D_i is the diameter of heat source [m]. This correlation is valid for circular, rectangular and annular channels, pressure of 0.1 to 20.6 MPa, velocity between 1 to 16 m/s and hydraulic diameter of 0.36 to 1.7 cm.

4. Results

4.1 Fuel temperature

Before beginning the experiments, the calibration of the thermal power released by the core were performed, and a power of 265 kW was found when the neutronic linear channel was indicating the power of 250 kW.

Figure 12 shows the radial power profile (neutron flux) calculated by Dalle et al. (2002) using the TRIGPOW code, the experimental fuel radial temperature profile, and the inlet/outlet coolant temperatures in the channel closest to the instrumented element. The theoretical results, for the IPR-R1 TRIGA, calculated by Veloso (2005) using the PANTERA code and the experimental results found in the ITU TRIGA Mark II reactor at the Istanbul University were also plotted (Özkul & Durmayaz, 2000).. All data are for the power of 265 kW.

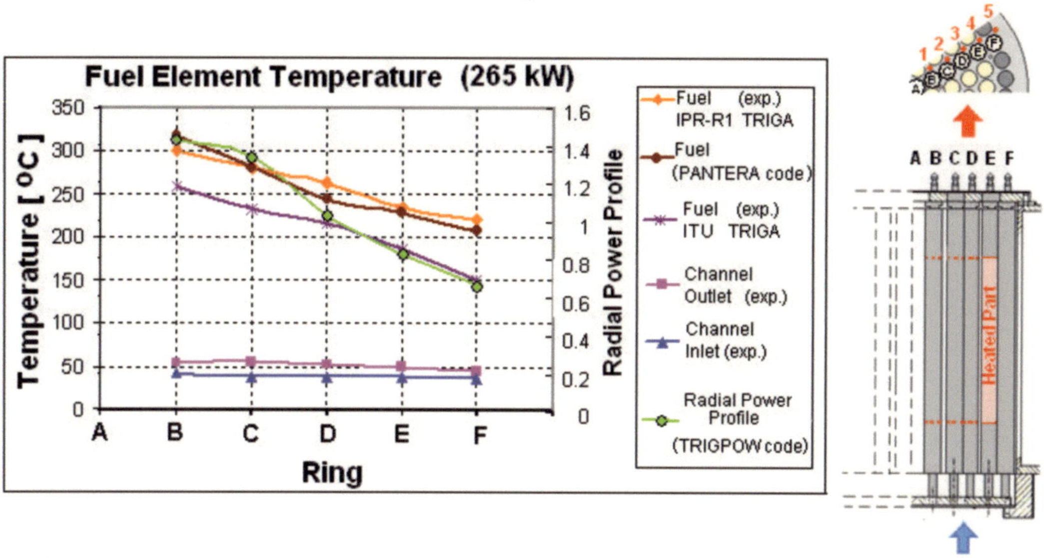

Fig. 12. Core temperature radial profile at 265 kW thermal power

Figure 13 shows the results of fuel temperature versus reactor thermal power. In the experiment the instrumented fuel element was positioned in each core ring.

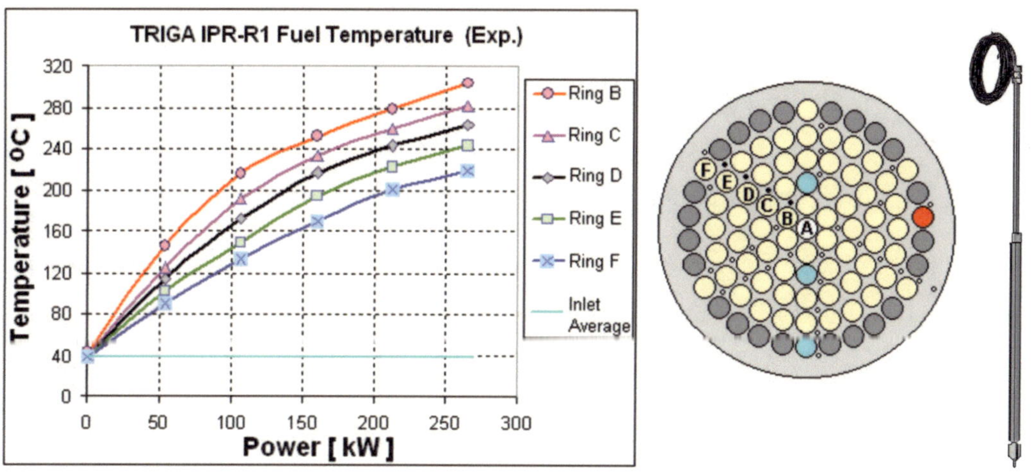

Fig. 13. Fuel temperature as function of the reactor power in all core rings

4.2 Core temperature

4.2.1 Outlet coolant temperature as function of the thermal power

The experimental coolant exit temperature for each core ring is shown in Fig 14 as a function of the reactor power. The aluminum probe with thermocouple was inserted in each hole at top grid plate, and the coolant inlet temperature was about 38 °C in all measurements.

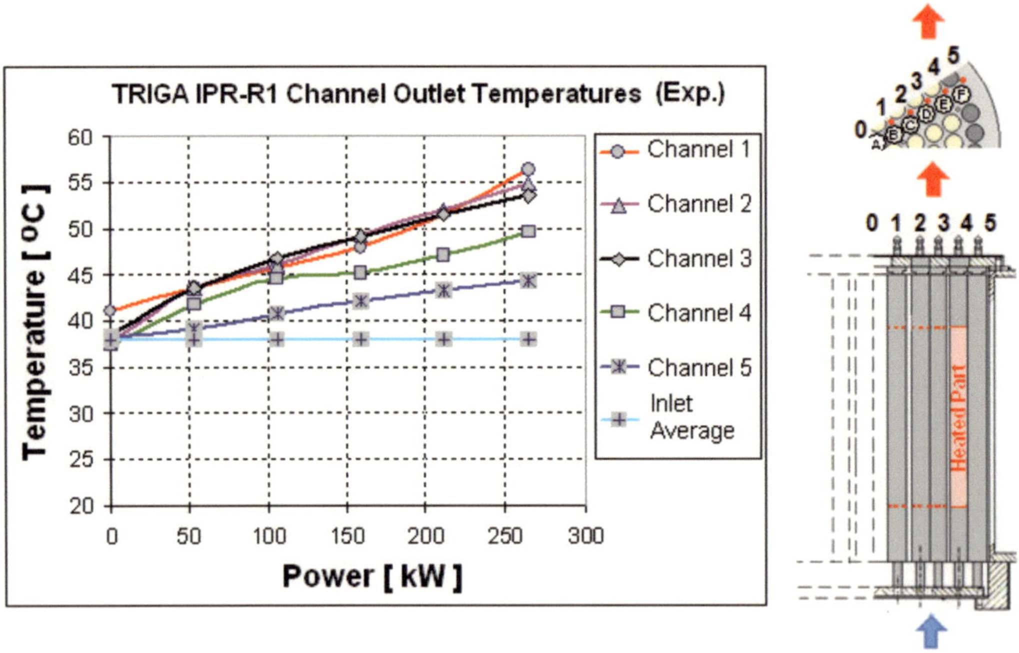

Fig. 14. Outlet coolant temperature as function of the thermal power

4.2.2 Radial temperature profile along the core coolant channels

Figure 15 shows the radial core coolant temperature profiles (inlet/outlet channel temperatures) at 265 kW. Theoretical results using the PANTERA code are also shown in the figure (Veloso, 2005).

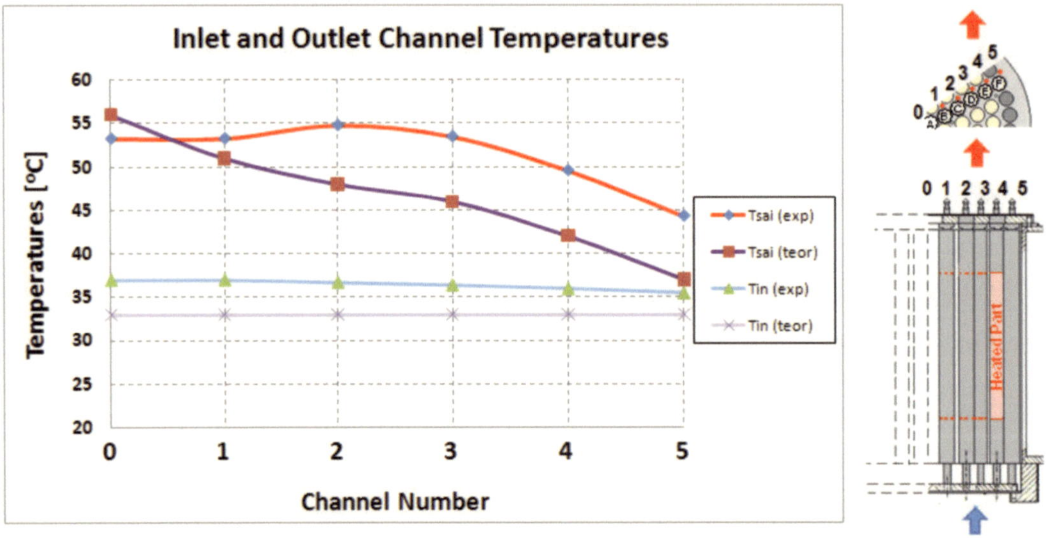

Fig. 15. Radial temperature profile in the core coolant channels at 265 kW thermal power

4.2.3 Axial temperature profile in the hot channel

The experimental bulk coolant temperatures profile in Channel 1 is shown in Fig. 16 as a function of the axial position, for the powers of 265 kW and 106 kW. Figure 16 shows also the curve predicted from the theoretical model using the PANTERA code at 265 kW (Veloso, 2005). The figure shows also the experimental results for other TRIGA reactors Bärs & Vaurio, 1966; Haag, 1971) and (Büke & Yavuz, 2000).

The experimental temperature profile along the coolant is different from that predicted from the theoretical model. Ideally, the coolant temperature would increase along the entire length of the channel, because heat is being added to the water by all fuel regions in the channel. Experimentally, the water temperature reaches a maximum near the middle length and then decreases along the remaining channel. The shape of the experimental curves is similar to the axial power distribution within the fuel rod as shown in Fig. 9. Although Channel 1 is located beside the control rod, the axial temperature profile was not influenced by a possible deformation of the neutron flux caused by this rod, because it was in its upper position, i.e. outside the core. The actual coolant flow is quite different probably, because of the inflow of water from the core sides (colder than its centre).

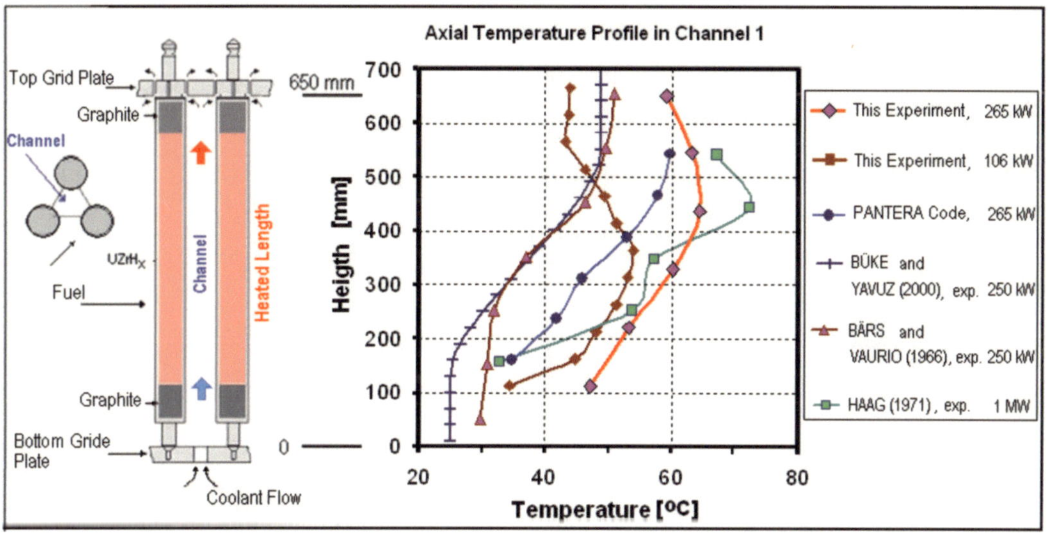

Fig. 16. Axial bulk coolant temperature profile along the Channel 1

4.2.4 Thermal hydraulic parameters of coolant channels

The pertinent parameters required for the analysis of coolant channels are tabulated in Table 3. Figure 17 shows the power dissipate and the temperature increase in each channel at 265 kW reactor total power. This power was the results of the thermal power calibration (Mesquita et al. 2007). The profile of the mass flow rate and velocity in the core is shown in the graphs of Figure 18. Figure 19 compares experimental and theoretical profile of mass flux G in the core coolant channels. The theoretical values were calculated using PANTERA code (Veloso, 2005). As it can see by the Reynolds number the flow regime is turbulent in channels near the core centre.

Channel	Channel Power q [kW]	$T_{out} - T_{in}$ ΔT [°C]	Flow Rate $\dot{m}$ [kg/s]	Area [cm²]	Mass Flux G [kg/m²s]	Velocity u [m/s]	Reynolds Number Re -
0	2.65	15.5	0.041	1.574	260.48	0.26	3228
1	9.81	15.5	0.151	8.214	183.83	0.18	5285
2	5.70	17.1	0.080	5.779	138.44	0.14	5181
3	4.85	16.3	0.071	5.735	123.79	0.12	4184
4	3.00	12.1	0.059	5.694	103.62	0.10	2525
5	0.93	7.7	0.029	3.969	73.06	0.07	549

[1]Specific heat (c_p) = 4.1809 [kJ/kgK], water density (ρ) 995 kg/m³ and dynamic viscosity(μ) = 0.620 10^{-3} kg/ms at 45 °C.

Table 3. Properties of the coolant channel at the power of 265 kW[1]

As can be seen in Figure 18 and Figure 19 the velocity and mass flux in each channel are proportional to power dissipated in the channel.

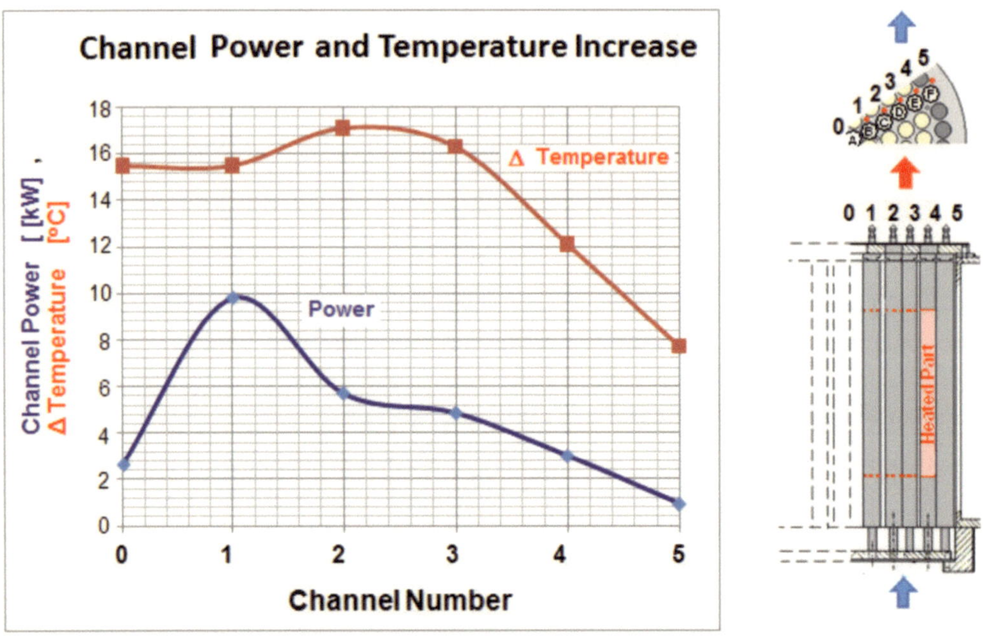

Fig. 17. Power and temperature increase in coolant channels at 265 kW

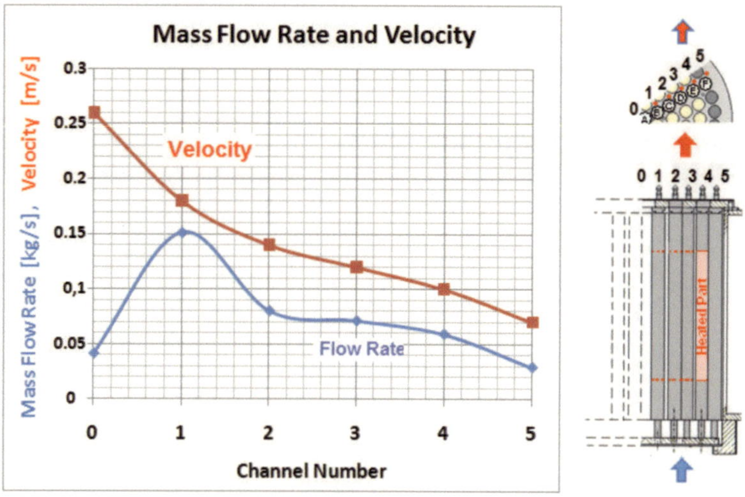

Fig. 18. Mass flow rate and velocity in coolant channels at 265 kW

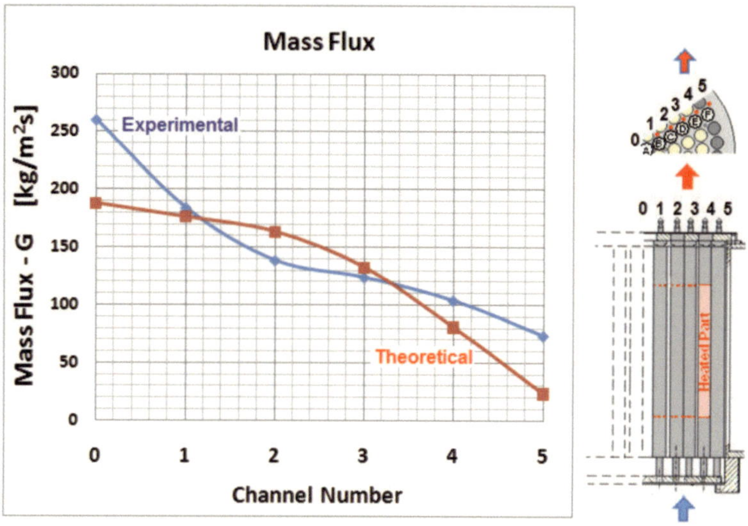

Fig. 19. Mass flux in coolant channels at 265 kW

4.3 Pool temperature

Figure 20 shows the water temperatures evolution at the reactor pool, and the inlet and outlet coolant temperature in the core's hottest channel until the beginning of steady state. The results showed that the thermocouples positioned 143 mm over the top grid plate (Inf 7) measure a temperature level higher than all the other thermocouples positioned over the reactor core. The temperature measurements above the core showed that thorough mixing of water occurs within the first centimeters above core top resulting in a uniform water temperature. It means that the chimney effect is not much high, less than 400 mm above the reactor core, in agreement with similar experiments reported by Rao et al. (1988). The chimney effect is considered as an unheated extension of the core. The chimney height is the distance between the channel exit and the fluid isotherm plan above the core and it depends of the reactor power.

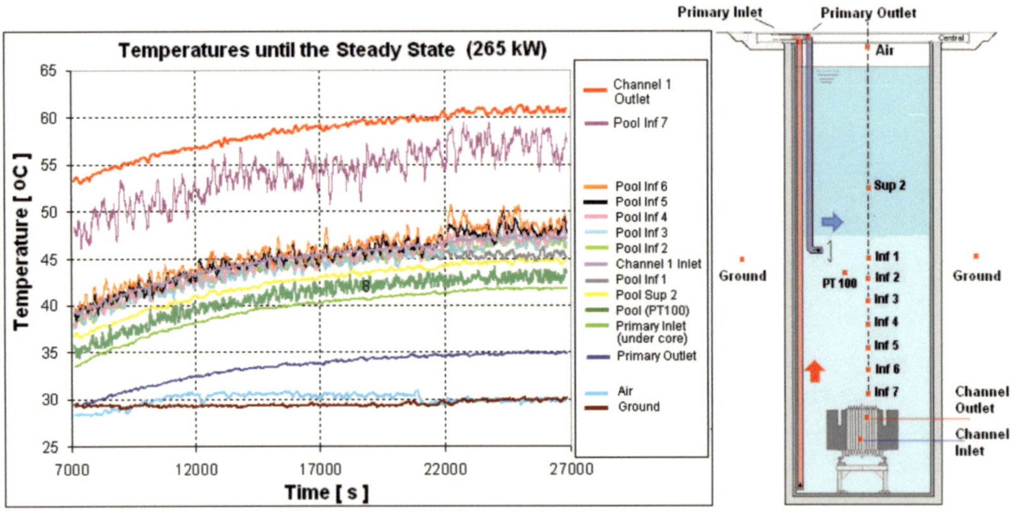

Fig. 20. Temperatures patterns in the reactor pool at 265 kW thermal power

4.4 Temperatures with the forced cooling system tuned off

Figure 21 shows the behavior of fuel element, channel outlet, reactor pool, and specimen rack temperatures at various operation powers, with the forced cooling system turned off.

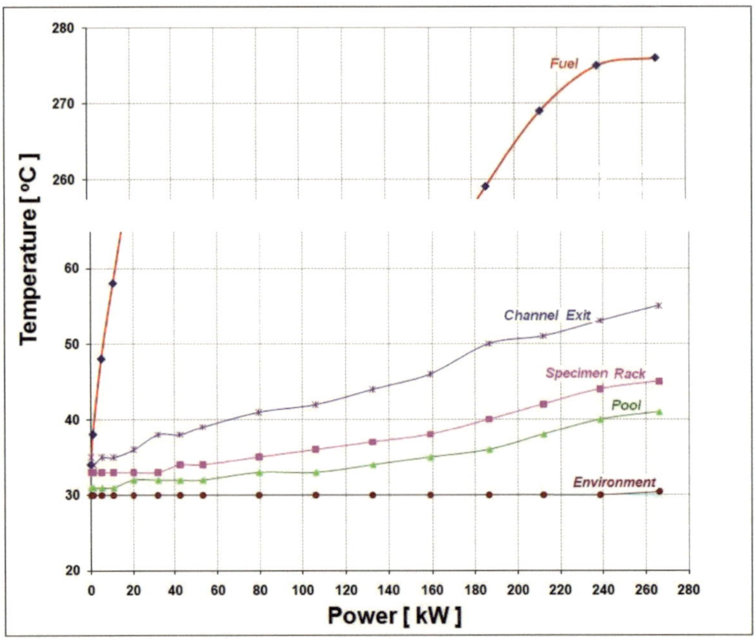

Fig. 21. Temperature evolution as a function of power with the forced cooling system off

4.5 Critical Heat Flux and DNBR

The heat generated by fission in the fuel material is conducted through the fuel, through the fuel-cladding interface, and across the cladding to the coolant. The thermal and hydrodynamic purpose of the design is to safely remove the heat generated in the fuel without producing excessive fuel temperatures or steam void formations and without closely approaching the hidrodynamic Critical Heat Flux (CHF) (Huda et al. 2001). As the IPR-R1 TRIGA reactor core power is increased, the heat transfer regime from the fuel cladding to the coolant changes from the single phase natural convection regime to subcooled nucleate boiling. The hottest temperature measured in the core channel was 65 °C (Channel 1'), below 111.4 °C, the water saturation temperature for the pressure of 1.5 bar. Therefore, the saturated nucleate boiling regime is not reached. Channel 1' is the closest channel to the centre of the reactor where it is possible to measure the water entrance and exit temperatures. The hottest channel is Channel 0, closer to the centre. With the measured temperature values in the Channel 1', the value of critical flow was evaluated in these two channels. The Bernath correlation was used (Eq. 5) for the calculation of the critical heat flux. With the reactor power of 265 kW operating in steady state, the core inlet temperature was 47°C. The critical flow for the Channel 0 is about 1.6 MW/m^2, giving a Departure from Nucleate Boiling Ratio (DNBR[2] of 8.5. Figure 22 and Figure 23 show the values of critical flow and DNBR for the two channels. The theoretical values for reactor TRIGA of the University of New York and calculated with the PANTERA code for the IPR-R1 are also shown (General Atomic, 1970) and (Veloso, 2005). The two theoretical calculations gave smaller results than the experiments. These differences are due to the core inlet temperature used in the models.

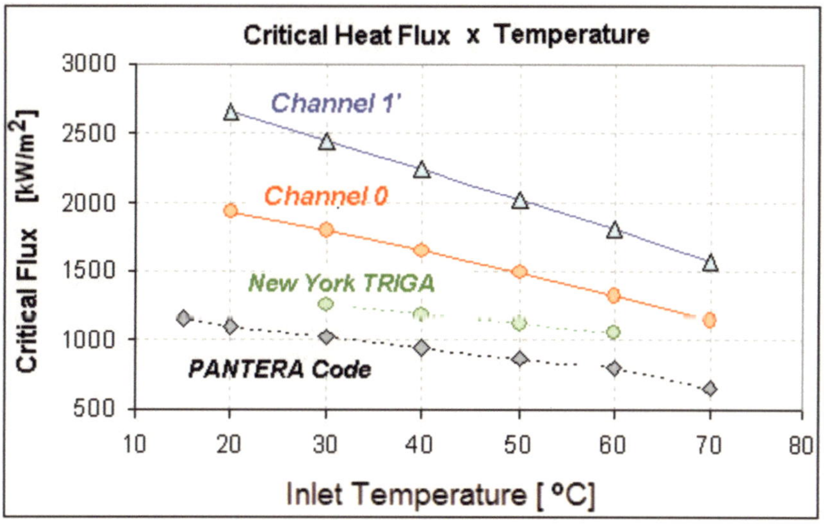

[2]CHF/actual local heat flux

Fig. 22. Critical heat flux as a function of the coolant inlet temperature

The minimum DNBR for IPR-R1 TRIGA (DNBR=8.5) is much larger than other TRIGA reactors. The 2 MW McClellen TRIGA calculated by Jensen and Newell (1998) had a DNBR=2.5 and the 3 MW Bangladesh TRIGA has a DNBR=2.8 (Huda and Rahman, 2004).

The power reactors are projected for a minimum DNBR of 1.3. In routine operation they operated with a DNBR close to 2. The IPR-R1 reactor operates with a great margin of safety at its present power of 250 kW, the maximum heat flux in the hottest fuel is about 8 times lesser than the critical heat flux that would take the hydrodynamic crisis in the fuel cladding. This investigation indicates that the reactor would have an appropriate heat transfer if the reactor operated at a power of about 1 MW.

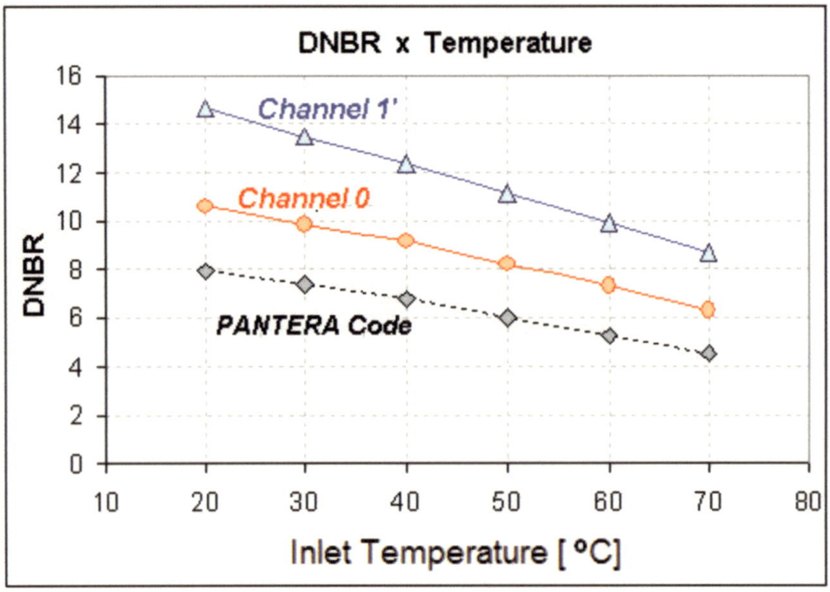

Fig. 23. DNBR as a function of the coolant inlet temperature

5. Conclusion

Experiments to understand the behavior of the nuclear reactors operational parameters allow improve model predictions, contributing to their safety. Developments and innovations used for research reactors can be later applied to larger power reactors. Their relatively low cost allows research reactors to provide an excellent testing ground for the reactors of tomorrow.

The experiments described here confirm the efficiency of natural convection in removing the heat produced in the reactor core by nuclear fission. The data taken during the experiments provides an excellent picture of the thermal performance of the IPR-R1 reactor core. The IPR-R1 TRIGA core design accommodates sufficient natural convective flow to maintain continuous flow of water throughout the core, which thereby avoids significant bubbles formation and restricts possible steam bubbles to the vicinity of the fuel element surface. The spacing between adjoining fuel elements was selected not only from neutronic considerations but also from thermohydrodynamic considerations. The experimental data also provides information, which allows the computation of other parameters, such as the fuel cladding heat transfer coefficient (Mesquita & Rezende, 2007). The theoretical temperatures and mass flux were determined under ideal conditions. There is a considerable coolant crossflow throughout the channels. Note that the natural convection flow is turbulent in all channels near the centre. The temperature measurements above the IPR-R1 core showed that water mixing occurs within the first few centimeters above the top of the core, resulting in an almost uniform water temperature. The temperature at the primary loop suction point at the pool bottom, as shown in Fig. 3, has been found as the lowest temperature in the reactor pool. Pool temperature depends on reactor power, as well as on the external temperature because it affects the heat dissipation rate in the cooling tower. The results can be considered as typical of pool-type research reactor. Further research could be done in the area of boiling heat transfer by using a simulated fuel element heated by electrical current (mock-up). The mock fuel element would eliminate the radiation hazard and allow further thermocouple instrumentation. By using a thermocouple near the fuel element surface, the surface temperature could be measured as a function of the heat flux.

It is suggested to repeat the experiments reported here, by placing a hollow cylinder over the core, with the same diameter of it, to verify the improvement of the mass flow rate by the chimney effect. These experiments can help the designers of the Brazilian research Multipurpose Reactor (RBM), which will be a pool reactor equipped with a chimney to improve the heat removal from the core (CDTN/CNEN, 2009).

6. Acknowledgment

This research project is supported by the following Brazilian institutions: Nuclear Technology Development Centre (CDTN), Brazilian Nuclear Energy Commission (CNEN), Research Support Foundation of the State of Minas Gerais (FAPEMIG), Brazilian Council for Scientific and Technological Development (CNPq) and Coordination for the Improvement of Higher Education Personnel (CAPES).

7. References

Bärs, B. & Vaurio, J. (1966). Power Increasing Experiments on a TRIGA Reactor. Technical University of Helsinki, Department of Technical Physics. Report No. 445. Otaniemi Filand.

Büke, T & Yavuz, H. (2000). Thermal-Hydraulic Analysis of the ITU TRIGA Mark-II Reactor. *Proceeding of 1st Eurasia Conference on Nuclear Science and its Application.* Izmir, Turquia, October 23-27, 2000.

CDTN/CNEN - Nuclear Technology Development Centre/Brazilian Nuclear Energy Commission. (2009). Brazilian Multipurpose Reactor (RMB), Preliminary Report of Reactor Engineering Group, General Characteristics and Reactors Reference". (in Portuguese).

Dalle, H.M., Pereira, C., Souza, R.M.G.P. (2002). Neutronic Calculation to the TRIGA IPR-R1 reactor using the WIMSD4 and CITATION codes. *Annals of Nuclear Energy*, Vol. 29, No. 8, (May 2002) , pp. 901–912, ISSN 0306-4549.

General Atomic. (1970). Safeguards Summary Report for the New York University TRIGA Mark I Reactor. (GA-9864). San Diego.

General Atomics. (June 2011). TRIGA. In: *TRIGA® Nuclear Reactors,* 07.06.2011. Available from: http://www.ga-esi.com/triga/.

Gulf General Atomic. (1972). 15" SST Fuel Element Assembly Instrumented Core. Drawing Number TOS210J220. San Diego, CA.

Haag, J.A. (1971). Thermal Analysis of the Pennsylvania State TRIGA Reactor. MSc Dissertation, Pennsylvania State University, Pennsylvania.

Huda, M.Q. & Rahman, M. (2004). Thermo-Hydrodynamic Design and Safety Parameter Studies of the TRIGA Mark II Research Reactor. *Annals of Nuclear Energy*, Vol. 31, (July 2004), pp.1102–1118, ISSN 0306-4549.

Huda, M.Q. et al. (2001). Thermal-hydraulic analysis of the 3 MW TRIGA Mark-II research reactor under steady-state and transient conditions. *Nuclear Tecnology*, Vol. 135, No. 1, (July 2001), pp. 51-66, ISSN 0029-5450.

Jensen, R.T. & Newell, D.L. (1998). Thermal Hydraulic Calculations to Support Increase in Operating Power in Mcclellen Nuclear Radiation Centre (MNRC) TRIGA Reactor. *Proceedings of RELAP5 International User´s Seminar*. College Station, Texas.

Lamarsh, J.R. & Baratta, A.J. (2001). *Introduction to Nuclear Engineering.* (3° ed.), Prendice Hall Inc., Upper Saddle River, ISBN 0-201-82498-1, New Jersey.

Marcum, W.R. (2008). Thermal Hydraulic Analysis of the Oregon State TRIGA® Reactor Using RELAP5-3D. MSc Dissertation. Oregon State University. Oregon.

Mesquita, A.Z., Costa, A.L., Pereira, C., Veloso, M.A.F. & Reis, P.A.L. (2011). Experimental Investigation of the Onset of Subcooled Nucleate Boiling in an Open-Pool Nuclear Research Reactor. *Journal of ASTM International,* Vol. 8, No. 6, (June 2011), pp. 12-20, 2011, ISSN 1546-962X.

Mesquita, A.Z. & Rezende, H. C. (2007). Experimental Determination of Heat Transfer Coefficients in Uranium Zirconium Hydride Fuel Rod. *International Journal of Nuclear Energy, Science and Technology*, Vol. 3, No. 2, (April 2007), pp. 170-179, ISSN 1741-6361.

Mesquita, A.Z. & Rezende, H.C. (2010). Thermal Methods for On-line Power Monitoring of the IPR-R1 TRIGA Reactor. *Progress in Nuclear Energy*, Vol. 52, (Abril 2010), pp. 268-272, ISSN: 0149-1970.

Mesquita, A.Z. & Souza, R.M.G.P. (2008). The Operational Parameter Electronic Database of the IPR-R1 TRIGA Research Reactor. *Proceedings of 4th World TRIGA Users Conference*, Lyon, September 8-9, 2008.

Mesquita, A.Z. & Souza, R.M.G.P. (2010). On-line Monitoring of the IPR-R1 TRIGA Reactor Neutronic Parameters. *Progress in Nuclear Energy*, Vol. 52, (Abril 2010), pp. 292-297, ISSN: 0149-1970.

Mesquita, A.Z. (2005). Experimental Investigation on Temperatures Distribuitions in the IPR-R1 TRIGA Nuclear Research Reactor, ScD Thesis, Universidade Estadual de Campinas, São Paulo. (in Portuguese).

Mesquita, A.Z., Rezende, H.C. & Tambourgi, E.B. (2007). Power Calibration of the TRIGA Mark I Nuclear Research Reactor, *Journal of the Brazilian Society of Mechanical Sciences*, Vol. 29, № 3, (July 2007), pp. 240-245, ISSN 1678-5878.

Özkul, E.H. & Durmayaz, A. (2000). A Parametric Thermal-Hydraulic Analysis of ITU TRIGA Mark II Reactor. *Proceedings of 16th European TRIGA Conference*, pp. 3.23-3.42, Institute for Nuclear Research, Pitesti, Romania.

Veloso, M.A. (2005). Thermal–Hydraulic Analysis of the IPR-R1 TRIGA Reactor in 250 kW, CDTN/CNEN, NI-EC3-05/05, Belo Horizonte, (in Portuguese).

Wagner, W & Kruse, A. (1998). Properties of Water and Steam – The Industrial Standard IAPWS-IF97 For The Thermodynamics Properties. Springer, Berlin

Rao, D.V., El-Genk, M.S., Rubio, R.A., Bryson, J.W., & Foushee, F.C. (1988). Thermal Hydraulics Model for Sandia's Annular Core Research Reactor, *Proceeding of Eleventh Biennial U.S. TRIGA Users's Conference*, pp. 4.89-4.113, Washington, USA, April 9-13, 1988.

2

Flow Instability in Material Testing Reactors

Salah El-Din El-Morshedy
Reactors Department, Nuclear Research Center, Atomic Energy Authority
Egypt

1. Introduction

Research reactors with power between 1 MW and 50 MW especially materials testing reactors (MTR), cooled and moderated by water at low pressures, are limited, from the thermal point of view, by the onset of flow instability phenomenon. The flow instability is characterized by a flow excursion, when the flow rate and the heat flux are relatively high; a small increase in heat flux in some cases causes a sudden large decrease in flow rate. The decrease in flow rate occurs in a non-recurrent manner leading to a burnout. The burnout heat flux occurring under unstable flow conditions is well below the burnout heat flux for the same channel under stable flow conditions. Therefore, for plate type fuel design purposes, the critical heat flux leads to the onset of the flow instability (OFI) may be more limiting than that of stable burnout. Besides, the phenomenon of two-phase flow instability is of interest in the design and operation of many industrial systems and equipments, such as steam generators, therefore, heat exchangers, thermo-siphons, boilers, refrigeration plants and some chemical processing systems. In particular, the investigation of flow instability is an important consideration in the design of nuclear reactors due to the possibility of flow excursion during postulated accident. OFI occurs when the slope of the channel demand pressure drop-flow rate curve becomes algebraically smaller than or equal to the slope of the loop supply pressure drop-flow rate curve. The typical demand pressure drop-flow rate curves for subcooled boiling of water are shown in Fig. 1 (IAEA-TECDOC-233, 1980). With channel power input S_2, operation at point d is stable, while operation at point b is unstable since a slight decrease in flow rate will cause a spontaneous shift to point a. For a given system, there is a channel power input S_c (Fig. 1) such that the demand curve is tangent to the supply curve. The conditions at the tangent point c correspond to the threshold conditions for the flow excursive instability. At this point any slight increase in power input or decrease in flow rate will cause the operating point to spontaneously shift from point c to point a, and the flow rate drops abruptly from M to M_c. For MTR reactors using plate-type fuel, each channel is surrounded by many channels in parallel. The supply characteristic with respect to flow perturbations in a channel (say, the peak power channel) is essentially horizontal, and independent of the pump characteristics. Thus, the criterion of zero slope of the channel demand pressure drop-flow curve is a good approximation for assessing OFI, i.e.

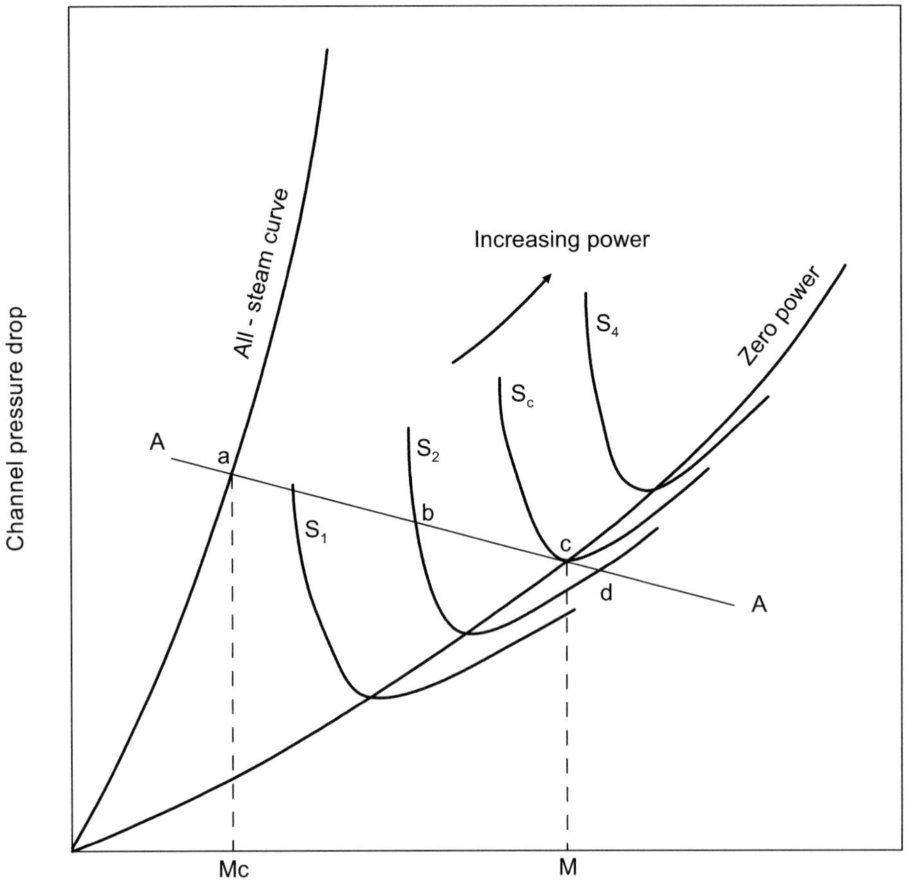

Fig. 1. Typical S-curves to illustrate OFI, (IAEA-TECDOC-233, 1980)

$$\frac{\partial(\Delta P)_{channel}}{\partial G} = 0 \tag{1}$$

Functionally, the channel pressure drop-flow curve depends on the channel geometry, inlet and exit resistances, flow direction, subcooled vapor void fraction, and heat flux distribution along the channel.

2. Background

There is a lot of research work in the literature related to flow instability phenomenon in two-phase flow systems. (Ledinegg, M., 1938) was the first successfully described the thermal-hydraulic instability phenomenon later named Ledinegg instability. It is the most common type of static oscillations and is associated with a sudden change in flow rate. (Whittle & Forgan, 1967) and (Dougherty et al., 1991) were performed an experimental investigations to obtain OFI data in a systematic methodology for various combination of operating conditions and geometrical considerations under subcooled flow boiling. (Saha et al., 1976) and (Saha & Zuber, 1976) carried out an experimental and analytical analysis on the onset of thermally induced two-phase flow oscillations in uniformly heated boiling channels. (Mishima & Nishihara, 1985) performed an experiment with water flowing in round tube at atmospheric pressure to study the critical heat flux, CHF due to flow instability, they found that, unstable-flow CHF was remarkably lower than stable-flow CHF and the lower boundary of unstable-flow CHF corresponds to the annular-flow boundary or flooding CHF. (Chatoorgoon, 1986) developed a simple code, called SPORTS for two-phase stability studies in which a novel method of solution of the finite difference equations was devised and incorporated. (Duffey & Hughes, 1990) developed a theoretical model for predicting OFI in vertical up flow and down flow of a boiling fluid under constant pressure drop, their model was based on momentum and energy balance equations with an algebraic modeling of two-phase velocity-slip effects. (Lee & Bankoff, 1993) developed a mechanistic model to predict the OFI in transient sub-cooled flow boiling. The model is based upon the influence on vapor bubble departure of the single-phase temperature. The model was then employed in a transient analysis of OFI for vertical down-wards turbulent flow to predict whether onset of flow instability takes place. (Chang & Chapman, 1996) performed flow experiments and analysis to determine the flow instability condition in a single thin vertical rectangular flow channel which represents one of the Advanced Test Reactor's (ATR) inner coolant channels between fuel plates. (Nair et al., 1996) carried out a stability analysis of a flow boiling two-phase low pressure and down flow relative to the occurrence of CHF, their results of analysis were useful in determining the region of stable operation for down flow in the Westinghouse Savannah River Site reactor and in avoiding the OFI and density wave oscillations. (Chang et al., 1996) derived a mechanistic CHF model and correlation for water based on flow excursion criterion and the simplified two-phase homogenous model. (Stelling et al., 1996) developed and evaluated a simple analytical model to predict OFI in vertical channels under down flow conditions, they found a parameter, the ratio between the surface heat flux and the heat flux required to achieve saturation at the channel exit for a given flow rate, is to be very accurate indicator of the minimum point velocity. (Kennedy et al., 2000) investigated experimentally OFI in uniformly heated micro channels with subcooled water flow using 22 cm tubular test sections, they generated demand curves and utilized for the specification of OFI points. (Babelli & Ishii, 2001) presented a procedure for predicting the OFI in down ward flows at low-pressure and low-flow conditions. (Hainoun & Schaffrath, 2001) developed a model permitting a description of the steam formation in the subcooled boiling regime and implemented it in ATHLET code to extend the code's range of application to simulate the subcooled flow instability in research reactors. (Li et al., 2004) presented a three dimensional two-fluid model to investigate the static flow instability in subcooled boiling flow at low-pressure. (Dilla et al., 2006) incorporated a model for low-pressure subcooled boiling flow into the safety reactor code RELAP5/Mod 3.2 to enhance the performance of the reactor code to predict the occurrence of the Ledinegg instability in two-phase flows. (Khater et al., 2007a, 2007b) developed a predictive model for OFI in MTR reactors and applied the model on ETRR-2 for both steady and transient states. (Hamidouche et al., 2009) developed a simple model based on steady-state equations adjusted with drift-flux correlations to determine OFI in research reactor conditions; they used RELAP/Mod 3 to draw the pressure drop characteristic curves and to establish the conditions of Ledinegg instability in a uniformly heated channel subject to constant outlet pressure. From the thermal-hydraulic point of view, the onset of significant void (OSV) leads to OFI phenomena and experimental evidence shows also that OSV is very close to OFI (Lee & Bankoff, 1993; Gehrke & Bankoff, 1993). Therefore, the prediction of OFI becomes the problem of predicting OSV. The first study that addressed the OSV issue was performed by (Griffith et al., 1958), they were the first to propose the idea that boiling in the channel could be divided into two distinct regions: a highly subcooled boiling region followed by a slightly subcooled region, they defined the OSV point as the location where the heat transfer coefficient was five times the single-phase heat transfer coefficient. A few years later, (Bowring, 1962) introduced the idea that OSV was related to the detachment of the bubbles from the heated surface and the beginning of the slightly subcooled region was fixed at the OSV point. (Saha & Zuber, 1974) developed an empirical model based on the argument that OSV occurs only when both thermal and hydrodynamic constraints are satisfied, where a general correlation is developed to determine OSV based on the Peclet and Stanton numbers. (Staub, 1968) postulated that OSV occurs when steam bubbles detach from the wall and assumed a simple force balance on a single bubble with buoyancy and wall shear stress acting on detach the bubble with surface tension force tending to hold it on the wall. He also postulated that the bubble could grow and detach only if the liquid temperature at the bubble tip was at least equal to the saturation temperature. (Unal, 1977) carried out a semi-empirical approach to determine and obtain a correlation of OSV point for subcooled water flow boiling. (Rogers et al., 1986; Chatoorgoon et al., 1992) developed a predictive model which relates the OSV to the location where the bubble first detaches assuming that bubble grow and collapse on the wall in the highly sub-cooled region. (Zeiton & Shoukri, 1996, 1997) used a high-speed video system to visualize the sub-cooled flow-boiling phenomenon to obtain a correlation for the mean bubble diameter as a function of the local subcooling, heat flux, and mass flux. (Qi Sun et al., 2003) performed a predictive model of the OSV for low flow sub-cooled boiling. The OSV established in their model meets both thermodynamic and hydrodynamic conditions. Several coefficients involved in the model were identified by Freon-12 experimental data.

It is clear that, there are several predictive models for OSV and OFI have been derived from theoretical and experimental analysis in the literature. However, their predictions in vertical thin rectangular channels still have relatively high deviation from the experimental data. Therefore, the objective of the present work is to develop a new empirical correlation with lower deviation from the experimental data in order to predict more accurately the OFI phenomenon as well as void fraction and pressure drop in MTR reactors under both steady and transient states.

3. Mathematical model

3.1 Correlation development

Experimental evidence shows that, the onset of significant voids, OSV is very close to the onset of flow instability, OFI (Lee & Bankoff, 1993; Gehrke & Bankoff, 1993). Therefore, the prediction of OFI in the present work becomes the problem of predicting OSV. Due to the complicated nature of the subcooled nucleate boiling phenomenon, it is often convenient to predict OSV by means of empirical correlations. In the present work, an empirical correlation to predict the onset of significant void is proposed takes into account almost all the related affecting parameters. The proposed correlation is represented best in terms of the following dimensionless groupings form:

$$\frac{\Delta T_{OSV}}{\Delta T_{sub,in}} = k_1 Bo^{k2} \Pr^{k3} \left(L/d_h\right)^{k4} \tag{2}$$

Where ΔT_{OSV} is the subcooling at OSV = $T_{sat} - T_{OSV}$

$\Delta T_{sub,in}$ is the inlet subcooling = $T_{sat} - T_{in}$ and

Bo is the boiling number = $\dfrac{\phi}{\rho_g U_g I_{fg}}$ where U_g is the rise velocity of the bubbles in the bubbly regime (Hari & Hassan, 2002)

$$U_g = 1.53 \left[\frac{\sigma g\left(\rho_f - \rho_g\right)}{\rho_f^{\,2}} \right]^{1/4} \tag{3}$$

By taking the logarithmic transformation of equation (2) and applying the least squares method, the constants k_1, k_2, k_3 and k_4 are evaluated as 1, 0.0094, 1.606 and -0.533 respectively. So the developed correlation takes the following form:

$$\frac{\Delta T_{OSV}}{\Delta T_{sub,in}} = Bo^{0.0094} \Pr^{1.606} \Big/ \left(L/d_h\right)^{0.533} \tag{4}$$

with all water physical properties calculated at the local bulk temperature. This correlation is valid for low pressures at heat flux ranges from 0.42 to 3.48 MW/m^2 and L/d_h ratios from 83 to 191.

3.2 Bubble detachment parameter

A parameter, η (the bubble detachment parameter) which indicates the flow stability is defined as follows (Bergisch Gladbach, 1992):

$$\eta = \frac{U \times \Delta T_{sub}}{\phi} \tag{5}$$

where U is the local velocity, ΔT_{sub} is the local subcooling and ϕ is the local heat flux. The physical meaning of η is that it controls the behavior of the steam bubbles formed at active sides of the heating surface. If η decreases below a certain value (η_{OFI}), the steam bubble will detach from the wall, otherwise it will stay there. In order to be sure of the maximum power channels are protected against the occurrence of excursive flow instability, the parameter η must be higher than η_{OFI} by a considerable safety margin. Based on the developed correlation, η_{OFI} can be determined by:

$$\eta_{OFI} = \frac{U \times \Delta T_{sub,in}}{\phi} \times Bo^{0.0094} \Pr^{1.606} \Big/ \left(L/d_h\right)^{0.533} \tag{6}$$

3.3 Void fraction modeling

The ability to predict accurately the void fraction in subcooled boiling is of considerable interest to nuclear reactor technology. Both the steady-state performance and the dynamic response of the reactor depend on the void fraction. Studies of the dynamic behavior of a two-phase flow have revealed that, the stability of the system depends to a great extent upon the power density and the void behavior in the subcooled boiling region. It is assumed that the void fraction in partially developed region between onset of nucleate boiling (ONB) and the OSV equal to 0 and in the fully devolved boiling region from the OSV up to saturation, the void fraction is estmated by the slip-ratio model as:

$$\alpha = 1 \Big/ \left[1 + \{(1-x)/x\} S \rho_g / \rho_f \right] \tag{7}$$

Where the slip, S is given by Ahmad, 1970 empirical relationship as:

$$S = \left(\frac{\rho_l}{\rho_g} \right)^{0.205} \left(\frac{G d_e}{\mu_l} \right) \tag{8}$$

The true vapor quality is calculated in terms of the thermodynamic equilibrium quality using empirical relationship from the earlier work of (Zuber et al., 1966; Kroeger & Zuber,1968) as:

$$x = \frac{x_{eq} - x_{eq,OSV} \exp\left(\frac{x_{eq}}{x_{eq,OSV}} - 1 \right)}{1 - x_{eq,OSV} \exp\left(\frac{x_{eq}}{x_{eq,OSV}} - 1 \right)} \tag{9}$$

Where the thermodynamic equilibrium quality, x_{eq} is given by:

$$x_{eq} = \frac{I_l - I_f}{I_{fg}} \tag{10}$$

and the thermodynamic equilibrium quality at OSV, $x_{eq,OSV}$ is given by:

$$x_{eq,OSV} = \frac{I_{l,OSV} - I_f}{I_{fg}} \tag{11}$$

3.4 Pressure drop modeling

Pressure drop may be the most important consideration in designing heat removal systems utilizing high heat flux subcooled boiling such as nuclear reactors. The conditions in which the pressure drop begins to increase during the transient from forced convection heat transfer to subcooled flow boiling are related to the OSV. The pressure drop is a summation of three terms namely; friction, acceleration and gravity terms.

3.4.1 Pressure drop in single-phase liquid

The pressure drop terms for single-phase liquid regime are given by:

$$\Delta P\big|_{friction} = \frac{2\, f\, G^2\, \Delta z}{\rho_l\, d_e} \tag{12}$$

where f is the Darcy friction factor for single-phase liquid. It is calculated for rectangular channels as:

for laminar flow (White, 1991)

$$f = 12 \Big/ \frac{G\, d}{\mu_l} \tag{13}$$

for turbulent flow (White, 1991)

$$\frac{1}{f^{1/2}} = 2.0 \log\left(\mathrm{Re}\, f^{1/2}\right) - 1.19 \tag{14}$$

$$\Delta P\big|_{acceleration} = (1/\rho_l - 1/\rho_{li})G^2 \tag{15}$$

$$\Delta P\big|_{gravity} = \mp \rho_l\, g\, \Delta z \tag{16}$$

3.4.2 Pressure drop in subcooled boiling

The pressure drop terms for subcooled boiling regime are given by:

$$\Delta P\big|_{friction} = \frac{f\, G^2}{2\, \rho_l\, d_e} \int_0^z \phi^2(z)\, dz \tag{17}$$

where $\phi^2(z)$ is the two-phase friction multiplier and is obtained from (Levy, 1960) correlation as:

$$\phi^2(z) = \left[\frac{1 - x(z)}{1 - \alpha(z)}\right]^{2-m} \tag{18}$$

where m is 0.25 as suggested by (Lahey & Moody, 1979)

$$\Delta P\big|_{acceleration} = G^2 \int_0^z \frac{d}{dz}\left[\frac{(1-x)^2}{(1-\alpha)\rho_l} + \frac{x^2}{\alpha_g\, \rho_g}\right] dz \tag{19}$$

$$\Delta P\big|_{gravity} = g \int_0^z \left[\alpha\, \rho_g + (1-\alpha)\rho_l\right] dz \tag{20}$$

3.5 Prediction of OFI during transients

In order to apply the present correlation on transient analysis, both the momentum and energy equations are solved by finite difference scheme to obtain the velocity variation and temperature distribution during transient. The conservation of momentum for unsteady flow through a vertical rectangular channel of length L and gap thickness *d* and heated from both sides is:

$$\rho \frac{dU}{d\tau} = \frac{dP}{dz}(\tau) - \frac{\tau_w}{d} \tag{21}$$

with the initial condition $U = U_0$ *at* $\tau = 0$.

where the wall shear stress, is defined by:

$$\tau_w = \frac{f \rho U^2}{8} \tag{22}$$

and the friction factor, *f* is given by Blasius equation as:

$$f = 0.316\,\mathrm{Re}^{-0.25} \tag{23}$$

The conservation of energy for unsteady state one-dimensional flow is:

$$\rho C_P \left(\frac{\partial T}{\partial \tau} + U(\tau)\frac{\partial T}{\partial z} \right) = \frac{\phi(t)}{d} \tag{24}$$

with the boundary condition $T = T_i$ *at* $z = 0$ and Initial condition $T = T_0(z)$ *at* $\tau = 0$.

The initial steady-state coolant temperature distribution is calculated from a simple heat balance up to the distance z from the channel inlet taking into account that, the channel is heated from both sides.

- for uniform heat flux distribution:

$$T_0(z) = T_{in} + \frac{\phi\, z}{G\, C_p\, d} \tag{25}$$

- for chopped cosine heat flux distribution:

$$T_0(z) = T_{in} + \frac{2 W_h L_p \phi_0}{\pi G C p W d} \times \left[\sin \frac{\pi (z - L/2)}{L_p} + \sin \frac{\pi L}{2 L_p} \right] \tag{26}$$

where the axial heat flux distribution is given by:

$$\phi(z) = \phi_0 \cos\left(\frac{\pi (z - L/2)}{L_p} \right) \tag{27}$$

Where:

Lp: is the extrapolated length, $L_P = L + 2e$,

e : is the extrapolated distance and

ϕ_0 : is the maximum axial heat flux in the channel, $\phi_0 = \bar{\phi} \times PPF$

Where . $\bar{\phi}$ is the average surface heat flux and *PPF* is the power peaking factor.

The coolant temperature distribution during transient resulted from the solution of equation (24) by finite difference method is:

$$T_j^{p+1} = \frac{K_1 \times T_{j-1}^{p+1} + T_j^p + K_2}{1 + K_1} \tag{28}$$

$$\text{where } K_1 = U^{p+1} \times \frac{\Delta\tau}{\Delta z} \text{ and}$$

- for uniform heat flux distribution:

$$K_2 = \frac{2\phi^p \Delta\tau}{\rho C_p d} \tag{29}$$

- for chopped cosine heat flux distribution:

$$K_2 = \frac{2\phi_0^p \Delta\tau L_p}{\pi \rho C_p d \Delta z} \times \left[\sin\left(\frac{\pi\left(z_j - L/2\right)}{L_p} \right) - \sin\left(\frac{\pi\left(z_{j-1} - L/2\right)}{L_p} \right) \right] \tag{30}$$

4. Results and discussion

4.1 Assessment of the developed correlation

The subcooling at OSV is evaluated by the present correlation and the previous correlations described in table 1 for (Whittle & Forgan, 1967) experiments. All the results and experimental data are plotted in Fig. 2. The solid line is a reference with the slope of one is drawn on the plot to give the relation between the predicted and measured data. The present correlation shows a good agreement with the experimental data, it gives only 6.6 % relative standard deviation from the experimental data while the others gives 20.2 %, 26.4 %, 27.4 % and 35.0 % for Khater et al., Lee & Bankoff, Sun et al. and Saha & Zuber correlations respectively as shown table 1.

The experimental data of (Whittle & Forgan, 1967) on light water cover the following operating conditions:

- Rectangular channel with hydraulic diameter from 2.6 to 6.4 mm.
- Pressure from 1.10 to 1.7 bar.
- Heat flux from 0.66 to 3.4 MW/m^2.
- Inlet temperature from 35 to 75°C.
- Velocity from 0.6096 to 9.144 m/s.

Correlation	Description
Khater et al., 2007	$\Delta T_{OSV} = \dfrac{\sqrt{1+16\left(\dfrac{\rho_f}{\rho_g}\right)\left(\dfrac{C_P\phi}{0.172\,h_i I_{fg}}\right)d_h} - 1}{2\left(\dfrac{\rho_f}{\rho_g}\right)\left(\dfrac{C_P}{I_{fg}}\right)}$
Lee & Bankoff, 1993	Approximated by: $St = 0.076 Pe^{-0.2}$
Sun et al., 2003	$\Delta T_{OSV} = \dfrac{\sqrt{1+16\left(\dfrac{\rho_f}{\rho_g}\right)\left(\dfrac{C_P\phi}{A_i h_i d_h I_{fg}}\right)} - 1}{2\left(\dfrac{\rho_f}{\rho_g}\right)\left(\dfrac{C_P}{I_{fg}}\right)}$ with $h_i = C1\dfrac{k}{d_b}\mathrm{Re}_{fg}^{C2}\,\mathrm{Pr}^{C3}\left(\dfrac{\rho_f}{\rho_g}\right)^{C4}$
Saha & Zuber, 1976	$Nu = \dfrac{\phi d_h}{k\Delta T_{OSV}} = 455$ for $Pe \le 70000$ $St = \dfrac{\phi}{GC_P\Delta T_{OSV}} = 0.0065$ for $Pe > 70000$

Table 1. Previous correlations used in comparison

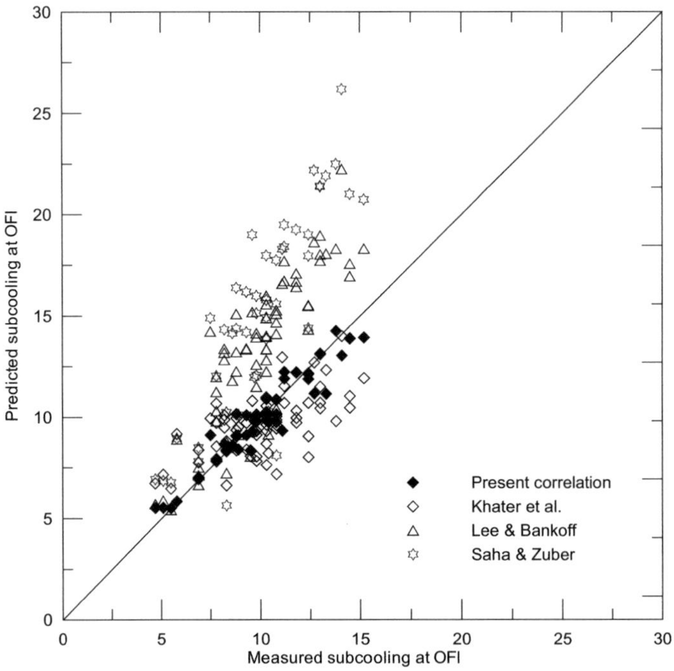

Fig. 2. Comparison of the present correlation with previous models

Correlation	Relative standard deviation
Present correlation	0.066
Khater et al.	0.202
Lee & Bankoff	0.264
Sun et al	0.274
Saha & Zuber	0.350

Table 2. Relative standard deviation from experimental data for subcooling at OSV

4.2 Prediction of S-curves

The pressure drop for Whittle & Forgan experimental conditions is determined and depicted in Figs 3 and 4 against the experimental data. The present model predicts the S-curves with a good agreement achieved with the experimental data. A well defined minimum occurred in all the S-curves. The change in slope from positive to negative was always abrupt and the pressure drop at the condition of the minimum was always approximately equal to that for zero-power condition. As subcooled liquid heat ups along the wall of a heated channel, its viscosity decreases. Increasing the wall heat flux causes further reduction in liquid viscosity. Therefore, pressure drop associated with pure liquid flow decreases with increasing wall heat flux. The trend changes significantly when bubbles begins to form. Here, increasing wall heat flux increases both the two-phase frictional and accelerational gradients of pressure drop. Pressure drop therefore begins to increase with increasing heat flux.

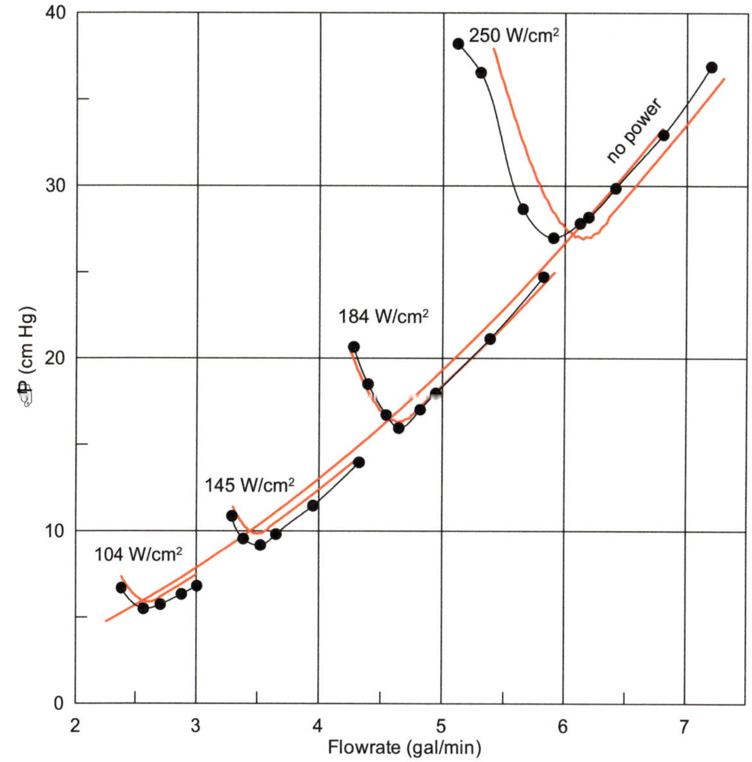

Fig. 3. S-curves prediction for (Whittle & Forgan, 1967) experiments (No. 1 test section)

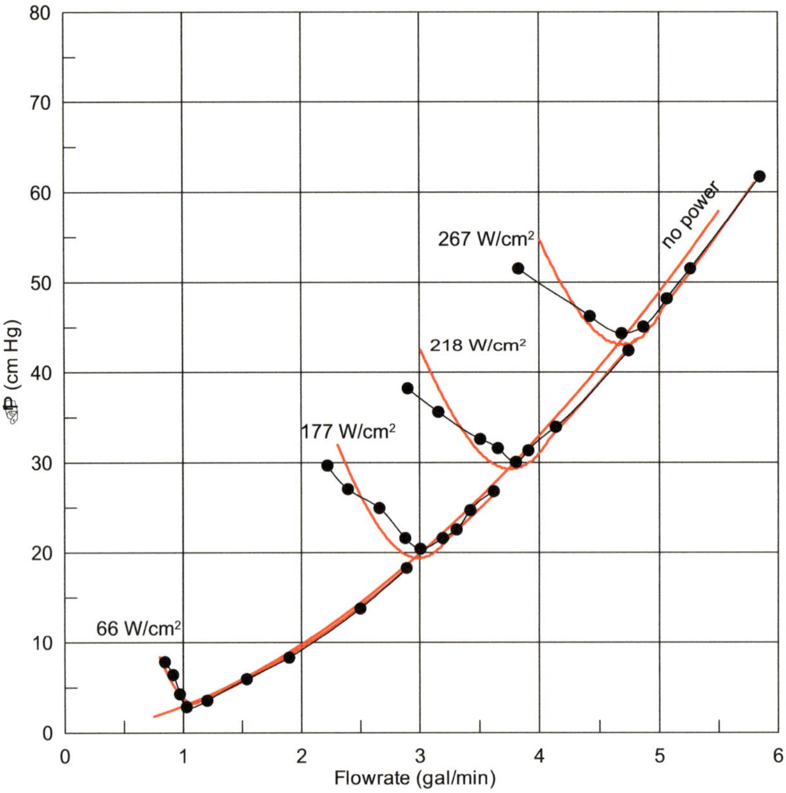

Fig. 4. S-curves prediction for (Whittle & Forgan, 1967) experiments (No. 3 test section)

4.3 Prediction of OFI during transients

The present model is used to predict the OFI phenomenon for the IAEA 10 MW MTR generic reactor (Matos et al., 1992) under loss of flow transient. The reactor active core geometry is 5×6 positions where both standard and control fuel elements are placed with a total of 551 fuel plates. A summary of the key features of the IAEA generic 10 MW reactor with LEU fuel are shown in Table 3 (IAEA-TECDOC-233, 1980). The pump coast-down is initiated at a power of 12 MW with nominal flow rate of 1000 m^3/h and reduced as $e^{\tau/T}$, with T = 1 and 25 seconds for fast and slow loss-of-flow transients respectively. The reactor is shutting down with Scram at 85 % of the normal flow. The pressure gradient is proportional to mass flux to the power 2. Therefore, the pressure gradient during transient is considered exponential and reduced as $e^{2\tau/T}$, with T = 1 and 25 seconds for fast and slow loss-of-flow transients respectively with steady-state pressure gradient, $\left.\frac{dP}{dz}\right|_{\tau=0.0}$ = 40.0. The calculation is performed on the hot channel where the axial heat flux is considered chopped cosine distribution of a total power peaking factor equal to 2.52 with the extrapolated length equal to 8.0 cm.

Coolant	Light Water
Coolant flow direction	Downward
Fuel thermal conductivity (W/cm K)	1.58
Cladding thermal conductivity (W/cm K)	1.80
Fuel specific heat (J/g K)	0.728
Cladding specific heat (J/g K)	0.892
Fuel density (g/cm^3)	0.68
Cladding density (g/cm^3)	2.7
Radial peaking factor	1.4
Axial peaking factor	1.5
Engineering peaking factor	1.2
Inlet coolant temperature	38.0
Operating pressure (bar)	1.7
Length (cm)	8.0
Width (cm)	7.6
Height (cm)	60.0
Number of fuel elements SFE/SCE	21/4
Number of plates SFE/SCE	23/17
Plate meat thickness (mm)	0.51
Width (cm) active/total	6.3/6.65
Height (cm)	60.0
Water channel thickness (mm)	2.23
Plate clad thickness (mm)	0.38

Table 3. IAEA 10 MW generic reactor specifications

Figures 5, 6, and 7 show the OFI locus on graphs of the flow velocity, the exit bulk temperature and the bubble detachment parameter as a function of time for fast loss-of-flow transient. The pressure gradient reduced exponentially from 40 kPa/m as $e^{2\tau}$, while the average heat flux is maintained at a constant value. The transient time is 0.16 second which represents the period from steady-state to the time of 85% of the normal flow (just before Scram). The flow velocity decreases, the bulk temperature increases, and the bubble detachment parameter decreases. Figure 5 shows slight changes of the velocity variation depending on the magnitude of the heat added from both plates. In this figure OFI is reached at end of each initial heat flux curve. Figure 6 shows that, OFI is always predicted at exit bulk temperature greater than 104°C while, Fig. 7 shows that, OFI phenomenon is always predicted at bubble detachment parameter value lower than 22. In case of slow loss-of-flow transient, the pressure gradient reduced exponentially from 40 kPa/m as $e^{-0.08\tau}$, the transient time is 4.0 seconds which represents the period from steady-state to the time just before Scram at 85% of the normal flow.

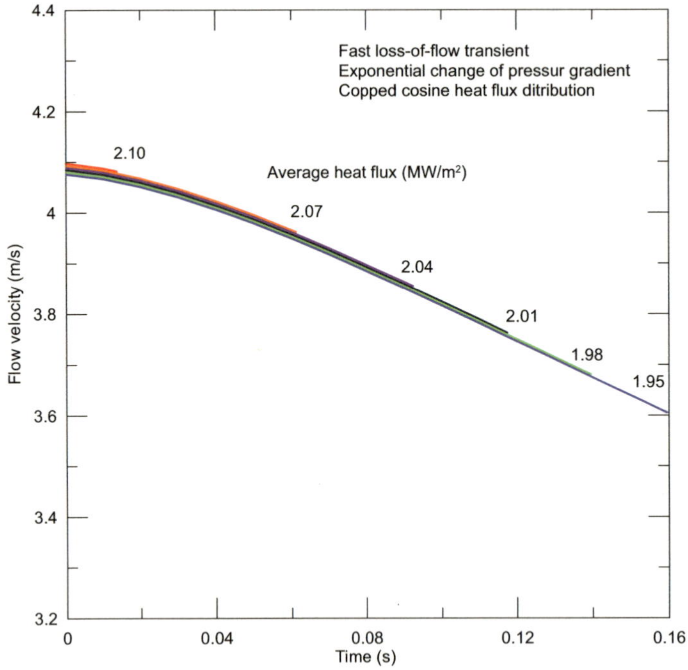

Fig. 5. Flow velocity variations for various heat fluxes under fast loss-of-flow transient, OFI reached at the end of each curve

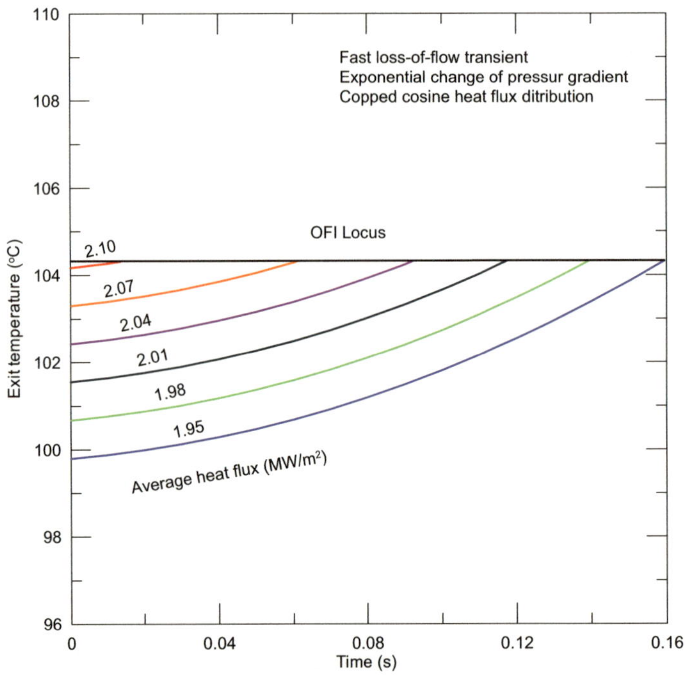

Fig. 6. Exit bulk temperature variations for various heat fluxes under fast loss-of-flow transient

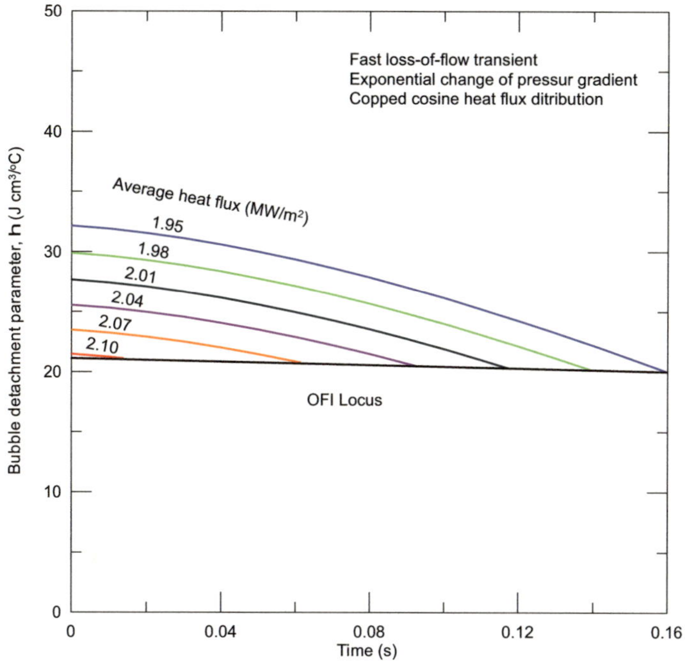

Fig. 7. Bubble detachment parameter variations for various heat fluxes under fast loss-of-flow transient

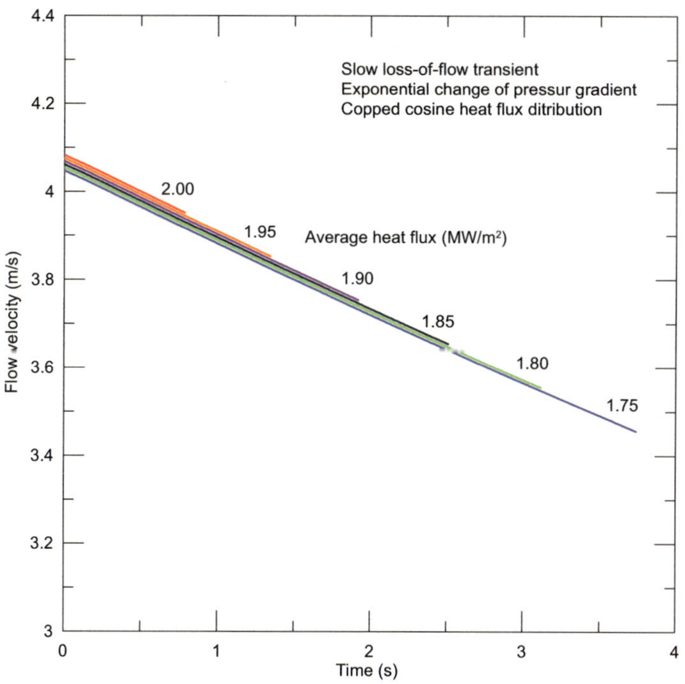

Fig. 8. Flow velocity variations for various heat fluxes under slow-of-flow transient, OFI reached at the end of each curve

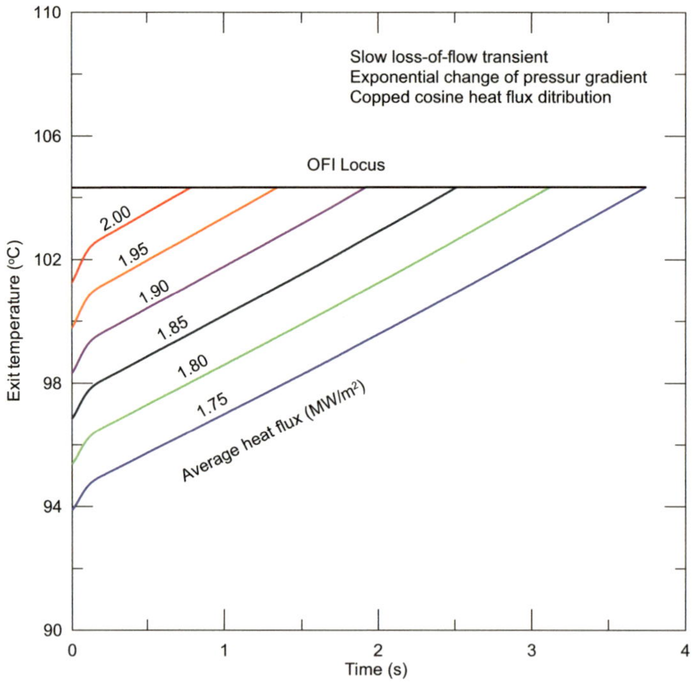

Fig. 9. Exit bulk temperature variations for various heat fluxes under slow-of-flow transient

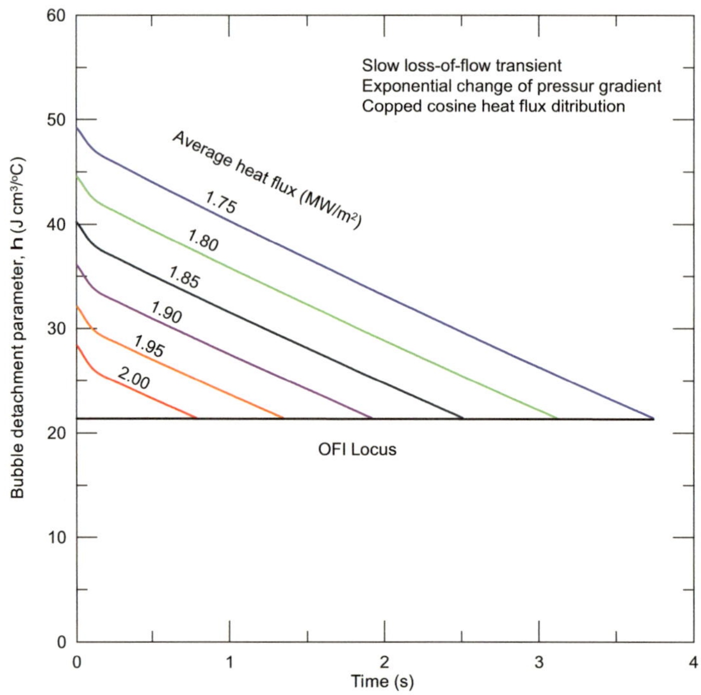

Fig. 10. Bubble detachment parameter variations for various heat fluxes under slow-of-flow transient

Figures 8, 9, and 10 show the OFI locus on graphs of the flow velocity, the exit bulk temperature and the bubble detachment parameter as a function of time for slow-of-flow transient. The graphs trends are same as for fast loss-of-flow-transient except that, OFI phenomenon could predicted at lower heat fluxes. Figures 9 and 10 show that, OFI phenomenon is always predicted at exit bulk temperature greater than 104°C and bubble detachment parameter value lower than 22 (the same values obtained for fast loss-of-flow-transient).

4.3 Safety margins evaluation

The safety margin for OFI phenomenon is defined as the ratio between the power to attain the OFI phenomenon within the core channel, and the hot channel power, this means that, OFI margin is equal to the ratio of the minimum average heat flux leads to OFI in the core channels and the average heat flux in the hot channel. It is found that, the OFI phenomenon occurs at an average heat flux of 2.1048 MW/m^2 for steady-state operation ($\tau = 0.0s$), and 1.7294 MW/m^2 just before Scram ($\tau = 4.0s$). Thus, these values can be regarded as the maximum possible heat fluxes to avoid OFI under steady-state operation and just before Scram respectively. The maximum hot channel heat flux is determined using the data of table 3 as 0.72595 MW/m^2 with an average value of 0.5648 MW/m^2. This means that, the reactor has vast safety margins for OFI phenomenon of 3.73 for steady-state operation, 3.45 and 3.06 just before Scram for both fast and low loss-of-flow transient respectively. Table 4 gives the estimated heat flux leading to OFI and the safety margin values for both the steady and transient states.

Description	Steady-state τ = 0.0 s	Transient τ = 0.16 s	Transient τ = 4.0 s
OFI heat flux (MW/m^2)	2.1048	1.9491	1.7294
Safety margin for OFI	3.73	3.45	3.06

Table 4. Reactor safety margins for OFI phenomenon.

5. Conclusion

Flow instability is an important consideration in the design of nuclear reactors due to the possibility of flow excursion during postulated accident. In MTR, the safety criteria will be determined for the maximum allowable power and the subsequent analysis will therefore restrict to the calculations of the flow instability margin. In the present work, a new empirical correlation to predict the subcooling at the onset of flow instability in vertical narrow rectangular channels simulating coolant channels of MTR was developed. The developed correlation involves almost all parameters affecting the phenomenon in a dimensionless form and the coefficients involved in the correlation are identified by the experimental data of Whittle and Forgan that covers the wide range of MTR operating conditions. The correlation predictions for subcooling at OSV were compared with predictions of some previous correlations where the present correlation gives much better agreement with the experimental data of Whittle and Forgan with relative standard deviation of only 6.6%. The bubble detachment parameter was also estimated based on the present correlation. The present correlation was then utilized in a model predicting the void fraction and pressure drop in subcooled boiling under low pressure. The pressure drop model predicted the S-curves representing the two-phase instability of Whittle and Forgan with good accuracy. The present correlation was also incorporated in the safety analysis of the IAEA 10 MW MTR generic reactor in order to predict the OFI phenomenon under both fast and slow loss-of-flow transient. The OFI locus for the reactor coolant channels was predicted and plotted against flow velocity, exit temperature and bubble detachment parameter for various heat flux values. It was found that the reactor has vast safety margins for OFI phenomenon under both steady and transient states.

6. Nomenclature

Cp : specific heat, J/kg°C
d : gap thickness, m
db : bubble diameter, m
dh : heated diameter, m
de : hydraulic diameter, m
g : acceleration of gravity, m/s^2
G : mass flux, kg/ms^2
I : enthalpy, J/kg
Ifg : latent heat of vaporization, J/kg
k : thermal conductivity, W/m°C
L : active length, m
Nu : Nusselt number, = $h d_e/k$
P : pressure, Pa
Pe : Peclet number, = $\mathrm{Re}\,\mathrm{Pr}$
Pr : Prantdel number, = $\mu Cp/k$
Re : Reynolds number = $G d_e/\mu$
St : Stanton number, = Nu/Pe
T : temperature, °C
U : coolant velocity, m/s
W : channel width, m
x : steam quality
z : distance in axial direction, m

Greek Letters

α : void fraction, dimensionless

ΔP : pressure drop, Pa
ϕ : heat flux W/m^2
μ : dynamic viscosity, kg/m s
ρ : density, kg/m^3
σ : surface tension, N/m
τ : time, s
τ_w : wall shear stress, N/m^2

Subscripts

f : liquid phase,
fg : difference of liquid and vapor,
g : vapor phase,
h : heated
in : inlet
OFI : onset of flow instability,
OSV : onset of significant void,
s : saturation,
w : wall.

7. References

Ahmad, S. Y. (1970). Axial distribution of bulk temperature and void fraction in a heater channel with inlet subcooling, Journal of Heat Transfer, Vol. 92, pp. 595-609.

Babelli, I. & Ishii, M. (2001). Flow excursion instability in downward flow systems Part I. Single-phase instability, Nuclear Engineering and Design, Vol. 206 pp. 91-96.

Bergisch Gladbach (April 1992). Safety Analyses for the IAEA Generic 10 MW Reactor, IAEA-TECDOC-643, Vol. 2, Appendix A.

Bowering, R. W. (1962). Physical model based on bubble detachment and calculation of steam voidage in the subcooled region of a heated channel", HPR-10, Institute for Atomenergi, Halden, Norway.

Chang, H. OH & Chapman, J. C. (1996). Two-Phase Flow Insatiability for Low-Flow Boiling in Vertical Uniformly Heated Thin Rectangular Channels, Nuclear Technology, Vol. 113, pp.327-337.

Chatoorgooon, V.; Dimmick, G. R.; Carver, M. B.; Selander, W. N. & Shoukri, M. (1992). Application of Generation and Condensation Models to Predict Subcooled Boiling Void at Low Pressures", Nuclear Technology, Vol. 98, pp.366-378.

Dilla, E. M.; Yeoh, G. H. & Tu, J. Y. (2006). Flow instability prediction in low-pressure subcooled boiling flows using computational fluid dynamics code, ANZIAM Jornal, Vol. 46, pp. C1336-C1352.

Doughherty, T.; Fighetti, C.; McAssey, E.; Reddy, G.; Yang, B.; Chen, K. & Qureshi, Z. (1991). Flow Instability in Vertical Channels, ASME HTD-Vol. 159, pp. 177-186.

Duffey, R. B. & Hughes, E. D. (1990). Static flow instability onset in tubes, channels, annuli and rod bundles, International Journal of Heat and Mass Transfer, Vol. 34, No. 10, pp. 1483-2496.

Gehrke, V. & Bankoff, S. G. (June 1993). Stability of Forced Convection Sub-cooled Boiling in Steady-State and Transient Annular Flow, NRTSC, WSRC-TR-93-406.

Griffith, P.; Clark, J. A. & Rohsenow, W. M. (1958). Void volumes in subcooled boiling, ASME Paper 58-HT-19, U.S. national heat transfer conference, Chicago.

Hainoun, A. & Schaffrath (2001). Simulation of subcooled flow instability for high flux research reactors using the extended code ATHLET, Nuclear Engineering and Design, Vol. 207, pp. 163-180.

Hamidouche, T.; Rassoul, N., Si-Ahmed, E.; El-Hadjen, H. & Bousbia, A. (2009). Simplified numerical model for defining Ledinegg flow instability margins in MTR research reactor, Progress in Nuclear Energy, Vol. 51, pp. 485-495.

IAEA-TECDOC-233 (1980). IAEA Research Reactor Core Conversion from the use of high-enriched uranium to the use of low enriched uranium fuels Guidebook.

Kaichiro Mishima & Hiedeaki Nishihara (1985). Boiling burnout and flow instabilities for water flowing in a round tube under atmospheric pressure, International Journal of Heat and Mass Transfer, Vol. 28, No. 6, pp. 1115-1129.

Kennedy, J. E.; Roach, G. M.; Dowling, M. F.; Abdel-Kalik, S. I.; Ghiaasiaan, S. M.; Jeter, S. M. & Quershi, Z. H. (2000). The onset of flow instability in uniformly heated horizontal microchannels, ASME Journal of Heat Transfer, Vol. 122, pp. 118-125.

Khater, H. A.; El-Morshedy, S. E. & Ibrahim, M. (2007). Thermal-Hydraulic Modeling of the Onset of Flow Instability in MTR Reactors, Annals of Nuclear Energy, vol. 34, issue 3, pp. 194-200.

Khater, H. A.; El-Morshedy, S. E. & Ibrahim, M. (2007). Prediction of the Onset of Flow Instability in ETRR-2 Research Reactor under Loss of Flow Accident, KERNTECHNIK, Carl Hanser Verlag, vol. 72, issue 1-2, pp. 53-58.

Kroeger, P. G. & Zuber, N. (1968). An analysis of the effects of various parameters on the average void fraction in subcooled boiling, International Journal of Heat Transfer, Vol. 11, pp. 211-233.

Lahey, R. T. & Moody, Jr. (1979). The thermal-hydraulics of a boiling water nuclear reactor, American Nuclear Society, La Grange Park, Illinois.

Ledinegg, M. (1938). Instability of flow during natural and forced circulation, Die Wärme, Vol. 61 (8), pp. 891-898.

Lee, S. C. & Bankoff, S. G. (1993). Prediction of the Onset of Flow Instability in Transient Sub-cooled Flow Boiling, Nuclear Eng. and Design, Vol. 139, pp. 149-159.

Levy, S., M. (1960). Steam slip-theoretical prediction from momentum model, Journal of Heat and Mass Transfer, Vol. 82, p. 113.

Li, Y., Yeoh; G. H. & Tu, J. Y. (2004). Numerical investigation of static flow instability in a low-pressure subcooled boiling channel, Heat and Mass Transfer, Vol. 40, pp. 355-364.

Matos, J. E.; Pennington, E. M.; Freese, K. E. & Woodruff, W. L. (April 1992). Safety-Related Benchmark Calculations for MTR-Type Reactors with HEU, MEU and LEU Fuels, IAEA-TECDOC-643, Vol. 3, Appendix G.

Nair, S.; Lele, S.; Ishii, M. & Revankar, S. T. (1996). Analysis of flow instabilities and their role on critical heat flux for two-phase down-flow and low pressure systems, International Journal of Heat and Mass Transfer, Vol. 39, No. 1, pp. 39-48.

Qi Sun ; Yang, R. & Zhao, H. (2003). Predictive Study of the Incipient of Net Vapor Generation in Low Flow Sub-cooled Boiling, Nuclear Eng. and Design, Vol. 225, pp.294-256.

Rogers,J. T.; Salcudean, M.; Abdullah, Z.; McLead, D. & Poirier, D. (1986). The Onset of Significant Void in Up-Flow Boiling of Water at Low Pressure and Velocities", Int. J. Heat and Mass Transfer, Vol. 30, pp. 2247-2260.

Saha, P.; Ishii, M. & Zuber, N. (1976). An Experimental Investigation of the Thermally Induced Flow Oscillations in Two-Phase Systems, ASME Journal of Heat Transfer, Trans. ASME, Vol. 98, pp. 616-622.

Saha, P. & Zuber, N. (1976). An Analytical Study of the Thermally Induced Two-Phase Flow Instabilities Including the Effects of Thermally Non-Equilibrium, International Journal of. Heat and Mass Transfer, Vol. 21, pp. 415-426.

Saha, P. & Zuber, N. (1974). Point of net vapor generation and vapor void fraction in subcooled boiling, Proceeding of the 5th international heat transfer conference, Vol. 4, pp. 175-179, Tokyo, Japan.

Soon Heung; Yun Il Kim & Won-Pil Beak (1996). Derivation of mechanistic critical heat flux model for water based on flow instabilities, International communications in Heat and Mass Transfer, Vol. 23, No. 8, pp. 1109-1119.

Sridhar Hari and Yassin A. Hassan (2002). Improvement of the subcooled boiling model for low-pressure conditions in thermal-hydraulic codes, Nuclear Engineering and Design, Vol. 216, pp. 139-152.

Staub, F. W. (1968). The Void Fraction in Sub-Cooled Boiling-Prediction of the Initial Point of Net Vapor Generation, J. Heat Transfer, Trans. ASME, Vol. 90, pp. 151-157.

Stelling, R.; McAssey, E. V.; Dougherty, T. & Yang, B. W. (1996). The onset of flow instability for downward flow in vertical channels, ASME Journal of Heat Transfer, Vol. 118, pp. 709-714.

Unal, H. C. (1977). Void Fraction and Incipient Point of Boiling During the Subcooled Nucleate Flow Boiling of Water, International Journal of Heat and Mass Transfer, Vol.20, pp. 409-419.

Vijay Chatoorgoon (1986). SPORTS A simple non linear thermal hydraulic stability code, Nuclear Engineering and Design, Vol. 93, pp. 51-67.

White, F. M. (1974), Viscous Fluid Flow, Copyright © 1991, McGraw-Hill, Inc.

Whittle, R. H. & Forgan, R. (1967). A Correlation for the Minima in the Pressure Drop Versus Flow-Rate Curves for Sub-cooled Water Flowing in Narrow Heated Channels, Nuclear Engineering and Design, Vol. 6, pp. 89-99.

Zeiton, O.& Shoukri, M. (1996). Bubble Behavior and Mean Diameter in Sub-cooled Flow Boiling, J. Heat Transfer, Trans. ASME, Vol. 118, pp. 110-116.

Zeiton, O. & Shoukri, M. (1997). Axial Void Fraction Profile in Low Pressure Sub-cooled Flow Boiling, J. Heat Transfer, Trans. ASME, Vol. 40, pp. 869-879.

Zuber, N.; Staub, F. W. & Bijwaard, G. (1966). Vapour void fraction in subcooled boiling systems, Proceeding of the third International heat transfer conference, Vol. 5, pp. 24-38, Chicago.

3

Herium-Air Exchange Flow Rate Measurement Through a Narrow Flow Path

Motoo Fumizawa
Shonan Institute of Technology
Japan

1. Introduction

Buoyancy-driven exchange flows of helium-air were investigated through horizontal and inclined small openings. Exchange flows may occur following a window opening as ventilation, fire in the room, over the escalator in the underground shopping center as well as a pipe rupture accident in a modular high temperature gas-cooled nuclear reactor. Fuel loading pipe is located in the inclined position in the pebble bed reactor such as Modular reactor (Fumizawa, 2005, Kiso, 1999) and AVR(El-Wakil, 1982, Juni-1965, 1965).

In safety studies of High Temperature Gas-Cooled Reactor (HTGR), a failure of a standpipe at the top of the reactor vessel or a fuel loading pipe may be one of the most critical design-base accidents. Once the pipe rupture accident occurs, helium blows up through the breach immediately. After the pressure between the inside and outside of he pressure vessel has balanced, helium flows upward and air flows downward through he breach into the pressure vessel. This means that buoyancy-driven exchange flow occurs through the breach, caused by density difference of the gases in the unstably stratified field. Since an air stream corrodes graphite structures in the reactor, it is important to evaluate and reduce the air ingress flow rate during the standpipe rupture accident.

Some studies have been performed so far on the exchange flow of two fluids with different densities through vertical and inclined short tubes. Epstein(Epstein, 1988) experimentally and theoretically studied the exchange flow of water and brine through the various vertical tubes. Mercer et al. (Mercer, 1975) experimentally studied an exchange flow through inclined tubes with water and brine. He performed the experiment in the range of $3.5 < L/D < 18$ and $0 \text{ deg} < \theta < 90 \text{ deg}$, and pointed out that the length-to-diameter ratio L/D, and the inclination angle $\theta$of the tube are the important parameter for the exchange flow rate. Most of these studies were performed on the exchange flow with a relatively small difference of the densities of the two fluids (up to 10 per cent). However, in the case of HTGR standpipe rupture accident, the density of the outside gas is at least three times larger than that of the gas inside the pressure vessel. Few studies have been performed so far in such a large range of density difference. Kang et al. (Kang, 1992) studied experimentally the exchange flow through a round tube with a partition plate. Although we may think that the partitioned plate, a kind of obstacle in the tube, decrease the exchange flow rate, he found that the exchange flow rate is increased by the partition plate because of separation of an upward and downward flow.

The objectives of the present study are to investigate the behavior of the exchange flow, i.e., exchange flow through the round long tube by several flow visualization method, then to evaluate the exchange flow rate by the PTV and PIV methods and mass increment with helium-air system. Therefore the following methods are investigated in the present study.

1. Smoke wire method
2. The optical system of the Mach-Zehnder interferometer
3. The method of the mass increment

2. Smoke wire method

2.1 Experimental apparatus and procedure

The smoke wire method was used for the present investigations. Figure 1 shows a typical sketch of the apparatus. It consists of a smoke pulse generator, thin Nichrome wire with oil and a test chamber. This figure also shows the high-speed camera system, and it transfers the visual digital data to the personal computer for data acquisition. The experimental procedure is as follows. The test chamber is filled with pure helium. By removing the cover plate placed on the top of the tube, exchange flow, i.e., exchange flow of helium and air is initiated. At such condition, the smoke pulse generator ignites the high voltage. Immediately a smoke appeared and visualized the helium up flow and the air down flow in the flow path in the long tube. Test chamber diameter and height are 350 mm, the long tube on the test chamber diameter(=D) are 17.4 and 20mm, length of it (=L) are 200 and 319mm. They denote L/D=10 and L/D=18.3 respectively. It simulates a typical long tube. The inclination angle θis 30 deg. The smoke wire conditions are as follows. The voltage is around 250 (V), current duration is 30 m sec, and the oil of thin wire is CRC-556. The high-speed camera system using D-file records the visual data up to 1600 flames in a second. Upward flows peak velocity measured by PTV method.

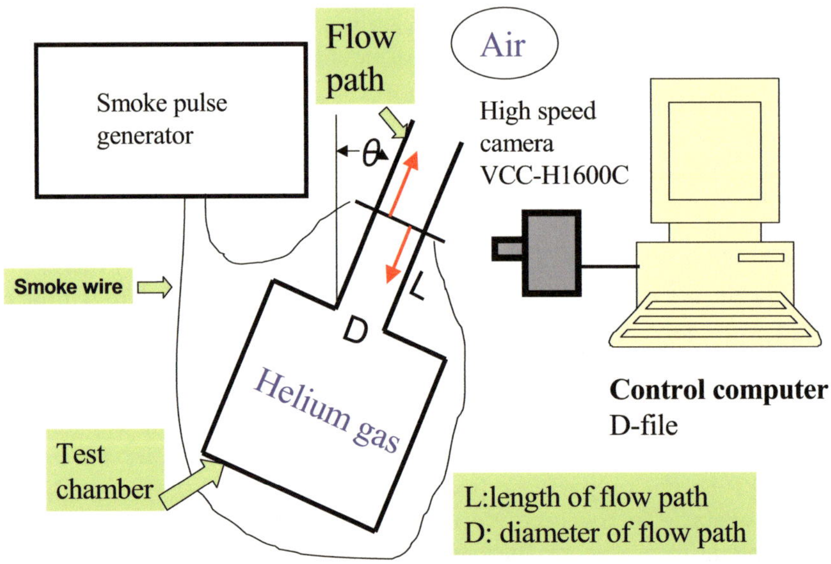

Fig. 1. Exchange Flow Apparatus of Smoke Wire Method High speed camera system

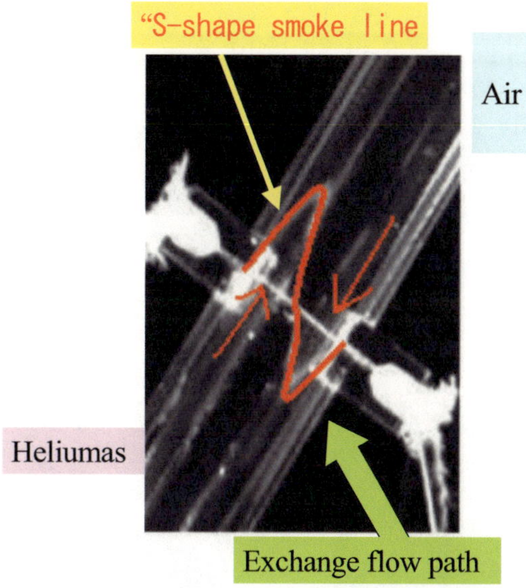

Fig. 2. Mechanism of exchange flow

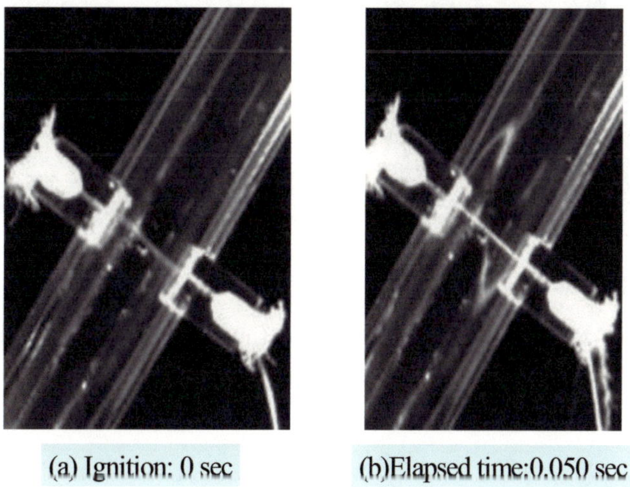

Fig. 3. The visualized data listed along the elapsed time (L/D=18.3, flame rate=200 f/s, θ =30 deg)

2.2 Results

The typical exchange flow in the tube was visualized in the figure 2. The visualized exchange flow resembles to the S-shape. The flames are detected 200 and 500 flames in a second by the high-speed camera. The visualized data listed along the elapsed time in the figures 3 and figure 4, respectively. It is clearly visualized that helium up-flow along left hand side and air down-flow along right hand side. Figure 5 shows time history of the upward flow peak velocity measured by PTV. In the case of L/D=18.3, the average velocity value $\boldsymbol{U_0}$ is evaluated as 0.315 m/s. In the case of L/d=10, the average velocity value $\boldsymbol{U_0}$ is evaluated as 0.662 m/s. It means that high exchange velocity detects in low L/D ratio. Figue.6 shows the visualization data of PIV measurement of 0.050sec after ignition (L/d=18.3 and flame rate=200 f/s). The adopted mesh size is 40x10 points in the flow area. The upward peak velocity was measured to 0.517 m/s. It means the value is higher than average velocity in this case. The exchange flow rate ***Q*** is derived as follows assuming parabolic flow profile, where ***r*** is the radius of flow path of the horizontal direction.

$$Q = \frac{2\pi U_0 r^2}{9} \tag{1}$$

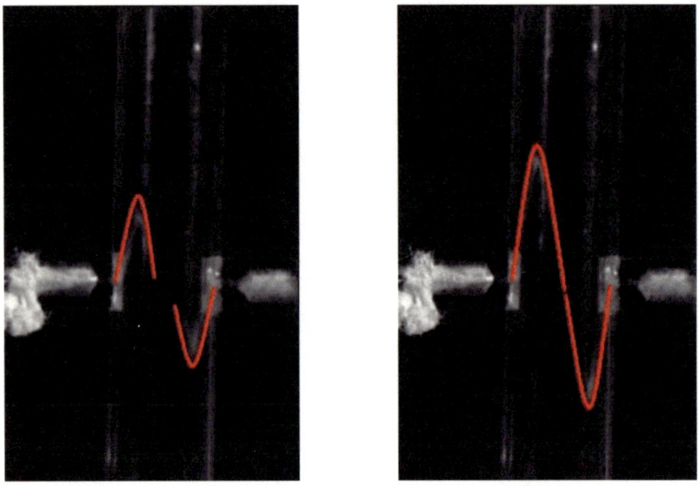

(a) Elapsed time 0.062 sec (b) Elapsed time 0.082 sec

Fig. 4. The visualized data listed along the elapsed time (L/D=10, flame rate=500 f/s, θ =30 deg)

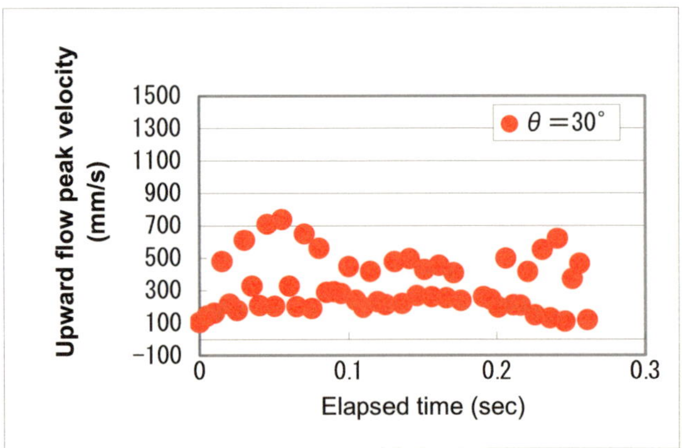

Fig. 5. Upward flow peak velocity measured by PTV method (L/D=18.3)

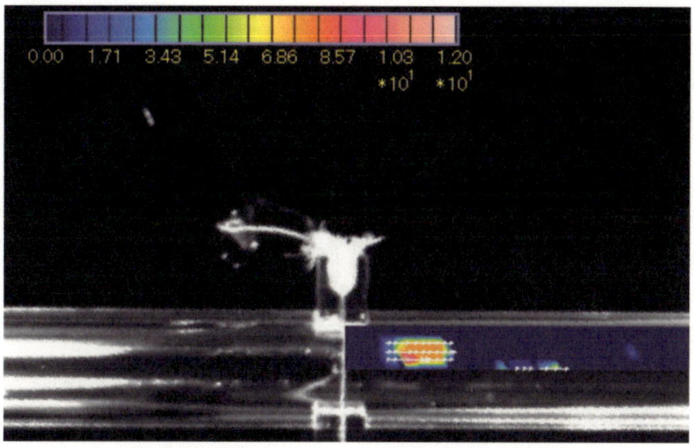

Fig. 6. The visualization data of PIV measurement of 0.050sec after ignition (L/D=18.3 and flame rate=200 f/s, θ =30 deg)

In PTV method, the exchange flow rate Q is calculated as $1.47x10^{-5}$ m^3/s under the conditon of . L/D=18.3. Therefore, the densimetric Froude number Fr is evaluated as 0.202, in this condition Reynords number Re is 79.2. In the condition of L/D=10, the densimetric Froude number Fr is evaluated as 0.287.

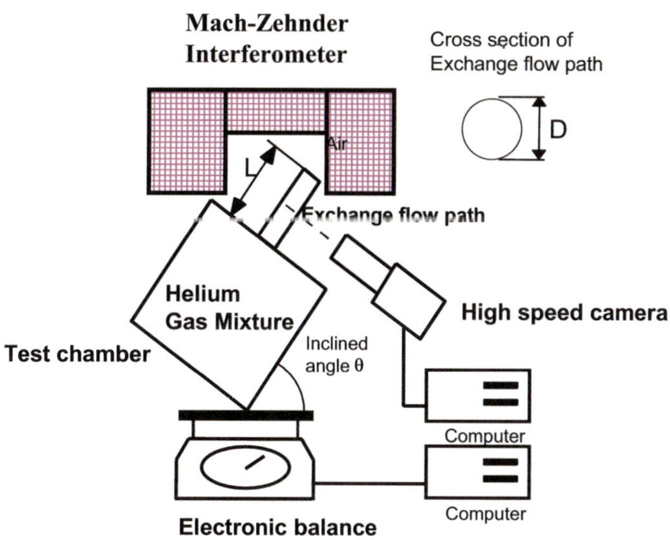

Fig. 7. Experimental apparatus of optical system and mass increment

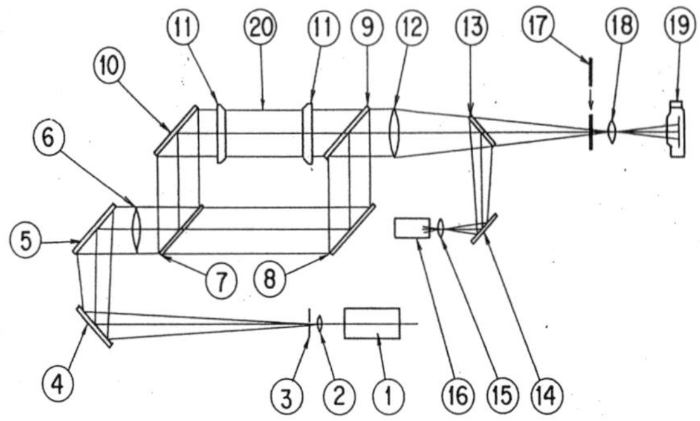

Fig. 8. The mechanism of optical system of the Mach-Zehnder interferometer, MZC-60S

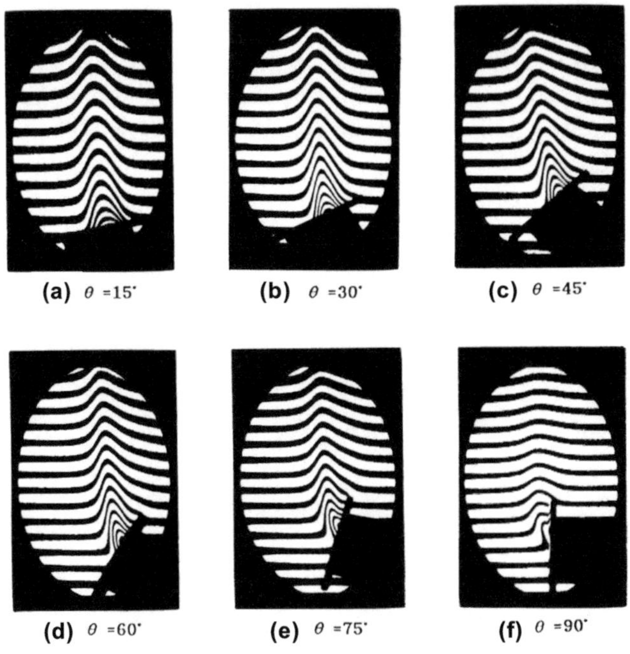

Fig. 9. The typical interference fringes for the inclined long round tube (L/D=5)

3. Optical system of mach-zehnderinterferometer

3.1 Experimental apparatus and procedure

The optical system of the Mach-Zehnder interferometer, MZC-60S is shown in figure 7 and figure 8 to visualize the exchange flow. After being rejoined behind the splitter, the test and reference laser beams interfere, and the pattern of interference fringes appears on the screen.

If the density of the test section is homogeneous, the interference fringes are parallel and equidistant (Keulegan, 1958). If it is not homogeneous, the interference fringes are curved. An inhomogeneity in the test section produces a certain disturbance of the non-flow fringe pattern. The digital camera and high-speed camera using D-file is able to attach to the interferometer.

3.2 Results

Figure 9 shows the typical interference fringes for the inclined long round tube (L/D=5). The curved interference fringes indicate that the lighter helium flows in the upper passage of the tube. The straight fringes indicate that the heavier air flows in the bottom of the tube. It is clearly visualized that the exchange flows take place smoothly and stable in the separated passages of the tube. This leads to less resistance for the exchange flow in the inclined tubes compared to he vertical ones. In the case of 30 deg, the curvature of the interference fringes is larger than that at other angles, indicating that the exchange flow rate and the densimetric Froude number are the largest at 30 deg.

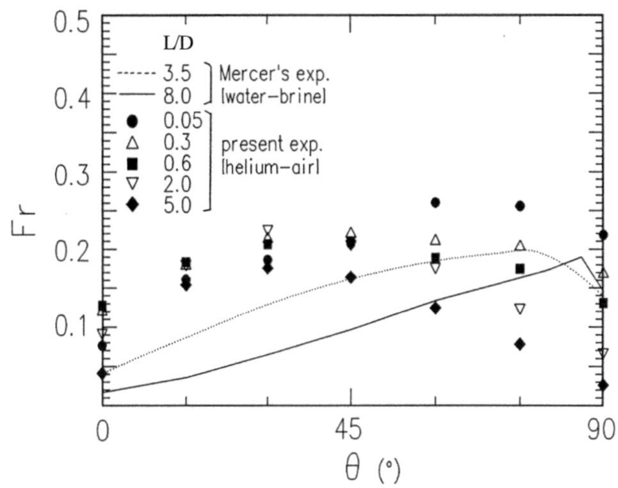

Fig. 10. The relationship between Fr and inclination angle θ with L/D as a parameter

4. Method of mass increment

4.1 Experimental apparatus and procedure

The method of the mass increment was used for the investigations. Figure 7 shows a rough sketch of the apparatus. It consists of a test chamber, an electronic balance and a personal computer for data acquisition. The experimental procedure is mentioned in Sec. 2.1. Air enters the test chamber and the mass of the gas mixture in the test chamber increases.

The mass increment Δ m is automatically measured by the electronic balance with high accuracy. From mass increment data, the density increment of the gas mixture $\Delta \rho_L = \Delta m/V$ is calculated. The density increment means the difference of densities of the gas mixture from the density of pure helium in the test chamber. Then, volumetric exchange flow rate is evaluated by the following equation:

$$Q = \frac{V}{\rho_H - \rho_L} \cdot \frac{d(\Delta\rho_L)}{dt} \tag{2}$$

The densimetric Froude number is defined by the following equation derived from the dimensional analysis suggested by Keulegan (Merzkirch, 1974):

$$Fr = \frac{Q}{A}\sqrt{\frac{\rho}{gD\Delta\rho}} \tag{3}$$

In the above equations, V is the volume of test chamber, ρ_H the density of air, ρ_L the density of gas mixture in the test chamber, $\Delta\ \rho_L$ ($=\rho_H - \rho_{He}$) = the density increment of the gas mixture, t the elapsed time, U(=Q/A) the exchange-velocity, $\rho(=\rho_H + \rho_{He})/2$, D the diameter and g the acceleration of gravity. The experiments are performed under atmospheric pressure and room temperature using the vertical and inclined round tubes, and using the vertical annular tube. The density of the gas mixture is close to that of helium in the present experiment. The sizes of the tubes are as follows. The diameter of the round tube D is 20 mm, which is much smaller than that of the test chamber. The inclination angle θ ranges from 15 to 90 deg and the height L ranges from 0.5 to 200 mm.

4.2 Results and discussion

It is already known that it is regarded as constant within a time duration when the gas in the upward flow can be assumed helium (Fumizawa, 1989). Figure 10 shows the relationship between Fr and inclination angle θ with L/D as a parameter. For inclined tubes, Fr is larger than that for vertical tubes. The black circles show the experimental data for the orifice (i.e. L/D =0.05) and the black rhombuses for the long tube (i.e. L/D = 5). Densimetric Froude number reaches the maximum at 60 deg for the orifice and 30 deg for the long tube. It is found that the angle for the maximum Fr decreases with increasing L/D in the helium-air system. On the other hand, Mercer's experiments with water and brine indicated that the inclination angle for the maximum Fr was about 80 deg in the several long tubes investigated. It may depend on the difference of dynamic viscosity between the gas and the liquid.

5. Conclusion

1. Flow visualization results indicate that the exchange flows through the inclined round tube take place smoothly and stable in the separated passages of the tube.
2. The visualized inclined exchange flow resembles to the S-shape.
3. In the inclined round long tube, the inclination angle for the maximum densimetric Froude number decreases with increasing length-to-diameter ratio for the helium-air system. On the other hand, this angle remains almost constant for the water-brine system.

6. Acknowledgements

The authors are deeply indebted to Dr. Makoto Hishida, who is professor of Chiba University in Japan, and Mr. Akira Furumoto who is manager of Digimo CO.,LTD for their unfailing interest and many helpful corporations to this study.

7. Nomenclature

A: flow passage area (m^2)

D: inner diameter of the tube of the flow path (m)

D_c: inner diameter of test chamber (m)

Fr: densimetric Froude number defined by eq.(3)

g :acceleration of gravity (m/s)

H_c: inner height of test chamber (m)

L: height of the tube of the flow path (m)

Q :volumetric exchange flow rate defined by eq.(l) (m/s)

r: radius of flow path of the horizontal direction (m)

T: elapsed time (s)

U: exchange-velocity (=Q/ A) (m/s)

U_0: maximum exchange-velocity (m/s)

V: volume of test chamber (m)

Greek

m: mass increment in test chamber (kg)

$\Delta\ \rho_L$: density increment (= Δ m/V) (kg/m^3)

θ: inclination angle if flow path from perpendicular line (deg)

ρ: mean density (=ρ_H +ρ_{He})/ 2 (kg/m^3)

Subscripts

L : lighter fluid (gas mixture)

H : heavier fluid (air)

He: helium

8. References

El-Wakil, M.M., Nuclear Energy Conversion, Thomas Y. Crowell Company Inc., USA (El-Wakil, 1982)

Epstein,M., Trans. ASME J. Heat Transfer, 110, pp885 -893 (Epstein, 1988)

Fumizawa,M. et. al., J. At. Energy Soc. Japan, Vol.31, pp1127-1128 (Fumizawa, 1989)

Fumizawa,M.; Proc. HT2005 ASME Summer Heat Transfer Conference, HT2005-72131, pp.1-7, Track 1-7-1 (Fumizawa, 2005)

Juni-1965, "Sicherheitsbericht fuer das Atom-Versuchskraftwerk Juelich", Arbeitsgemeinschaft Versuchs-Reactor AVR (Juni-1965, 1965)

Kang,T. et al., NURETH-5, pp541-546 (Kang, 1992)

Keulegan,G.H., U.S.N.B.S.Report 5831 (Keulegan, 1958)
Kiso et.al.; JSME Annual MTG, pp.339-340 (Kiso, 1999)
Mercer, A. and Thompson.H., J. Br. Nucl. Energy Soc., 14, pp327-340 (Mercer, 1975)
Merzkirch,W., "Flow Visualization", Academic Press (Merzkirch, 1974)

4

New Coolant from Lead Enriched with the Isotope Lead-208 and Possibility of Its Acquisition from Thorium Ores and Minerals for Nuclear Energy Needs

Georgy L. Khorasanov[1], Anatoly I. Blokhin[1] and Anton A. Valter[2]
[1]*Institute for Physics and Power Engineering Named After A.I. Leypunsky, Obninsk*
[2]*Institute for Applied Physics, Sumy*
[1]*Russian Federation*
[2]*Ukraine*

1. Introduction

In critical and subcritical fast reactors functional materials fulfill various tasks, including:

- heat transfer from pins to heat exchangers,
- heat removal from ADS target under dissipation of high energy intensive proton beam in the liquid metal – a source of spallation neutrons.

As such of materials molten heavy metals – mercury, lead, eutectic of lead (45%) and bismuth (55%) and others are using or to be used in future.

Heavy metals posses acceptable for FRs and ADSs neutron and physical characteristics while due to some their properties, for example chemical passivity to water, high boiling temperature, they are better as coolant in comparison to liquid light metal which is now used in sodium cooled FRs such as BN-600 and BOR-60 in Russia.

One of important parameters of functional material considered is a value of neutron absorption in coolant because it is desirable the neutron losses in the core of FR and ADS blanket have to be minimized.

Ways of minimization of neutron absorption in FR are well-known: it is offered to use wrapper less fuel assemblies, low neutron absorbing nitrogen isotope ^{15}N in nitride fuel contents, structural materials with low cross section of neutron capture, etc.

The authors of this paper are pointing out on one more possibility of reducing the neutron losses in the core cooled with lead: it is connected with enrichment of lead isotope, lead-208, from its value in the natural lead isotope mix, equal to 52.3%, up to the value of 99.0% [1-9]. Lead-208 as a twice magic nucleus possesses a very low cross section of neutron radiation capture. This unique feature leads to economy of neutrons in the core and other profitable factors which are listed in the Part I of this paper.

The limiting factor of usage highly enriched ^{208}Pb as the coolant is its high price in the world market. In the ISTC #2573 project [10], executed in the RF, the opportunity of creation of the plant for separation of lead isotopes using selective photoreactions was considered. The complex of calculations and theoretical works were carried out, the outline sketch of the separation installation was developed, and economic and technical estimations of industrial production of highly enriched ^{208}Pb were made. Developers of the ISTC #2573 project expect that at the scale of manufacture equal to 150-300 kg of ^{208}Pb per year its price will be of US $200/kg [11]. But these theoretical predictions have not been confirmed experimentally yet.

Presently lead isotopes are separated in gaseous centrifuges in using tetra methyl of lead $Pb(CH_3)_4$ as a working substance. According to estimations given in Ref. 12 the price of lead-208 with enrichment of 99.0% will be about 1000-2000 US $/kg, which is relatively high for nuclear power plants. For comparison, another heavy metal coolant, Pb-Bi costs approximately 50 US $/kg.

Meanwhile, in nature besides lead of usual isotopic content: 1.48% Pb-204, 23.6% Pb-206, 22.6% Pb-207, 52.32% Pb-208, it can be found lead with higher enrichment of lead-208. Such type of lead can be found in ores and placers containing thorium. Lead-208 is a final product of decay the radioactive nucleus Th-232 and that is why such type of lead is called as radiogenic lead. The period of half decay of Th-232 nucleus is $1.4 \cdot 10^{10}$ year. In ancient ores ($\sim 3 \cdot 10^{9}$ year) the total content of thorium of 3-5 wt% is usual. In this case concentration of radiogenic lead reaches approximately 0.3 wt%. The enrichment of lead-208 in radiogenic lead is about 85-93%, depending on uranium content in ores and minerals. Uranium-238 produces in isotope mix the input of lead-206 which is product of uranium-238 radioactive decay.

As known, thorium containing ores and minerals can be found in India, Brazil, Australia, Ukraine, Russia and other countries. In Part 2 of this paper the possibility of reprocessing thorium containing ores and minerals for production of thorium-232 and lead-208 for nuclear engineering needs is discussed.

2. Advantages of using lead enriched with lead-208 as coolant of FR and ADS

2.1 Reducing of neutron absorption in cores of FRs and ADSs

In Fig. 1 microscopic cross sections of radiation neutron capture, s(n, g), by the lead isotopes ^{204}Pb, ^{206}Pb, ^{207}Pb, ^{208}Pb and the natural mix of lead isotopes ^{nat}Pb in the ABBN-93 (Abagian-Bazaziants-Bondarenko-Nikolaev of 1993 year) system of 28 neutron energy groups [13] are given. The cross sections are cited on the basis of files of the evaluated nuclear data for the ENDF/B-VII.0 version.

As can be seen, the microscopic cross sections of radiation neutron capture by the lead isotope ^{208}Pb for all of the 28 neutron energy groups of the ABBN-93 system are smaller than the cross sections of radiation neutron capture by the mix of lead isotopes ^{nat}Pb, and this difference is especially large, by 3-4 orders of magnitude, for intermediate and low energy neutrons, $E_n < 50$ keV.

In Fig. 2 microscopic cross sections of radiation neutron capture, s(n, g), by the lead isotope ^{208}Pb and the eutectic lead(45%) – bismuth (55%) in the ABBN-93 system of 28 neutron energy groups are given. The cross sections are cited on the basis of files of the evaluated nuclear data for the ENDF/B-VII.0 version.

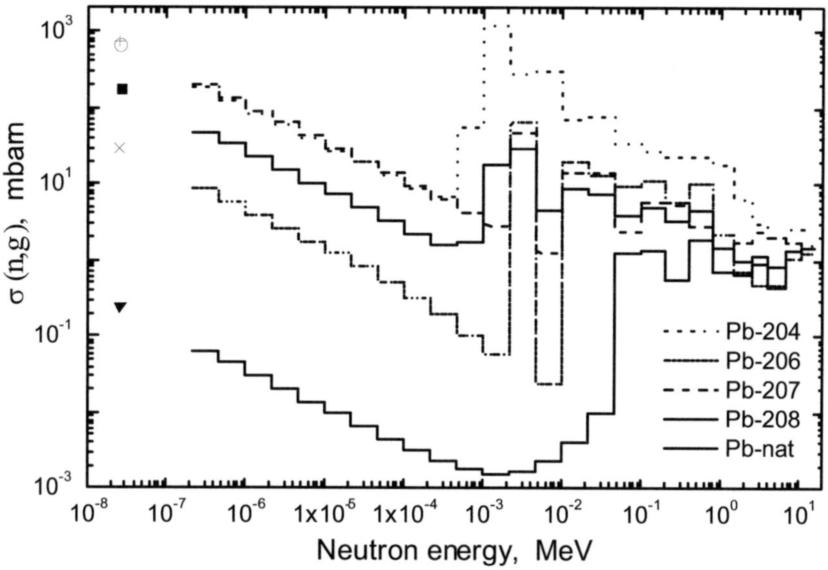

Fig. 1. Microscopic cross sections of radiation neutron capture s(n,g) by stable lead isotopes and by natural mix of lead isotopes taken from the ENDF/B-VII.0 library.
Cross sections are represented in the ABBN-93 system of 28 neutron energy groups.

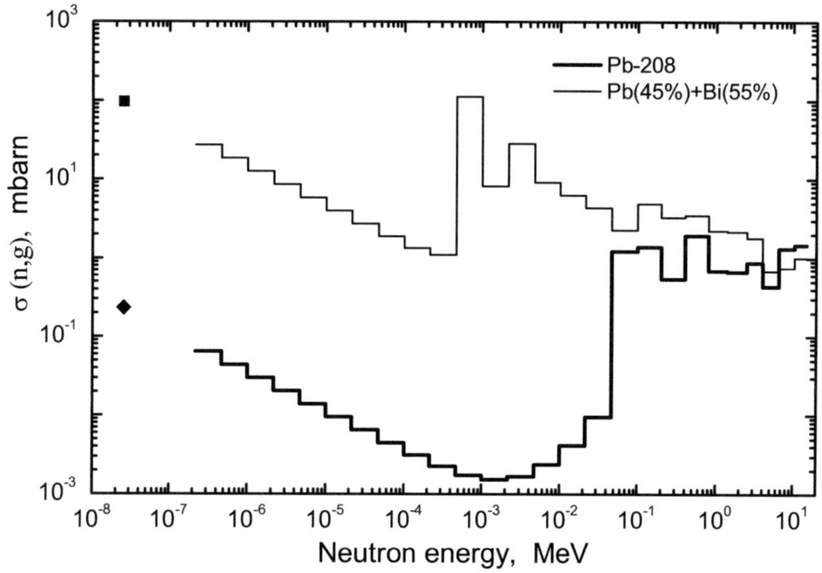

Fig. 2. Microscopic cross sections of radiation neutron capture s(n, g) by stable lead-208 isotope and by the eutectic Pb-nat(45%) - Bi (55%) taken from the ENDF/B-VII.0 library.
Cross sections are represented in the ABBN-93 system of 28 neutron energy groups.

As can be seen, the microscopic cross sections of radiation neutron capture by the lead isotope ^{208}Pb for all of the 28 neutron energy groups of the ABBN-93 system are smaller than the cross sections of radiation neutron capture by mix of lead ^{nat}Pb (45%) and bismuth, Bi

(55%), and this difference is especially large, by 3-5 orders of magnitude, for intermediate and low energy neutrons, $E_n < 50$ keV.

Share of neutrons with energies less than 50 keV, $E_n<50$ keV, usually is about 20-25% of all neutrons in FR or ADS cores and it increases in lateral and topical blankets of the core.

In Table 1 one-group cross sections of neutron radiation capture by two types of coolants - Pb-208 or the eutectic of Pb-Bi – in the lead-bismuth fast reactor project named as RBEC-M and designed in the Russian Kurchatov Institute [14] are given.

Reactor and its coolant	Core 1 with small enrichment of fuel	Core 2 with middle enrichment of fuel	Core 3 with large enrichment of fuel	Lateral blanket	Topical blanket under core 1	Topical blanket under core 2	Topical blanket under core 3
RBEC-M, Pb-Bi	**3.71190**	**3.62388**	**3.66404**	**4.82878**	**5.32383**	**5.22481**	**5.40967**
RBEC-M, Pb-208	**0.93296**	**0.94187**	**0.93931**	**0.86595**	**0.80867**	**0.81212**	**0.79005**

Table 1. One-group cross sections of radiation neutron capture by various coolants in the fast reactor RBEC-M core consisted from core 1, 2, and 3. Data are given for the standard lead-bismuth coolant, as it has been designed at the Kurchatov Institute, and for lead-208 coolants, proposed by authors of this paper.
Cross sections in millibarns are given.

From Table1 follows that the coolant from lead-208 is characterized with minimum one-group cross section, about <s>=0.93-0.94 millibarns. In standard lead-bismuth coolant the value of the same one-group cross section is by ~4 times bigger, about <s>=3.62-3.71 millibarns. In lateral and topical blankets one-group cross sections for Pb-208 by ~6-7 times are less than for Pb-Bi. The small values of one-group sections in RBEC-M cooled with lead-208 and corresponding excess of neutrons can be used for minimization of fuel load of the core, increasing fuel breeding and transmutation of long-lived fission products in lateral and topical blankets.

2.2 Hardening of neutron spectra in FRs and ADSs cooled with lead-208

In Fig.3 neutron spectra for the core 1(small enrichment of fuel) of the reactor RBEC-M cooled with its standard PB-Bi coolant and proposed Pb-208 coolant are given. Spectra were calculated for cases when their total fluxes were similar and the neutron multiplication factors were equal to 1 in using both of coolants. It can be seen that replacement of standard lead-bismuth coolant in RBEC-M leads to neutron hardening: the mean neutron energy increases from the value of 0.402 MeV to 0.428 NeV , i.e. on 6.5%.

In Fig.4 the ratio of neutron fluxes in the core 1 of RBEC-M in linear scale is represented. It is shown the increasing of share of fast neutrons ($E_n>0.4$ MeV) and the increasing of the very low share (less than 1%) of neutrons with energies $E_n < 100$ eV in the core cooled with lead-208. In whole the mean neutron increases on 6.5% as it has been mentioned above.

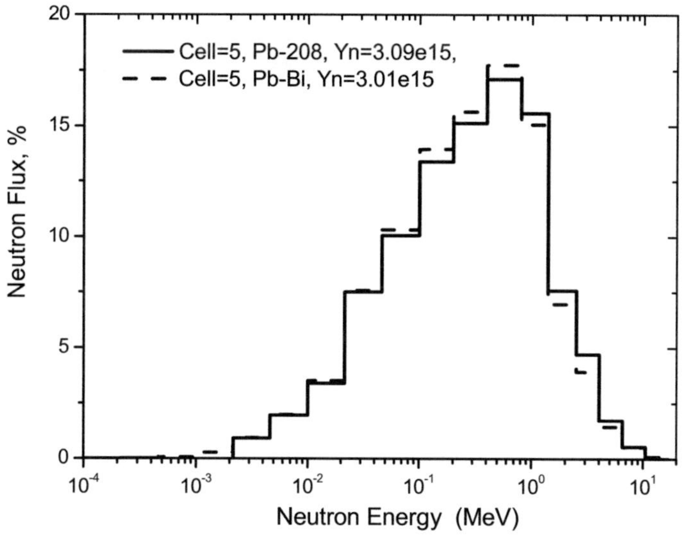

Fig. 3. Neutron spectra for the core1(small enrichment of fuel) of the reactor RBEC-M cooled with its standard Pb-Bi coolant (dash line) and Pb-208 coolant (solid line).
Yn- total flux of neutrons in core 1.

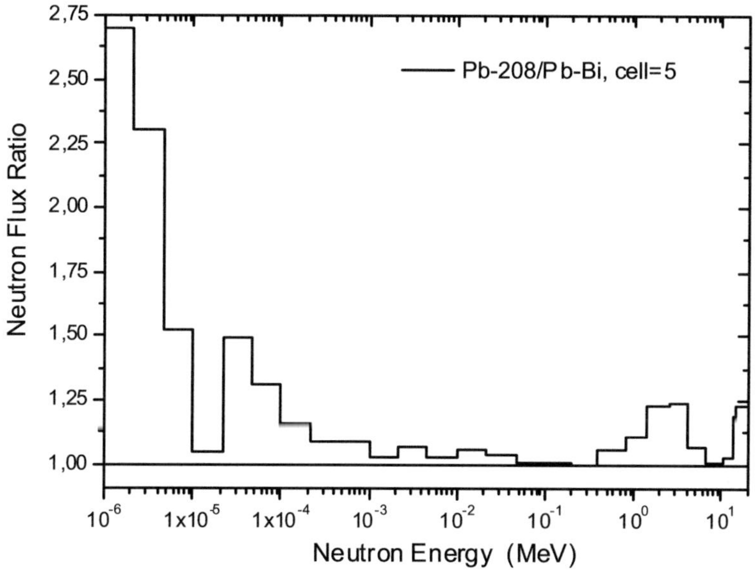

Fig. 4. The ratio of neutron fluxes in the core 1 of RBEC-M given in linear scale.The core is cooled with lead-208 leading to increasing the mean neutron energy on 6.5%.

2.3 Increasing effective neutron multiplication factor in FRs and ADSs cooled with lead-208

In the reactor RBEC-M in replacement its standard coolant to lead natural its effective neutron multiplication factor, K_{ef}, decreases from its standard value, K_{ef}=1.0096, to the value

K_{ef}=0.9815. But replacement of Pb-Bi to lead-208 leads to the value K_{ef}=1.0246, i.e. K_{ef} increases approximately on 1.5%. For reducing this increased value to the standard value, K_{ef}=1.0096, the plutonium enrichment must be decreased from its initial value equal to 13.7% as designed in lead-bismuth RBEC-M project to the value equal to 13.0%. It means that initial plutonium fuel loading must be decreased from 3595 kg to 3380 kg, i.e. on 215 kg. Thus, it means that economy of plutonium will be of 650 kg per 1 GW electrical power in using lead-208 as coolant instead of lead-bismuth in RBEC-M type reactors. It can be noted, that this quantity of power grade plutonium is comparable with the annual value of plutonium quantity, about 1 tone, which is now obtaining after reprocessing the spent fuel of Russian NPPs - VVER-440 and BN-600.

In the ADS with subcritical blanket of 80 MW thermal power [5] K_{ef} increases approximately on 1.7% in replacement lead natural as coolant to lead-208, from its value of K_{ef}=0.95289 for lead natural to K_{ef}=0.96997 for lead-208. In this case to liberate the nominal 80 MW thermal power in the blanket the power of the proton beam can be reduced from 2.59 MW to 1.68 MW, i.e. by 1.5 times.

2.4 Increasing the fuel breeding gain in FRs and ADSs cooled with lead-208

The excess of neutrons due to their small absorption in lead-208 can be used for fuel breeding and transmutation of long-lived radiotoxic fission products. Here, as an example, we assume the radiation capture of neutrons by uranium-238 leading to creation of plutonium-239. The affectivity of this process will be as large as the value of one-group cross section of radiation neutron capture by uranium-238 nucleus is large. In Fig.5 microscopic cross sections of radiation neutron capture by U-238 taken from ENDF/B-VII.0 library are given.

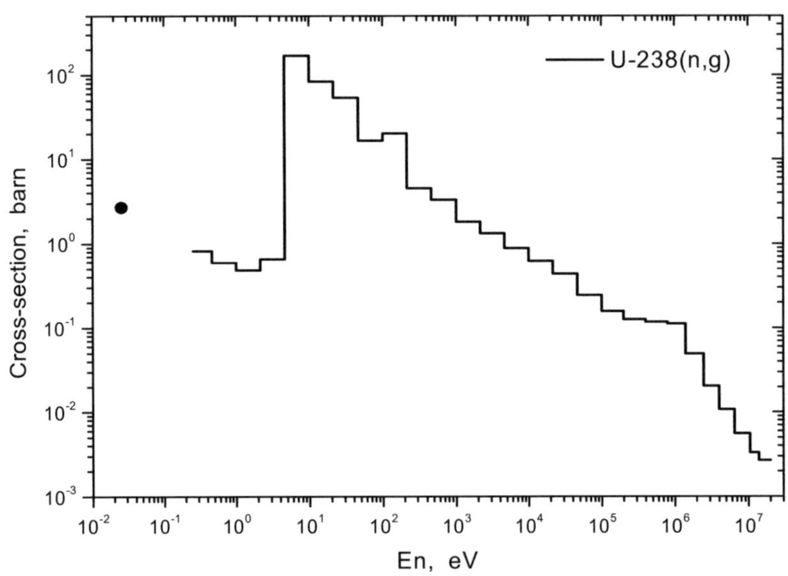

Fig. 5. Microscopic cross sections of radiation neutron capture by uranium-238 taken from ENDF/B-VII.0 library.

From Fig.5 it can be seen that at neutron energies near to E_n=5-10 eV these cross sections have maximum equal to 170 barns. That is why if the neutron spectra contains an increased share of neutrons of small and intermediate energies the corresponding one-group will be large enough.

In table 2 the one-group cross sections of radiation neutron capture by U-238 averaged over neutron spectra of the 80 MW ADS and various FRs (BREST, BN-600 and RBEC-M) are given.

Reactor	Coolant	One-group cross sections in barns
ADS-80 MW th.	Pb-208	**0.6393**
ADS-80 MW th.	Pb-nat	**0.4053**
BREST-300 MW el.	Pb-nat	**0.3089**
BN-600 MW el.	Na-23	**0.2965**
RBEC-M -340 MW el.	Pb-208	**0.1874**
RBEC-M-340 MW el.	Pb-Bi	**0.1886**

Table 2. One-group cross sections of radiation neutron capture by U-238 averaged over neutron spectra of the 80 MW ADS and various FRs (BREST, BN-600 and RBEC-M) cores. Cross sections in barns are given.

It can be noted that one-group cross section of neutron capture by uranium-238 in ADS spectrum is by 2.15 times bigger than for sodium reactor BN-600 spectrum and this fact indicates to the possibility of enhancing the breeding gain in the blanket of ADS 80 MW cooled with lead-208.

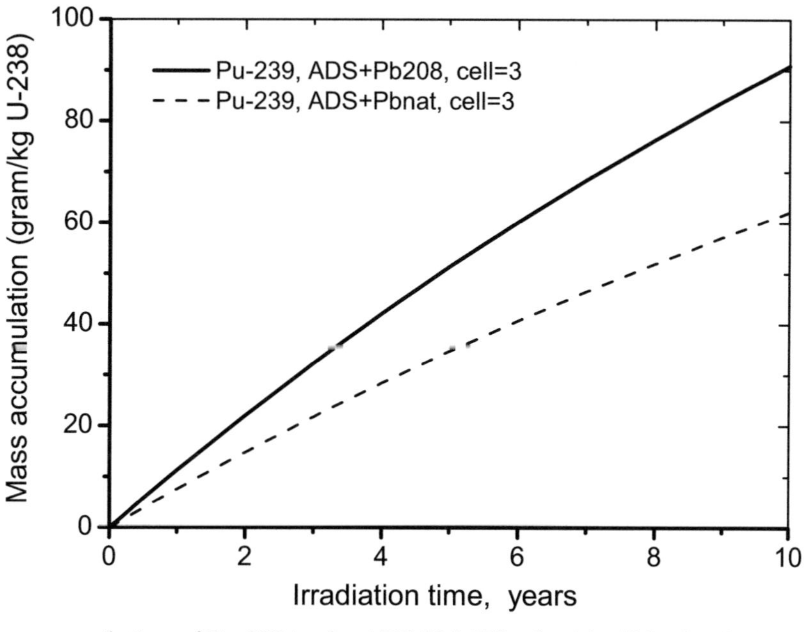

Fig. 6. Mass accumulation of Pu-239 in the ADS 80 MW subcritical blanket in inserting 1 kg of U-238. in the cell 3, near the blanket's far margin. The solid curve corresponds to the case, when the blanket is cooled with lead-208, the dash curve – to the case, when the blanket is cooled with lead natural.

As an illustration, in Fig. 6 and 7 the results of burning 1 kg of uranium-238 placed in the one part of ADS 80 MW subcritical blanket (cell 3 near the blanket far margin) and corresponding accumulation of plutonium-239 are given. Calculations have been performed on the basis of code ACDAM [15] developed at the IPPE Centre of nuclear data.

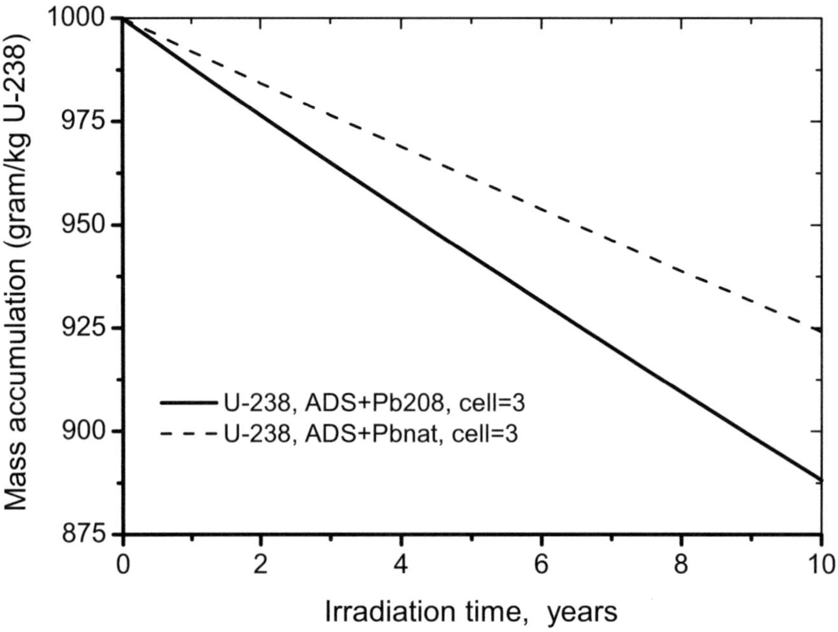

Fig. 7. Mass burning of 1 kg of U-238 in the neutron spectra of 80 MW ADS blanket, in the cell 3 which is near far blanket's margin. The solid curve corresponds to the case, when the blanket is cooled with lead-208, the dash curve – to the case, when the blanket is cooled with lead natural.

3. On the possibility of acquisition of radiogenic lead enriched with lead-208

3.1 On the sources of radiogenic lead enriched with lead-208 in Russia

The problem of acquisition of radiogenic lead enriched with lead-208 is coupled with perspectives of involving thorium into nuclear power engineering of Russia. As it is noted in Ref.16 to develop the thorium nuclear energetic it is necessary to obtain at least 10-13 thousand tones of thorium per year at the stage of 20-30 years of this century.

Content of lead-208 in thorium ores and minerals can reach 0.3-0.5% wt of thorium mass. In acquisition 10-13 thousand tones of thorium per year it will be possible to recover about 65 tones of radiogenic lead per year. This quantity of lead is insufficient to cover the needs in lead coolant of large scale nuclear power which requires approximately 2000 tones of lead per 1 GW of electrical power. But 65 tones of lead are sufficient to cool the blanket of 80 MW_{th} ADS. About 700 tones of lead can be enough to cool the reactor RBEC-M delivering 340 MW electrical.

As it is shown in Ref. 16, the main source of thorium in Russia is the Lovozerskoe deposit at Kola Peninsula. Estimations show that in reprocessing 2 mln tones of loparit ore per year 500-600 thousand tones of Ln_2O_3 and TiO_2, 100 thousand tones of Nb_2O_5, 10 thousand tones of Ta_2O_5, 13 thousand tones of ThO_2 and 65 tones of radiogenic lead can be produced. In Ref 16 the conclusion was made that is possible to extract in near future large quantities of thorium from the progress of industry and as co-product of rear metal raw.

The separate problem is the level of lead-208 enrichment of lead-208 in various deposits. It can be strongly different. For example, in Brazil monazites radiogenic lead is enriched by lead-208 up to 88.34% [17]. For FRs and ADSs it can be desirable the following isotopic composition of radiogenic lead: lead-208-93% and lead-206-6% with minimum content of lead-207 - the isotope with large cross section of neutron capture. In Ref. 18 the data concerning thorium-containing ores and monazites in the world scale are given. The authors of this paper pointed out that as a rule radiogenic lead contains very small quantities of lead-204 and lead-207–isotopes with large cross sections of neutron capture.

It can be noted that the advantages of lead-208 can be used, besides nuclear power plants, in other branches of nuclear science and technology. It seems that lead-208 as low moderating material will be preferable in the lead slowing down neutron spectrometers [19] and also in the spallation neutron sources to have the harder neutron spectra under interaction of high energy protons with liquid proton target from lead-208 [2, 20].

3.2 Prospects of ancient monazite from placers and bed-rock's deposits of Ukraine as the raw materials to produce highly enriched ^{208}Pb

Monazite is the phosphate containing mainly ceric rare earths ((Ce, La, Nd ..., Th) PO_4) and is the main natural concentrator of thorium. It is widely spread (though usually in small amounts) in rocks and some types of ores. Owing to chemical and mechanical durability monazite is accumulated in placers.

The crystal structure of monazite can be presented as three-dimensional construction of oxygen nine apex polyhedron with rare-earth center atoms and oxygen tetrahedrons with the central atom of phosphorus. Nine-fold coordination allows a wide occurrence of relatively large ions of the light rare earths and thorium in mineral structure. The total content of thorium in a mineral can reach 28 wt%, and concentration of 5-7 wt% is usual. Though there are no experimental data about the form of radiogenic lead presented in the monazite structure, the numerous data, summarized for example in work [21], argued for its good stability in a monazite crystal matrix that allows monazite to be used for isotope dating.

In Ukraine monazite contains in developed fine-grained titanium-zirconium placers. By the explored easily enriched titanium-zirconium ores Ukraine comes to the forefront in Europe and in the CIS. The resources of zirconium in Ukraine make more than 10% of world ones. Now the largest Malyshevsky (Samotkansky) placer is developed and the working off of the Volchansky placer has been started.

Owing to the marked paramagnetism monazite at existing capacity of mines can be taken in passing by working out of placers in quantity of about 100 tons per year that corresponds approximately to 3.5 tons of thorium and 0.5 tons of the lead enriched with 208 isotope. Now monazite is considered as a harmful radioactive impurity and it is not produced.

The composition of monazite from the Malyshevsky placer as to the amounts of U, Th and Pb for dating purposes is well studied in work [21] by means of X-ray-fluorescent technique specially developed for individual grain analysis. In Table 3 the data about the contents of thorium, uranium and about isotope contents of lead for monazite of the Malyshevsky deposit is cited. The average composition of lead is confirmed by direct mass spectrometry determinations.

Average values from 224 X-ray-fluorescent determinations according to [21] data, wt %.			Isotopic composition of lead by mass spectrometry analysis of average sample, relative %%.				Average value of 70 uranium depleted samples. Elements - mass %%, Lead isotopes -relative %%.					
										Lead isotopes		
Th	U	Pb	^{204}Pb	^{206}Pb	^{207}Pb	^{208}Pb	U	Th	Pb	^{206}Pb	^{207}Pb	^{208}Pb
3,52	0,23	0,30	0,04	13,11	1,43	85,42	0,06	3,63	0,33	3,8	0,4	95,7

Table 3. Contents of thorium, uranium, lead and isotopic composition of lead for monazite of the Malyshevsky placer (Ukraine)

As is seen from Table 3, enrichment by ^{208}Pb in the average for all monazite is insufficiently high. However, there is a probability of monazite separation by the flotation, magnetic or other characteristics with release of low uranium fraction of the mineral.

Extraction of total monazite concentrate by working out of the Malyshevsky placer scattering of an average almost won't demand additional costs and its price as at first approximation can be accepted as the equal to zircon concentrate, i.e. ~ 1 US $/ kg. Cost of hydrometallurgical emanation of lead from monazite by analogy with similar processes can be estimated as (24÷30 US $/ kg). The removal of differences with low U/Th ratio and the high content of ^{208}Pb from monazite concentrate will require additional researches and will cause some rise in price of a product.

In Ukraine there are insufficiently studied shows of monazite in ancient radical breeds, their barks of aeration and in placers, i.e. enriched ^{208}Pb. According to the available analytical data there is a possibility to detect monazite with highly enriched ^{208}Pb.

For extraction of thorium and the lead enriched with 208 isotope Russia has a great opportunities by preparation the fine-grained titanium-zirconium placers for development and by the extraction from raw materials in complex deposits.

4. Conclusions

The paper is dedicated to the proposal of using lead enriched with the stable isotope ^{208}Pb in FRs and ADSs instead of lead natural, ^{nat}Pb.

It seems that unique neutron features of ^{208}Pb make it as one of the best among the molten metal coolants now assumed for FRs and ADSs: sodium, lead-bismuth, lead natural and others.

The main advantage of ^{208}Pb is its low neutron absorption ability: for neutron energies E_n=0.1-20.0 MeV the microscopic cross sections of radiation neutron capture by ^{208}Pb are by 1.5-2.0 times smaller as compared with ^{nat}Pb, and for energies, E_n<50 keV, the difference in the cross section values reaches 3-4 orders of magnitude. Averaged over neutron spectra of the LFR or ADS the one-group cross sections for a coolant from ^{208}Pb are by 5-6 times smaller than those for the coolant consisted from ^{nat}Pb.

The second advantage of using ^{208}Pb consists in achievement the core neutron spectra hardening on 5-6% due to low energy losses. Low neutron absorbing and moderating features of ^{208}Pb permit to reach the gain in the multiplication factor K_{ef} on 2-3% for critical or subcritical core fueled with U-Pu mix. In this case to have the multiplication factor K_{ef} =1.01 for the LFR or K_{eff} =0.97 for the ADS, both cooled with lead-208, the enrichment of power grade Pu in the U-Pu fuel can be reduced approximately on 0.7-0.8%.

The third important advantage of using ^{208}Pb is coupled with increasing the small share of neutrons of low energies, 5-10 eV in spite of the neutron spectra hardening in whole. In this region of neutron energies the microscopic cross sections for such nuclides as ^{238}U and ^{99}Tc are maximum and very high, and the one-group cross sections for these nuclides averaged over neutron spectra of LFRs and ADSs cooled with lead-208 are equal to 0.6 and 0.8 barn respectively which are comparable with the one-group cross sections for typical breeders and transmutters.

The possibility of using ^{208}Pb as coolant in commercial fast critical or subcritical reactors requires a special considering but relatively high content of this isotope in natural lead, 52.3%, and perspectives of using high performance photochemical technique of lead isotope separation permit to expect obtaining in future such a material in large quantities and under economically acceptable price. In the paper it is shown that principal possibility of acquisition of radiogenic lead containing high enriched lead -208, up to 93%, exists. Nowadays in Russian Federation and Ukraine thorium- containing loparit ores and monazite minerals are reprocessed for production of rare metal raw. Thorium and lead are not required now and they are deposited in sludge. Nevertheless, the scales of future thorium and radiogenic lead production for innovative nuclear reactors have some prospects in near-term future. The conclusion is made that to obtain the minimum amount of required in future radiogenic lead (65 t/year) for small sized FRs and ADSs the very large quantities of loparit ores or monazite minerals must be reprocessed and acquisition of radiogenic lead-208 can be economically acceptable as a co-product of rare metal raw.

5. References

[1] G.L. Khorasanov, A.P. Ivanov, A.I. Blokhin Isotopic tailored materials for nuclear engineering. Issues of Atomic Science and Technology, Series: Nuclear Constants, 2006, Issue 1-2, p. 99-109, (in Russian), (ISSN 0207-3668).

[2] G.L. Khorasanov, A.P. Ivanov, A.I. Blokhin. Reduction of the induced radioactivity in an ADS target by changing the target material isotope composition. Nuclear Engineering and Design, 2006, v. 236, Nos. 14-16, p. 1606-1611, (ISSN 0029-5493).

[3] G.L. Khorasanov, A.P. Ivanov, A.I. Blokhin. Neutronic parameters of the ADS target using lead-208 as a coolant target material. Issues of Atomic Science and Technology, Series: Nuclear Constants, 2005, Issue 1-2, p. 94-100, (in Russian), (ISSN 0207-3668).

[4] G.L. Khorasanov, V.V. Korobeynikov, A.P. Ivanov, A.I. Blokhin. Minimization of an initial fast reactor uranium-plutonium load by using enriched lead-208 as a coolant. Nuclear Engineering and Design, v. 239, No 9, p. 1703-1707, 2009, (ISSN 0029-5493).

[5] G.L. Khorasanov and A. I. Blokhin. Macroscopic cross sections of neutron radiation capture by coolant, uranium-238 and technetium -99 in ADS subcritical core cooled with natural and enriched lead. Special issue of the journal "Perspective materials" #8, 2010, p.361-365, (ISSN 1028-978X).

[6] Georgy L. Khorasanov and Anatoly I. Blokhin. Macroscopic cross sections of neutron radiation capture by Pb-208, U-238 and Tc-99 nuclides in the accelerator driven subcritical core cooled with molten Pb-208. In CD-ROM Proceedings of the International Conference PHYSOR 2010 – Advances in Reactor Physics to Power the Nuclear Renaissance, Pittsburgh, Pennsylvania, USA, May 9-14, 2010, Paper #286 at the Session 5C "Advanced Reactors Design".

[7] G.L. Khorasanov and A.I. Blokhin. A low neutron absorbing coolant for fast reactors and accelerator driven systems.. Issues of Atomic Science and Technology, Series: Nuclear Constants, 2010, Issue 1, (in Russian), (ISSN 0207-3668).

[8] D.A. Blokhin, E.A. Zemskov, G.L. Khorasanov. The influence of the coolant from lead-208 on neutron characteristics of a fast reactor core. Issues of Atomic Science and Technology, Series: Nuclear Constants, 2010, Issue 1, (in Russian), (ISSN 0207-3668).

[9] G.L. Khorasanov and A.I. Blokhin. A low neutron absorbing coolant for fast reactors and accelerator driven systems. In the book "Cooling Systems", Editor Aaron L. Shanley, Chapter 5, Edition: Nova Science Publishers, Inc., USA, 2011, in press, (ISBN 978-1-61209-379-6).

[10] ISTC #2573 project: "Investigation of Processes of High - Performance Laser Separation of Lead Isotopes by Selective Photoreactions for Development of Environmentally Clean Perspective Power Reactor Facilities", Project Manager: A.M. Yudin (Saint-Petersburg, Efremov Institute, NIIEFA), Project Submanagers: G.L. Khorasanov

(Obninsk, Leypunsky Institute, IPPE) and P.A. Bokhan (Novosibirsk, Institute for Semiconductor Physics, ISP), 2004-2005.

[11] A.L. Bortnyansky, V.L.Demidov, S.A. Motovilov, F.P. Podtikan, Yu.I. Savchenko, V.A. Usanov, A.M. Yudin, B.P. Yatsenko, 2005. Experimental Laser Complex for Lead Isotope Separation by means Selective Photochemical Reactions. Proc. of the X Int. Conf. "Physical and Chemical Processes on Selection of Atoms and Molecules", 3-7 October 2005. Moscow, TSNIIATOMINFORM, 76-82. (ISBN 5-85389-122-7).

[12] V.D. Borisevich, G.A. Sulaberidze, A.Yu. Smirnov. Production of highly enriched lead-208: separation problems. Paper presented at the Russian-Chinese Bilateral Workshop "Possibility of using stable isotope lead-208 in nuclear engineering and its acquisition", 12-13 October, 2010. Tsinghua University, Beijing, P.R. China.

[13] G.N Manturov, M.N. Nikolaev, A.M. Tsiboulia, Group constant system ABBN-93. Part 1: Nuclear constants for calculation of neutron and photon emission fields, Issues of Atomic Science and Technology, Series: Nuclear Constants, Issue 1, p. 59 (1996) (in Russian), (ISSN 0207-3668).

[14] Alekseev P.N., Mikityuk K.O., Vasiljev A.V., Fomichenko P.A., Shchepetina T.D., Subbotin S.A. Optimization of Conceptual Decisions for the Lead-Bismuth Cooled Fasdt Reactor RBEC-M. Atomnaya Energiya, 2004, v. 97, issue 2, pp.115-125.); http://www.iaea.org/NuclearPower/ SMR/crpi25001 /html/).

[15] A.I Blokhin, N.A. Demin, V.N. Manokhin et al. Code ACDAM for study nuclear and physical properties of materials under long-term neutron irradiation. Perspective materials 2010, No 2, p. 46-55, (ISSN 1028-978X).

[16] V.M. Decusar, B.Ya. Zil'berman, A.I. Nikolaev et al. Analysis of potential sources of satisfactory the near term requirement in thorium accounting the various scenarios of thorium involving in nuclear energetic of Russia. Preprint IPPE-3186, Obninsk, 2010. - 36p.

[17] J.A. Seneda , C.A.L.G. de O. Forbicini, C.A. da S. Queiroz, M.E. de Vasconcellos, S. Forbicini, S.M. da R. Rizzo, Vera L.R. Salvador and A. Abrão. Study on radiogenic lead recovery from residues in thorium facilities using ion exchange and electrochemical process. Progress in Nuclear Energy, 2010, v. 52, No 3, pp. 304-306.

[18] G.G. Kulikov, A.N. Shmelev, V.A. Apse, V.V. Artisyuk. On the possibility of using radiogenic lead in nuclear energetic. Yadernaya energia, 2011 (in Russian), (ISSN 0204-3327).

[19] A.A. Alekseev, A.A. Bergman, O.N. Goncharenko et al. Investigation of the neutron-fission processes on the lead neutron slowing down spectrometer of INR RAS. In Proc. XII Int. Sem. on Interaction of Neutrons with Nuclei, ISINN-12, Dubna, May 26-29, 2004, p.237.

[20] V.I. Yurevich. Production of Neutrons in Thick Targets by High-Energy Protons and Nuclei. Physics of elementary particles and atomic nuclei, 2010, vol.41, part 5, pp. 1451-1530, (ISSN 0367-2026).

[21] A.A. Andreev. Monazite age, geochemical peculiarities and possible sources of origin on the territory of Ukraine. Ph. D. Thesis's. Kiev, 2011, 190 p.

5

Decay Heat and Nuclear Data

A. Algora and J. L. Tain
Instituto de Fisica Corpuscular, CSIC-Univ. de Valencia, Valencia
Spain

1. Introduction

The recent incidents at the Fukushima Daiichi nuclear power plant, following the great tsunami in Japan, have shown publicly, in a dramatic way, the need for a full knowledge and proper handling of the decay heat in reactors and spent-fuel pools.

In this chapter, after a short introduction to decay heat from the historical perspective we will discuss, how the decay heat is calculated from available nuclear data, and how the quality of the available beta decay data plays a key role in the accuracy and predictive power of the calculations. We will present how conventional beta decay experiments are performed and how the deduced information from such conventional measurements can suffer from the so-called pandemonium effect. Then we will introduce the total absorption technique, a technique that can be used in beta decay experiments to avoid the pandemonium effect. Finally, we will present the impact of some recent measurements using the total absorption technique, performed by an international collaboration that we lead on decay heat summation calculations and future perspectives.

1.1 The discovery of nuclear fission

In 1934 Fermi bombarded an uranium target with neutrons slowed down in paraffin in an attempt to produce transuranic elements. The first impression after the experiment was that uranium did undergo neutron capture and the reaction product was beta radioactive. Subsequent investigation of this reaction showed that the final activity produced included a range of different half-lives. This puzzle triggered intensive research from 1935 to 1939.

The identification of one of the activities produced as the rare-earth lantanum, first by Curie and Savitch in 1938 and then by Hahn and Strassmann in 1939, started to shed light on the puzzle. Indeed it was this fact that lead Hahn and Strassman to interpret the experimental activities as barium, lanthanum and cerium instead of radium, actinium and thorium. Shortly afterwards Meitner and Frisch (1939) suggested that the uranium nucleus, after the absorption of a neutron, splits itself into two nuclei of roughly equal size. Because the resemblance with the biological process in a living cell, the process was called fission. A typical example of the splitting is represented in Equation 1. Later measurements established the asymetric character of the process, the large energy release ($\sim$ 200 MeV) and the emission of prompt neutrons, which could trigger new fission processes and produce a chain reaction.

$$^{235}_{92}U +^{1}_{0} n \rightarrow^{140}_{54} Xe +^{94}_{38} Sr + 2^{1}_{0}n + \gamma + \sim 200 MeV \tag{1}$$

The first self-sustained chain reaction was achieved by Fermi in 1942 at the University of Chicago, which marked the begining of the nuclear age.

Since then many types of reactor have been developed for research, military and civil applications. Some examples include the Gas-Cooled reactor (GCR), Light Water Reactor (LWR), Heavy Water Reactor (HWR), Boiling Water Reactor (BWR), Liquid Metal Cooled Fast Breeder Reactor (LMCFBR) and so on. Independently of the kind of reactor one is considering there are important design and operating criteria which require a knowledge of the energy released in the decay of the fission products. In Table 1 an approximate distribution of the energy released in the fission of ^{235}U is presented. This table shows that from the total energy released inside the reactor (190 MeV), approximately 7 % is due to the beta decay of the fission products in the form of gamma and beta radiation. This source of energy is commonly called *decay heat* and depends on the fuel used in the reactor.

Distribution	MeV
Kinetic energy of light fission fragments	100
Kinetic energy of heavy fission fragments	67
Energy of prompt neutrons	5
Energy of prompt gamma rays	5
Beta energy of fission products	7
Gamma energy of fission products	6
Subtotal	190
Energy taken by the neutrinos	11
Total	201

Table 1. Approximate distribution of the energy released in the fission process of ^{235}U.

1.2 Decay heat

Once the reactor is shutdown, the energy released in radioactive decay provides the main source of heating. Hence, the coolant needs to be maintained after the termination of the neutron-induced fission process in a reactor. The decay heat varies as a function of time after shutdown and can be determined theoretically from known nuclear data. Such computations are presently based on the inventory of nuclei created during the fission process and after reactor shutdown, and their radioactive decay characteristics:

$$f(t) = \sum_i (\overline{E}_{\beta,i} + \overline{E}_{\gamma,i} + \overline{E}_{\alpha,i})\lambda_i N_i(t) \quad (2)$$

where $f(t)$ is the power function, $\overline{E}_i$ is the mean decay energy of the ith nuclide (β, γ and α components), λ_i is the decay constant of the ith nuclide ($\lambda_i = ln(2)/T_{1/2,i}$), and $N_i(t)$ is the number of nuclide i at cooling time t. In the summation calculations 2, the first step is the determination of the inventory of nuclei $N_i(t)$, which can be obtained by solving a linear system of coupled first order differential equations that describe the build up and decay of fission products:

$$\frac{dN_i}{dt} = -(\lambda_i + \sigma_i \phi)N_i + \sum_j f_{j\rightarrow i}\lambda_j N_j + \sum_k \mu_{k\rightarrow i}\sigma_k \phi N_k + y_i F \quad (3)$$

where N_i represents the number of nuclides i, λ_i stands for the decay constant of nuclide i, σ_i is the average capture cross section of nuclide i, ϕ is the neutron flux, $f_{j\to i}$ is the branching ratio of the decay from nuclide j to i, $\mu_{k\to i}$ is the production rate of nuclide i per one neutron capture of nuclide k, y_i is the independent fission yield of nuclide i and F is the fission rate. These calculations require extensive libraries of cross sections, fission yields and decay data.

As mentioned earlier, an accurate assessment of the decay heat is highly relevant to the design of nuclear facilities. Consider for example the safety analysis of a hypothetical loss-of-coolant accident (LOCA) in a light water reactor or the cooling needs of a spent-fuel pool. Calculations of the decay heat are also important for the design of the shielding of discharged fuel, the design and transport of fuel-storage flasks and the management of the resulting radioactive waste. This assessment is obviously not only relevant to safety, it also has economic and legislative consequences. For example, the accuracy of the presently available decay data is still not high enough and this situation translates into higher safety margins implying greater economic costs. Reducing the uncertainty in available decay data is one of the main objectives of the work devoted to decay heat in present-day research.

Nowadays the most extended way to calculate the decay heat in reactors is the summation calculation method based on equations 2 and 3. As can be seen from 2 this method is simply the sum of the activities of the fission products produced during the fission process and after the reactor shutdown weighted by the mean decay energies. As a consequence, the summation calculation method relies heavily on the available nuclear data and this is the reason why in the early days of the nuclear age this method was not satisfactory. A different method, the so-called *statistical method* was the prefered way of evaluation of the decay heat at that time.

The statistical method was based on the work of Way and Wigner (Way & Wigner, 1948). They considered fission products as a sort of statistical assembly and relying on mean nuclear properties, they deduced empirical relations for the radioactive half-lives and atomic masses of fission products. With these relations the gamma (P_γ MeV/fission-s) and beta plus gamma (P_t MeV/fission-s) decay heat power functions in ^{235}U were determined as follows:

$$P_\gamma(t) = 1.26t^{-1.2} \tag{4}$$

$$P_t(t) = 2.66t^{-1.2} \tag{5}$$

These relations were considered valid for decay times t in seconds in the range of 10 seconds to 100 days. For many years this was the only available method for calculations of the decay heat. Other fissile materials were supposed to behave similarly to ^{235}U. This method is clearly simpler than the summation calculations, however it is incomplete by nature and less accurate for longer cooling times. In a natural way, with an increasing volume of nuclear data, the statistical method was gradually superseded by the summation calculations. Even though it is not used anymore, the work of Way and Wigner (Way & Wigner, 1948) was a seminal article, and the interested reader is encouraged to read it. Their predictions are approximatelly off by 60 % from todays accepted values, but the form of the parametrization of the decay heat as function $a \times t^{-b}$ is still used for benchmarks determination (benchmarks are obtained as a sum of functions of the form $a \times t^{-b}$ that cover the full cooling period, which are adjusted to experimental data).

It should be noted that in the beta part of equations 4 , 5 as well as in 2 the neutrino energy is not included.

2. Decay heat measurements

In the introduction we discussed how the decay heat can be calculated using two possible methods: the statistical method and via summation calculations, but any model for calculating the decay heat is only useful if it is able to describe properly experimental data. In this section we will discuss briefly how the decay heat can be measured and what benchmarks can be used to validate the calculations.

In general terms we can classify the decay heat measurements in two ways: a) radiation detection experiments and b) calorimetric experiments.

Radiation detection experiments consist of measuring the beta energy and the gamma energy coming from a small sample of fissile material that has been irradiated for a known length of time. In these measurements the aim is to increase the sensitivity of the setup for the particular goal of the study and reduce the sensitivity to any other type of radiation, which otherwise can lead to systematic errors. For example, in a beta energy measurement we will be prone to use a thin plastic detector, which has high efficiency for beta detection and a reduced efficiency for gamma rays. Conversely, in a gamma energy measurement one would use detectors of high efficiency for gamma rays and would try to avoid as much as possible the penetration of the betas in the gamma detection setup. Examples of these kinds of measurements can be seen in Refs. (Rudstam, 1990), (Dickens, 1981) and (Tasaka, 1988). These measurements have been also labelled in the past as "nuclear calorimetric" measurements, but the method itself is not truly calorimetric one, since it is only the detection of pure nuclear radiation with very high efficiency.

Real calorimetric experiments consist of absorbing the decay radiations and measuring the heat in the absorber. They follow well developed procedures that have been used extensively in studies of the energetics of chemical reactions. Many calorimetric measurements have been limited to intermediate cooling times, because of the difficulties of building an instrument that responds rapidly to changes in the power level, which is characteristic of short cooling times. Actually, for the application to the decay heat problem, the challenge is to build a calorimeter with a short time constant, or in other words to construct a setup that reacts quickly to the power release of the sample. Examples of such measurements can be found in the works of (Schrock, 1978) and (Yarnell, 1977).

There are several publications that summarize the efforts to improve the decay heat benchmarks covering different time periods. Some examples can be found in the reviews of Schrock (Schrock, 1979), Tobias (Tobias, 1980) and Tasaka (Tasaka, 1988). Nowadays the results of decay heat calculations are compared with the measurements of Akiyama *et al.* (Akiyama, 1982), Dickens *et al.* (Dickens, 1981), (Dickens, 1980) and Nguyen *et al.* (Nguyen, 1997). Decay heat benchmarchs can be found in the work of Tobias (Tobias, 1989), which is considered the standard in the field.

3. Nuclear data and beta decay

We will now concentrate on the data needs for summation calculations. Equation 3, which is the first step in the calculations requires a knowledge of decay constants, neutron capture cross-sections, decay branching ratios and independent fission yields. If this information is available, we can solve the coupled system of differential equations numerically or analytically and we will obtain the inventory of nuclei. The next step is to apply Eq. 2, which requires the inventory of nuclei previously determined from 3, a knowledge of the decay constants and the mean energies released per decay. In this chapter we will concentrate on this last part of the problem: how the mean beta and gamma energies are determined from experimental data and what is the best technique to determine them.

Most nuclear applications involving beta decay rely on data available from databases, see for example the Evaluated Nuclear Structure Data file (ENSDF) (ENSDF, n.d.). The compiled data are typically the result of the evaluation of different measurements, using different techniques, but until now they have been mainly based on the use of Ge detectors (the technique that uses Ge detectors is conventionally called the high resolution technique, since Ge detectors have a very good energy resolution ($\Delta E/E \sim 0.15\%$)). In such experiments the main goal is to determine the levels populated in the decay (feeding probability to the different nuclear levels), as well as the quantum numbers that characterize the levels, since these data provide the basic nuclear structure information. As part of the analysis the level scheme populated in the decay, based preferably on $\gamma\gamma$ coincidence relations, should be constructed and its consistency should be tested using the intensity balance of gamma rays that populate and de-excite the different levels. Depending on the case, these experiments can suffer from systematic uncertainties.

3.1 The pandemonium effect

One example of systematic uncertainty is the so called pandemounium effect (Hardy, 1977), introduced by Hardy and coworkers in 1977. Pandemonium is the seat of Satan, where chaos reigns, in the epic poem *Paradise Lost* by John Milton, but in nuclear physics it has a very different meaning. Here pandemonium stands for the problems we face when constructing a complex level scheme from high resolution data in a beta decay experiment. The main difficulty is related to the relatively poor efficiency of the Ge detectors. Since building the level scheme is based on the detection of the individual gamma rays and on the detection of coincidences between them, if the detectors have poor efficiency we may not detect some of the gamma rays emitted in the de-excitation process of the populated levels. This means that the resulting level scheme is incomplete and the feeding pattern is incorrectly determined. Fig. 1 shows the effect of the pandemonium effect in a simplified level scheme.

We can explain the effect using the picture in Fig. 1. The left panel depicts the "real" situation in this simplified case. A schematic beta decay that goes 100 % of the time to level 2 is shown. Following the beta decay, level 2 in the daughter nucleus is de-excited by a cascade of two gamma rays (γ_1 and γ_2). In a beta decay experiment using Ge detectors the aim will be to detect the gamma rays de-exciting the populated levels and from the intensity balance of the gamma rays populating and de-exciting the levels in the daughter nucleus one will infer the beta feeding ($F = I_{out} - I_{in}$). So, in an ideal experiment one will infer that the feeding to level $F_2 = I_{\gamma_2} - 0$ is 100 % and $F_1 = I_{\gamma_1} - I_{\gamma_2}$ is 0 %. In such an experiment the detection setup will have an efficiency for detecting the individual gamma rays of ϵ_1 and ϵ_2 respectively and the probability of detecting coincidences will be proportional to the product $\epsilon_1\epsilon_2$. What happens if in reality the efficiency for detecting γ_2 is very small and we do not see this transition in our spectra and(or) we miss the coincidence relationship? We will detect only γ_1 and we will assign the feeding probability to level 1 ($F_1 = I_{\gamma_1}$). This means that instead of assigning the feeding to level 2 (real situation), we will have an apparent feeding of 100 % to level 1 and the nuclear structure information deduced will be incorrect. If this happens, we say that the decay suffers from the pandemonium effect, but the real problem is that normally we do not know if the decay data suffers or not from the effect.

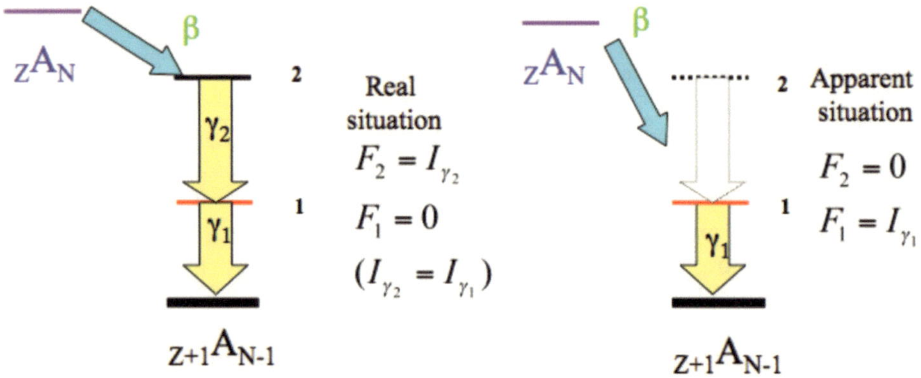

Fig. 1. Schematic decay to ilustrate the pandemonium effect. F represents the feeding (normalized to 100 %), which is determined from the intensity balance of gamma rays feeding and de-exciting the level and I_γ the gamma intensity.

Before entering into the details of how the pandemonium effect can be avoided it is worth to mention how the mean decay energies included in the summation calculations are determined. If we know the feeding probabilities ($f = F/100$) and the levels populated in the decay of nuclei i the mean γ and β energies released in the decay can be calculated according to the following:

$$\overline{E}_{\gamma,i} = \sum_j f_\beta(E_j)E_j \tag{6}$$

$$\overline{E}_{\beta,i} = \sum_j f_\beta(E_j) < E_{\beta j} > \tag{7}$$

where j runs over all levels fed in the daughter nucleus, $f_\beta(E_j)$ stands for the beta-feeding probability to level j, E_j is the excitation energy of level j in the daughter nucleus, and $< E_{\beta j} >$ is the mean energy of the beta particles emitted when feeding level j. This last quantity takes into account only the beta particle energies that feed the level j and does not include the energy taken away by the neutrinos. Since the betas have a continuous spectrum, $< E_{\beta j} >$ has to be calculated separately for each populated level j.

3.2 The total absorption technique

In the previous subsection it was assumed that to determine the feeding probability we depend on the detection of the gamma rays emitted following the beta decay. The main reason for this is that we cannot extract the feeding information easily from the detection of the beta particles. In a beta minus (plus) decay, the beta decay transition energy is shared between the beta minus (plus) particle, the recoiling nucleus and the anti-neutrino (neutrino). This means that if we measure the energy of the beta particles we do not have a discrete distribution, but a continuous one, in other words we are dealing with a three-body problem. Extracting information on the feeding probability from a continuous spectrum is not an easy task, which is why people prefer to work with the gamma-ray spectra associated with the beta decay. This situation is illustrated in Fig. 2 where again we show the ideal beta decay presented in Fig.

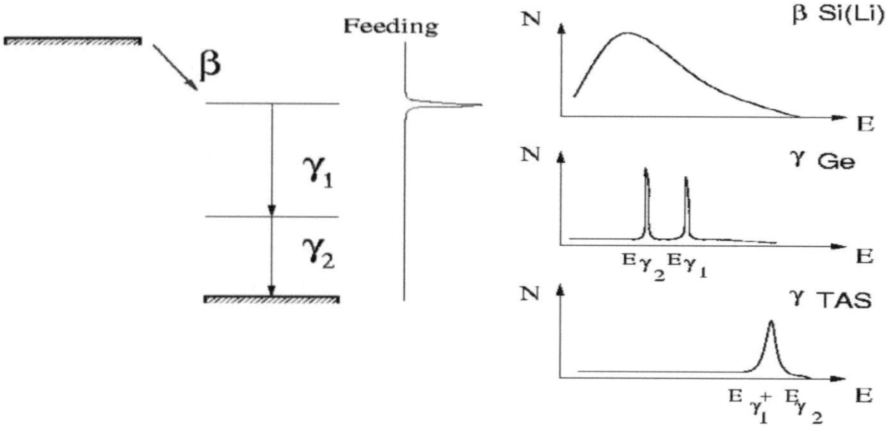

Fig. 2. Ideal beta decay seen by different detectors

1, but in this case we emphasize how the decay is seen by different detectors. The discrete nature of the gamma spectrum is what makes the problem tractable.

We have already discussed the pandemonium effect, which is related to the relatively low efficiency of the Ge detectors. If we have to rely on the detection of the gamma transitions then what can be done? It appears that the only solution is to increase drastically the gamma detection efficiency. In this way we arrive to the total absorption spectrometer (TAS) concept, which is presented in Fig. 3.

A TAS can be constructed using a large volume of detector material with high intrinsic efficiency for gamma rays, which surounds the radioactive source in a 4π geometry. In the left panel of Fig. 3 an ideal TAS is presented. There is also a change in the detection "philosophy" used in the TAS technique compared with high resolution (Ge detectors). Instead of detecting the individual gamma rays, the idea here is to detect the gamma cascades that follow the beta decay. So in principle, with an ideal TAS, the measured spectrum is proportional to the feeding probability of the decay. Now we can fully explain the rigth-hand panel of Fig. 2. What we see in the right panel is how the decay of the left panel is seen by different detectors. In the upper panel the betas detected by a Si detector are seen, which shows a continuous spectrum from which the beta feeding information is very difficult to extract if more levels are fed in the decay. The middle graph shows the individual gammas detected in an ideal high resolution (Ge) experiment, where the two individual gammas that follow the beta decay are detected. The lower graph shows how the decay is seen by an ideal TAS detector. In this case a spectrum is shown, which is proportional to the feeding pattern of the beta decay.

The left panel of Fig. 3 shows a photo of a real TAS, *Lucrecia*, which was installed by the Madrid-Strasbourg-Surrey-Valencia international collaboration at ISOLDE (CERN). It has a cylindrical geometry (∅=h=38 cm), with a longitudinal hole perpendicular to the symmetry axis. The light produced in the NaI scintillator material by the detected radiation is read by eight photomultipliers. This photograph also shows the main reason why an ideal 100% efficient TAS can not be constructed. For the measurements we need to place the sources in the centre of the crystal, and to achieve this we need the telescopic tube that is seen in the foreground of the photograph. The hole, the dead material of the tube and of the tape transport system needed to take the sources to the centre of the detector makes the detector less efficient. The second reason is that we would need very large detectors indeed to have a truly 100 % efficiency.

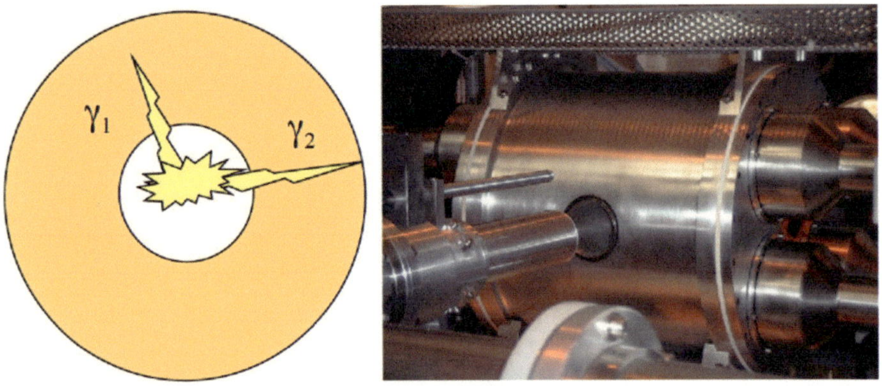

Fig. 3. Total absorption spectrometer. On the left panel an ideal TAS is presented, on the right a photo of the *Lucrecia* TAS installed at Isolde (CERN) can be seen.

The fact that we can not build a 100 % efficient detector makes the analysis of the TAS experiments difficult. To extract the feeding information we need to solve the following:

$$d = R(B)f \tag{8}$$

where d represents the measured TAS spectrum (free of contaminants), $R(B)$ is the response matrix of the detector and f is the feeding distribution we would like to determine. In this equation B represents the branching ratios of the levels populated in the beta decay. We have developed algorithms and techniques that allow us to solve this problem. Because of their complexity we will not discuss them here in detail and the interested reader is encoraged to look at Refs (Tain, 2007; Cano, 1999). An example of a recent application of the technique, where details of how the analysis is performed, is presented in (Estevez, 2011). This recent work has implications for neutrino physics.

3.3 Pandemonium and decay heat

As mentioned earlier, to calculate the mean energies the feeding probability to the different levels populated in the decay is needed (Eqs. 6, 7). Now we can understand why TAS measurements can have an impact and should be applied to decay heat studies. If we have a beta decay that suffers from the pandemonium effect, the mean gamma energies will be underestimated and the beta energies will be overestimated. This is shown schematically in Fig. 4. Since the application of the TAS technique is the only way to avoid the pandemonium effect, this technique should be used for studying decays that are important for the decay heat problem.

Our interest in the decay heat topic was triggered by the work of Yoshida and coworkers (Yoshida, 1999). At the begining of the 1990s one of the most successful data bases for summation calculations was the JNDC-V2 (Japanese Nuclear Data Committee version 2) database. In this database the gross theory of beta decay (Takahashi & Yamada, 1969) was used to supplement experimental data that might suffer from the pandemonium effect. For example in (Tasaka, 1988) it is mentioned that the mean energies of many decays having a beta

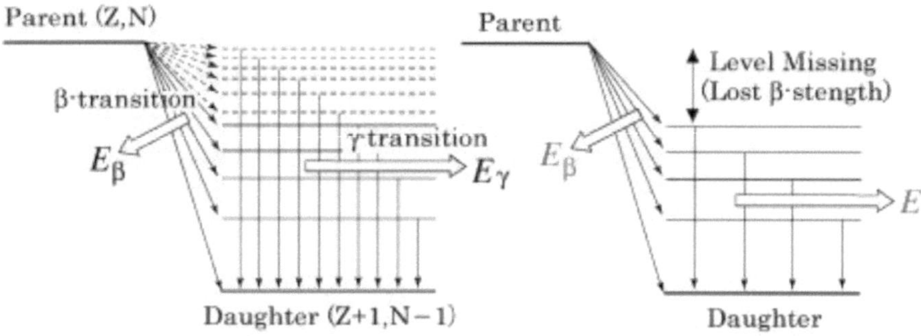

Fig. 4. Effect of missing levels in the mean energies.

decay Q value larger that 5 MeV were supplemented by theoretical values calculated using the gross theory. Even though the database in general worked well for the gamma component of the ^{239}Pu decay heat, there was a significant remaining discrepancy in the 300-3000 s cooling time. Yoshida studied the possible causes for the discrepancy and proposed several possible explanations. The most plausible was the possibility that the decay energies of some nuclides with half-lives in the range of 300-3000 s or with precursors with similar half-lives suffered from the pandemonium effect. After a careful evaluation of possible candidates he proposed that the decays of $^{102,104,105}Tc$ should be measured with the total absorption technique.

In the process of defining priorities for the TAS measurements related to decay heat, contact with specialists in the field was established and a series of meetings were held under the auspices of the International Atomic Energy Agency (IAEA). As a result of these meetings a list of nuclei that should be measured with the total absorption technique was defined (Nichols, 2007). This list included the Yoshida cases ($^{102,104,105}Tc$) and some additional nuclei (in total 37). The nuclei were identified based on their contribution to the decay heat in different fuels, and in order to reduce the discrepancies between the major international databases (JENDL (Shibata, 2011), JEFF (Kellett, 2009), ENDF (Chadwick, 2006)). Additionally, this list included some cases that deserve to be measured for other reasons. For example there are TAS measurements that were performed in the past by Greenwood and coworkers (Greenwood, 1992). These TAS measurements were analyzed using different procedures from those used nowadays. It is important to verify the results and compare them with new measurements using different analysis techniques to look for systematic uncertainties. Similarly the measurements of (Rudstam, 1990), obtained from direct spectral measurements can be checked against TAS measurements (Taín & Algora, 2006).

4. Experiments and experimental techniques

We have started a research programme aimed at the study of nuclei included in the list of the WPEC-25 group of the IAEA (Nichols, 2007). To perform a succesful experiment, the first step is to define the best facility to carry it out. In recent years there have been extensive developments in the methods used to produce radioactive beams. Indeed, we are living a period of renaissence and renewal for these facilities, since new ones have been constructed and others are under development, construction and upgrade (FAIR, HIE-ISOLDE, HRIBF, etc.). In essence these facilities are based on two main methods of producing radioactive beams, the ISOL method and the fragmentation or In-Flight method. In the ISOL technique an intense particle beam, impinges on a thick target. After diffusion and effusion, the radioactive nuclei produced are mass selected, ionized and re-accelerated in a post-accelerator. This method in general produces cleaner nuclear species, but requires specific developments of the ion sources, which involve chemistry and physics aspects. Its major constrain is related to the production of very short-lived isotopes and the production of beams of refractory elements that are difficult to extract from ion sources. The fragmentation technique, as the name implies, is based on the fragmentation of high energy projectiles on target nuclei and the subsequent separation and selection in-flight of the radioactive nuclei produced using magnetic spectrometers. With this technique it is possible to study very short-lived isotopes, but typically the nuclei are produced in a less clean enviroment. In this case the experiments rely on the identification of the nuclear species produced on an ion-by-ion basis.

Some of the nuclei included in the WPEC-25 list are refractory elements, so it was not possible to produce them in "conventional" ISOL facilities such as ISOLDE(CERN). So for our experiments we decided to use the Ion-Guide Isotope Separator On-Line (IGISOL) facility of the University of Jyväskylä (Äystö, 2001). In this facility the ion guide method was developed, which can be considered to be a "chemistry" independent ISOL method (Dendooven, 1997). The working principle of the ion guide method is that the radioctive nuclides are produced in a thin target after bombarment with the accelerator beam. The reaction products (recoils) fly out of the target and are transported by a differential pumping system to the first stage of the accelerator. The mean path of the recoils is optimized in such a way that they survive as singly charged ions. By this method we obtain a system, which is chemically insensitive and very fast (ms). A schematic picture of the ion-guide principle is presented in Fig. 5.

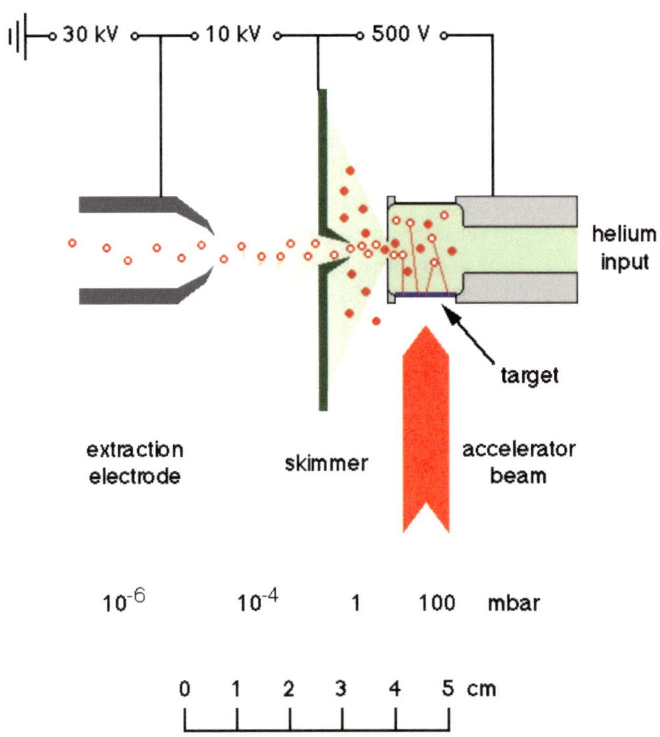

Fig. 5. The ion guide principle.

At the IGISOL facility several ion guides were developed. For our experiments we used a fission ion guide, which allows the extraction of fission products produced in proton-induced fission. This was the chosen method for the production of the radioactive nuclides of interest. Following the extraction, the ions can be separated using a dipol magnet. The magnetic field of the dipol bends the ions to different trajectories depending on their charge-to-mass ratio. Since most of the ions come out of the ion guide singly charged, mass separation is achieved. To characterize the separation quality, the mass resolving power of the system is used, which is a measure of how well species with different masses are separated ($\Delta m/m$). At IGISOL the mass resolving power typically varies from 200 to 500, depending on the experimental conditions. This is not enough to separate ions that have the same mass number (isobars), but it is adequate to separate isotopes of the same element. That is why the instrument is called an isotope separator. For the TAS measurements this separation is not enough, since you can have several isobars produced in fission that can not be separated with the isotope separator, and their decay will appear as contaminants in the TAS spectrum. An advantage of the IGISOL facility is that a Penning trap system (JYFLTRAP) (Kolhinen, 2004) can be used for further isobaric separation. Penning traps are devices for the storage of charged particles using a homogeneous static magnetic field and a spatially inhomogeneous static electric field. This kind of trap is particularly well suited for precision mass spectroscopy, but they can also be used as high resolution separators for "trap-assisted" spectroscopic studies as in our experiments.

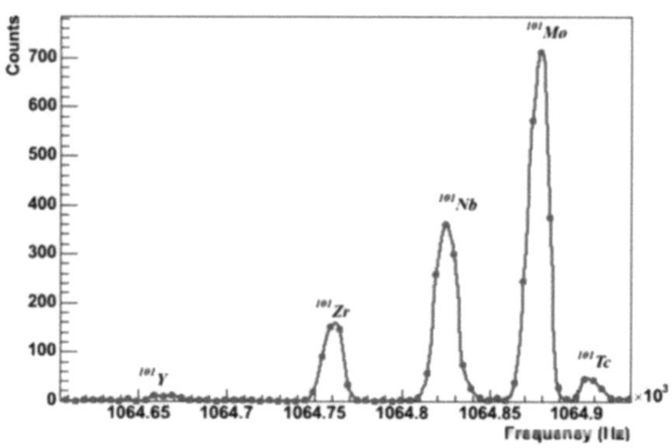

Fig. 6. Mass scan in the Penning trap for A=101 fission products. During the experiment the frequency corresponding to the isotope of interest is set in the trap, and then a very pure beam can be used for the measurements.

With a Penning trap system a mass resolving power of the order of 10^5 and even 10^6 can be achieved. The good separation is shown in Fig 6, where a frequency scan in the Penning trap for mass A= 101 prior to one of our experiments is presented. Once a frequency in the trap is set for a particular isobar, a very pure radioactive beam can be obtained. The only disadvantage of this system is the relatively low intensity of the ion beam, since the transmission of the trap is only a few percent.

Figure 7 shows an schematic picture of the setup used in our experiments in Jyväskylä. In this setup the radioactive beam coming from the trap is implanted in a tape system that allows us to transport the radioactive sources to the measuring position and to remove the undesired daughter activities. The cycles of the tape are optimized according to the half-life of the isotope of interest.

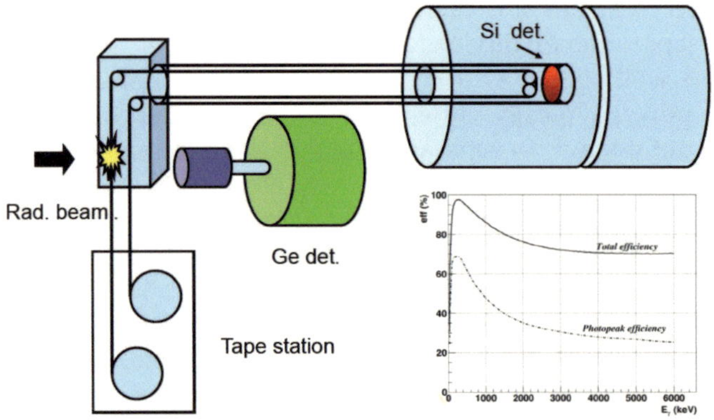

Fig. 7. Schematic picture of our experimental setup at IGISOL for TAS measurements. In the inset the peak and total efficiency of the used TAS is presented.

4.1 First trap assisted TAS experiment

At IGISOL we have performed two trap-assisted TAS experiments related to the decay heat problem. Here we will discuss briefly the first experiment and its impact. The second experiment is presently in the analysis phase. It is worth noting that in the second experiment we have used for the first time a segmented BaF_2 TAS detector which additionally provides information on the multiplicity of the gamma cascades following the beta decay. This extra information is useful for the analysis of the complex beta decay data.

The analysis of a TAS experiment is a lengthy procedure and requires several stages. The first phase requires a careful evaluation of the contaminants and distortions of the measured spectrum in order to determine *d*. Then the calibration of the experimental data in energy and width and a precise characterization of the TAS detector using Monte Carlo (MC) techniques is

Nuclide	$T_{1/2}$ s	$\overline{E_\gamma}$ ENDF	$\overline{E_\gamma}$ TAGS	$\overline{E_\beta}$ ENDF	$\overline{E_\beta}$ TAGS
^{101}Nb	7.1(3)	270(22)	445(279)	1966(307)	1797(133)
^{105}Mo	35.6(16)	552(24)	2407(93)	1922(122)	1049(44)
^{102}Tc	5.28(15)	81(5)	106(23)	1945(16)	1935(11)
^{104}Tc	1098(18)	1890(31)	3229(24)	1595(75)	931(10)
^{105}Tc	456(6)	668(19)	1825(174)	1310(205)	764(81)
^{106}Tc	35.6(6)	2191(51)	3132(70)	1906(67)	1457(30)
^{107}Tc	21.2(2)	515(11)	1822(450)	2054(254)	1263(212)

Table 2. Comparison of mean gamma and beta energies included in the ENDF/B-VII database with the results of the analysis of our measurements (in keV).

required. For the MC simulations the GEANT4 code (Agostinelli, 2007) is used. In this phase a careful characterization of the setup is performed until measurements with conventional radioactive sources like ^{24}Na, ^{137}Cs and ^{60}Co are very well reproduced by the MC code. Once this has been achieved, the response function of the detector to the decay of interest can be calculated. This requires the definition of the level scheme that may be populated in the decay (B or branching ratio matrix). To construct the branching ratio matrix we take into account known levels up to a certain excitation in the daughter (E_{cut}) and above that cut-off energy we use a statistical model to generate levels and their branchings. The information on the low-lying levels and their branchings is taken from conventional high resolution measurements, because this information is correct in general if available (known levels). For the statistical model we use a back shifted Fermi formula for the level density and gamma strength functions, which define the probabilities that gamma rays connect the different levels (known and unknown part). Once the B is determined, the $R(B)$ is calculated from the MC responses of the detector to the different γ and β transitions and 8 is solved. As part of the analysis, the cut-off energy, the accepted low-lying levels and the parameters of the statistical model are changed if necessary. The final result of the analysis is a feeding distribution, from which nuclear structure information can be obtained in the form of the beta strength distribution (S_β), and in the case of the decay heat application mean beta and gamma energies can be calculated (6 and 7).

The impact of our first experiment can be seen in Table 2, where the mean energies of the ENDF database are compared with the results of our measurements. From this table the relevance of performing experiments is also clear. Two nuclei (^{101}Nb, ^{102}Tc), that were suspected to suffer from the pandemonium effect did not, even though they have large Q_β decay values. The remaining nuclei all suffered from the effect (see for example the large increase in the mean gamma energy and the reduction in the mean beta energy with respect to high resolution measurements for ^{104,105}Tc). In Fig. 8 the results for the gamma component of ^{239}Pu are presented. They are compared with ENDF (ENSDF, n.d.) before and after the inclusion of our new data (Sonzogni, n.d.). Similar conclusions have been obtained recently using the JEFF database (Mills, n.d.). The new TAS results were published in (Algora, 2010).

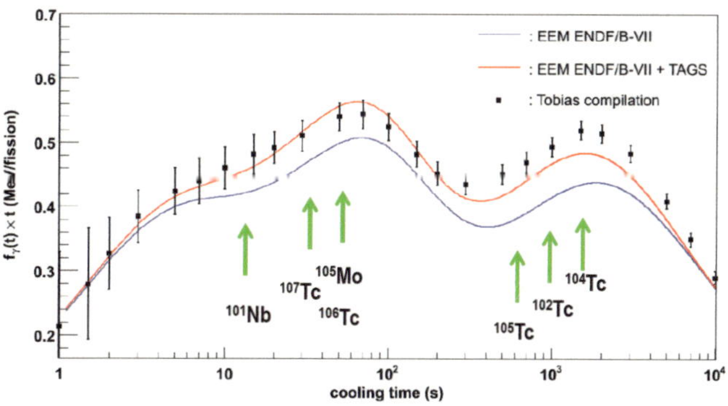

Fig. 8. Comparison of the summation calculations for the gamma component of the decay heat in ^{239}Pu. The experimental points with errors are taken from the Tobias compilation (Tobias, 1989). The blue line represent the results obtained using the ENDF data base without the inclusion of the new results (Algora, 2010). The red line represents the results after the inclusion of the new TAS measurements. The cooling time at which the contribution of the measured nuclei is maximal is represented by arrows.

As a result of our measurements, a large part of the discrepancy pointed out by (Yoshida, 1999) in the 300-3000 s cooling interval and additionally the discrepancy at low energies has been solved within the ENDF database. The new data were also used to perform summation calculations for the gamma component of ^{235}U. In this case the results were disappointing. Our new results had very little impact. This can be understood in terms of the cummulative fission yields of the nuclei in question. They sample approximatelly 33.8 % of the fission in ^{239}Pu, but only 13.5 % in ^{235}U. Additionnally from the 13.5 % in ^{235}U, ^{101}Nb, ^{102}Tc amounts to 9.2 %, which does not bring a large change in the mean energies. This explains why our measurements to date can only represent a relative change of approximatelly 4.3 % in ^{235}U, compared to the 22.6 % relative impact in ^{239}Pu with respect to the earlier values of the ENDF database.

5. Conclusions and outlook

In this chapter we have described how total absorption measurements can play an important role in improving the beta decay data necessary for summation calculations. We have discussed the technique and how its combination with IGISOL and the JYFL Penning trap has allowed us to perform measurements that had a large impact in the decay heat of ^{239}Pu. These measurements can also be relevant for other reasons. The beta feeding distributions can also be used to deduce the beta strength and test nuclear models. This region (nuclear mass $\sim$ 100) is interesting from the point of view of nuclear structure. For example it has been suggested that triaxial shapes play a role in this region of the nuclear chart (Möller, 2006). Most known nuclei have prolate (rugby ball shape) or spherical shapes in their ground state. If triaxiality plays a role in the structure of these nuclei, this will afect the distribution of the strength in the daughter and it may be studied using the TAS technique. Actually, we have previously used the TAS technique to infer shape effects in the A $\sim$ 70 region (Nacher, 2004) and have started recently a related research programme in the lead region (Algora, 2005).

We plan to continue to make measurements using the TAS to obtain data of relevance to decay heat, but it is important to mention that there also similar efforts ongoing in other facilities and by other groups. In Argonne National Laboratory (Chicago, USA), there is a new facility under construction *CARIBU*, that will allow for the production of neutron-rich species from the fission of ^{252}Cf. Here there are plans to use again the TAS detector employed in the measurements of Greenwood (Greenwood, 1992) and coworkers. Another example is the development of the MTAS detector by the group of Rykaczewski and coworkers, that will be used at the HRIBF facility at Oak Ridge (USA). These new facilities will contribute in the future to improving the quality of beta decay data for the decay heat application.

Our work had a large impact in ^{239}Pu, but there is still a large amount of work to be done for ^{235}U as was mentioned in the previous section. Additionally decays relevant for other fuels like ^{232}Th should be also studied. Recent work by Nichols and coworkers has identified which nuclei should be measured for the ^{232}Th fuel (Gupta, 2010).

Another aspect worth mentioning is the possible impact of these measurements in the prediction of the neutrino spectrum from reactors. In the same fashion as beta and gamma summation calculations are performed, neutrino summation calculations can be done for a working reactor. Because of the very small interaction cross section, neutrinos leave the reactor almost without interaction in the core. They carry information on the fuel composition and on the power level and their flux can not be shielded or controlled. Because of the small interaction cross section with matter they are difficult to detect ($\sim 10^{-43} cm^2$), but they are produced in very large numbers from the fission products. For example, approximately six antineutrinos are produced per fission, and a one GW_{el} reactor produces of the order of 10^{21} neutrinos every second. The precision of the neutrino spectrum measurements can be important for neutrino oscillation experiments in fundamental physics experiments like Double CHOOZ and for non-proliferation applications (Fallot, 2007). There is presently a working group of the IAEA, which studies the feasibility of building neutrino detectors, which if positioned outside and close to a nuclear reactor can be used to monitor the power level and the fuel composition of the reactor. These measurements, if they reach the necessary precision, can be used to indicate the fuel used and to monitor manipulations of the fuel in a non-intrusive way (Porta & Fallott, 2010). We plan future measurements to address this topic of research.

6. References

K. Way and E. Wigner, Phys. Rev. 73 1318 (1948)

G. Rudstam *et al.*, Atom. Dat. and Nucl. Dat. Tabl. 45, 239 (1990)

J. K. Dickens *et al.*, Nucl. Sci. Eng, 78, 126 (1981)

K. Tasaka, J. Katakura, T. Yoshida, Nuclear Data for Science and Technology (1988 MITO), p 819-826, 1988

V. E. Schrock *et al.* EPRI Report, NP616, Vol. 1, 1978

J. L. Yarnell and P. J. Bendt, Los Alamos Scientific Laboratory Report, LA-NUREG-6713, 1977

V. E. Schrock, Progress in Nucl. Energy, Vol. 3. pp. 125-156, 1979

A. Tobias, Progress in Nucl. Energy, Vol. 5. pp. 1-93, 1980

K. Tasaka, J. Katakura, T. Yoshida, Nuclear Data for Science and Technology (1988 MITO), p 819-826, 1988

M. Akiyama, S. An, Proc. Int. Conf. on Nuclear Data for Science and Technology, Antwerp, p. 237, 1982 and references therein

J. K. Dickens *et al.*, Nucl. Sci. Eng 74, 106, (1980)

H. V. Nguyen *et al.*, Proc. Int. Conf. on Nucl. Data for Science and Technology, Triestre, p. 835 (1997)

A. Tobias, CEGB Report No. RD/B/6210/R89, 1989

http://www.nndc.bnl.gov/ensdf

J. C. Hardy *et al*, Phys. Letts 71B (1977) 307.

J. L. Tain, D. Cano-Ott, Nucl. Instrum. and Meth. Phys. Res. A 571 (2007) 728 and 719.

D. Cano-Ott *et al.*, Nucl. Instrum. and Meth. Phys. Res. A 430 (1999) 488 and 333.

E. Estevez *et al.*, Phys. Rev. C 84 (2011) 034304

T. Yoshida *et al*, J. Nucl. Sci. and Tech. 36 (1999) 135.

K. Takahashi, M. Yamada, Prog. Theor. Phys. 41 (1969); K. Takahashi, Prog. Theor. Phys. 45 (1971) 1466; T. Tchibana, M. Yamada, Y. Yoshida, Prog. Theor. Phys. 84 (1990) 641

A. Nichols, NEA report NEA/WPEC-25 (2007) 1.

K. Shibata *et al.*, J. Nucl. Sci. and Tech. 48 (2011) 1; K. Shibata *et al.*, J. Nucl. Sci. and Tech. 39 (2002) 1125

M. A. Kellett, *et al.*, JEFF Report 20, OECD 2009, NEA No. 6287; A. J. Koning *et al.*, Proc. of the Int. Conf. on Nucl. Data for Science and Technology, Nice, 2007

M. B. Chadwick *et al.*, Nucl. Data Sheets 107 (2006) 2931

R. C. Greenwood *et al.*, Nucl. Instrum. and Meth. Phys. Res. A 390 (1997) 95, Nucl. Instrum. and Meth. Phys. Res. A 314 (1992) 514

J. L. Tain, A. Algora, IFIC-06-1 Report, 2006

J. Äystö, Nucl. Phys. A 693 (2001) 477.

P. Dendooven, Nucl. Instrum. Meth. Phys. Res. B 126 (1997) 182

V. Kolhinen et al., Nucl. Instrum. Methods Phys. Res., Sect. A 528 (2004) 776

S. Agostinelli *et al*, Nucl. Instrum. and Meth. Phys. Res. A 506 (2003) 250.

A. Sonzogni, private communication

A. Algora *et al*, Phys. Rev. Letts. 105 (2010) 202501; D. Jordan, PhD thesis, Valencia, 2010; D. Jordan *et al*, in preparation

R. W. Mills, private communication

P. Möller *et al*, Phys. Rev. Letts. 97, 162502 (2006)

E. Nacher *et al*, Phys. Rev. Lett. 92, 232501 (2004)

A. Algora *et al*, ISOLDE Experimental Proposal IS440, CERN-INTC-2005-027, INTC-P-199

M. Gupta *et al*, INDC(NDS)-0577

M. Fallot *et al*, Proceedings of the Int. Conf. on Nucl. Data for Science and Technology 2007, p. 1273.

A. Porta, M. Fallott, JEFF Meeting 2010 and private communication.

6

Transport of Interfacial Area Concentration in Two-Phase Flow

Isao Kataoka, Kenji Yoshida, Masanori Naitoh, Hidetoshi Okada and Tadashi Morii
Osaka University, The Institute of Applied Energy, Japan Nuclear Energy Safety Organization
Japan

1. Introduction

The accurate prediction of thermal hydraulic behavior of gas-liquid two-phase flow is quite important for the improvement of performance and safety of a nuclear reactor. In order to analyze two-phase flow phenomena, various models such as homogeneous model, slip model, drift flux model and two-fluid model have been proposed. Among these models, the two-fluid model (Ishii (1975), Delhaye (1968)) is considered the most accurate model because this model treats each phase separately considering the phase interactions at gas-liquid interfaces. Therefore, nowadays, two-fluid model is widely adopted in many best estimate codes of nuclear reactor safety. In two-fluid model, averaged conservation equations of mass, momentum and energy are formulated for each phase. The conservation equations of each phase are not independent each other and they are strongly coupled through interfacial transfer terms of mass, momentum and energy through gas-liquid interface. Interfacial transfer terms are characteristic terms in two-fluid model and are given in terms of interfacial area concentration (interfacial area per unit volume of two-phase flow) Therefore, the accurate knowledge of interfacial area concentration is quite essential to the accuracy of the prediction based on two-fluid model and a lot of experimental and analytical studies have been made on interfacial area concentration. In conventional codes based on two-fluid model, interfacial area concentration is given in constitutive equations in terms of Weber number of bubbles or droplets depending upon flow regime of two-phase flow (Ransom et al. (1985), Liles et al. (1984)). However, recently, more accurate and multidimensional predictions of two-phase flows are needed for advanced design of nuclear reactors. To meet such needs for improved prediction, it becomes necessary to give interfacial area concentration itself by solving the transport equation. Therefore, recently, intensive researches have been carried out on the models, analysis and experiments of interfacial area transport throughout the world

In view of above, in this chapter, intensive review on recent developments and present status of interfacial area concentration and its transport model will be carried out.

2. The definition and rigorous formulation of interfacial area concentration

Interfacial area concentration is defined as interfacial area per unit volume of two-phase flow. Therefore, the term "interfacial area concentration" is usually used in the meaning of volume averaged value and denoted by $\overline{a_i}^{V}$. For example, one considers the interfacial area concentration in bubbly flow as shown in Fig.1. In this figure, A_i is instantaneous interfacial area included in volume, V. The volume averaged interfacial area concentration is given by

$$\overline{a_i}^{V} = \frac{A_i}{V} \tag{1}$$

For simplicity, bubbles are sphere of which diameter is d_b, interfacial area concentration is given by

$$\overline{a_i}^{V} = \frac{\pi N d_b^{\,2}}{V} = \frac{6\alpha}{d_b} \tag{2}$$

Here, N is number of bubbles in volume V, and α is void fraction (volumetric fraction of bubbles in volume V).

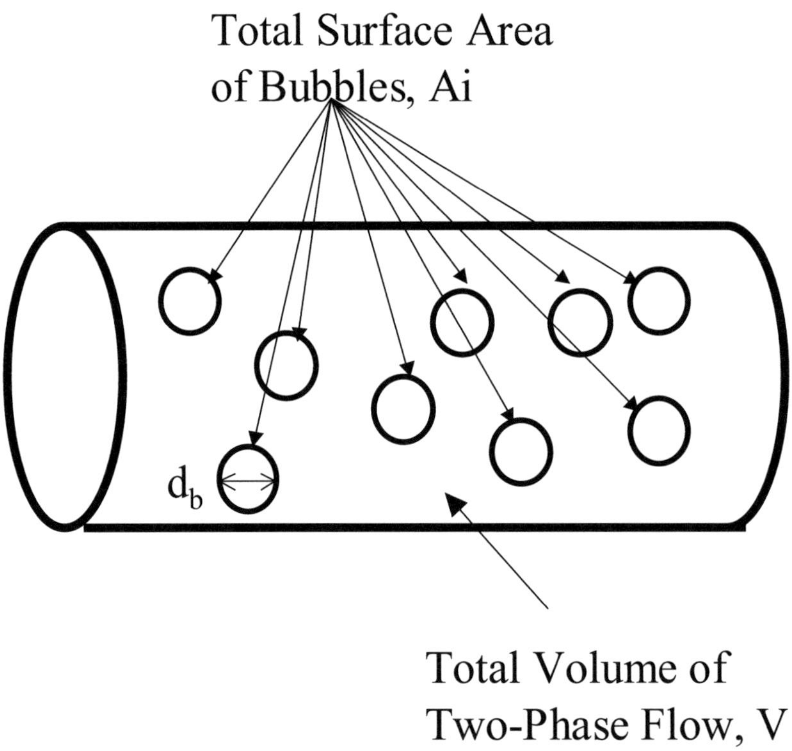

Fig. 1. Interfacial area in bubbly flow

Similarly, time averaged interfacial area concentration, $\overline{a_i}$ and statistical averaged interfacial area concentration, $\overline{a_i}^{A}$ can be defined. The transport equation of interfacial area concentration is usually given in time averaged form in terms of time averaged interfacial area concentration, $\overline{a_i}$. However, for the derivation of the transport equation, it is desirable to formulate interfacial area concentration and its transport equation in local instant form.

Kataoka et al. (1986), Kataoka (1986) and Morel (2007) derived the local instant formulation of interfacial area concentration as follows.

One considers the one dimensional case where only one plane interface exists at the position of $x=x_0$, as shown in Fig.2. In the control volume which encloses the interface in the width of Δx, as shown in Fig.2, average interfacial area concentration is given by

$$\overline{a_i} = \frac{1}{\Delta x} \tag{3}$$

When one takes the limit of $\Delta x \to 0$, local interfacial area concentration, a_i, is obtained. It takes the value of zero at $x \neq x_0$ and infinity at $x=x_0$. This local interfacial area concentration is given in term of delta function by

$$a_i = \delta(x - x_0) \tag{4}$$

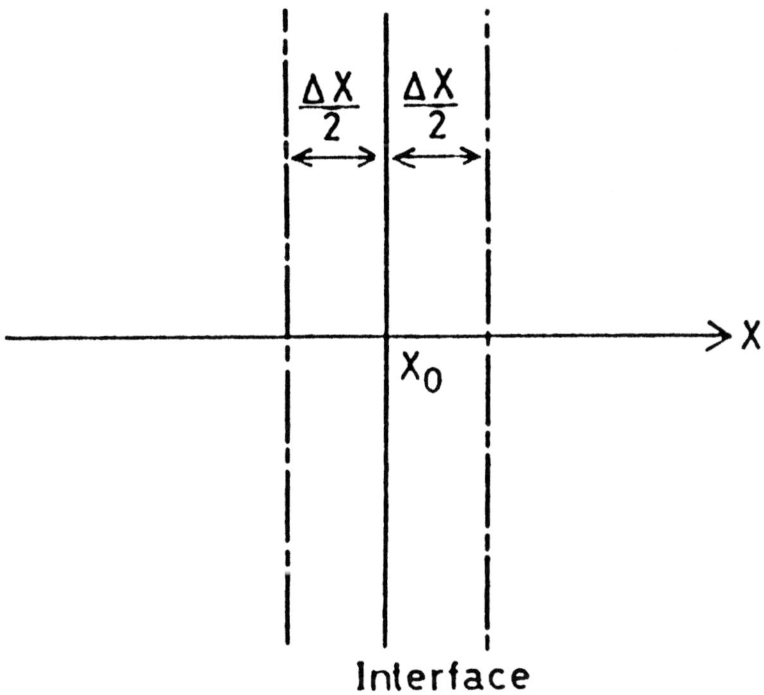

Fig. 2. Plane interface at $x=x_0$

This formulation can be easily extended to three dimensional case. As Shown in Fig.3, three dimensional interface of gas and liquid is mathematically given by

$$f(x,y,z,t)=0 \tag{5}$$

$$f(x,y,z,t)>0 \text{ (gas phase)}, \quad f(x,y,z,t)<0 \text{ (liquid phase)} \tag{6}$$

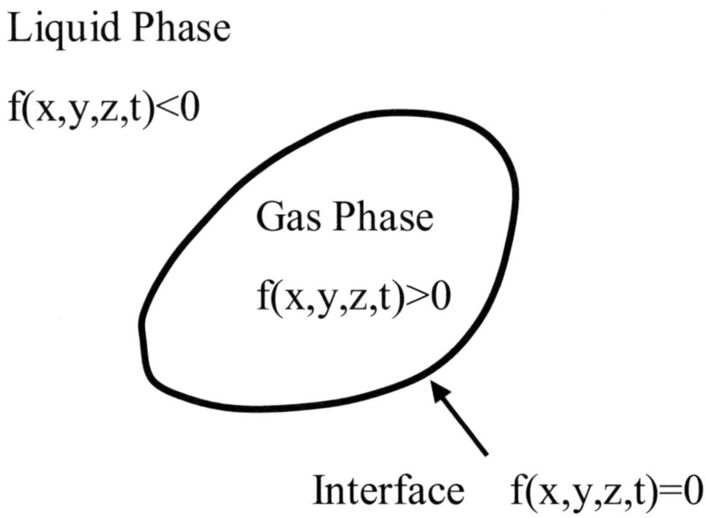

Fig. 3. Mathematical representation of three-dimensional interface

As shown in Fig.4, one considers the control volume which encloses the interface by following two surfaces.

$$f(x,y,z,t)=\Delta f/2 \tag{7}$$

$$f(x,y,z,t)=-\Delta f/2 \tag{8}$$

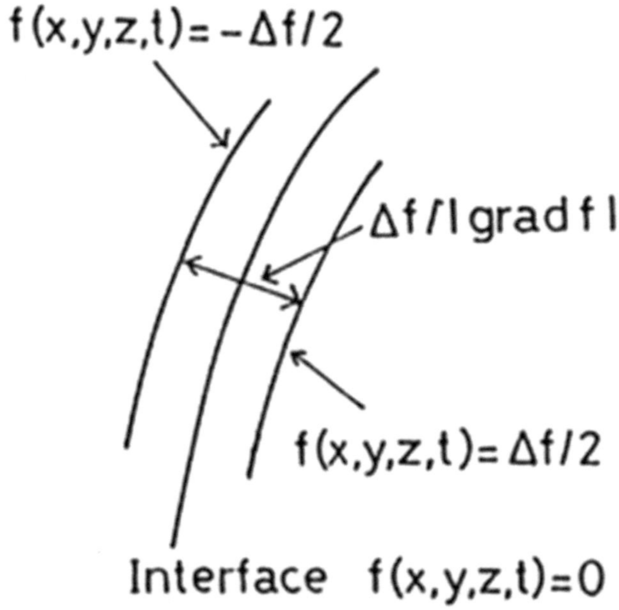

Fig. 4. Control volume enclosing three-dimensional interface

By the differential geometry, the width of the control volume is given by

$$\Delta f / |\text{grad } f(x,y,z,,t)|$$

Then, average interfacial area concentration in this control volume is given by

$$\overline{a_i} = |\text{grad } f(x,y,z,t)| / \Delta f \tag{9}$$

When one takes the limit of $\Delta f \to 0$, local interfacial area concentration, a_i, is obtained by

$$a_i = |\text{grad } f(x,y,z,t)|\, \delta(f(x,y,z,t)) \tag{10}$$

where $\delta(w)$ is the delta function which is defined by

$$\int_{-\infty}^{\infty} g(w)\delta(w - w_0)dw = g(w_0) \tag{11}$$

where $g(w)$ is an arbitrary continuous function..

In relation to local instant interfacial area concentration, characteristic function of each phase (denoted by ϕ_k) is defined by

$$\phi_G = h(f(x,y,z,t)) \quad \text{(gas phase)} \tag{12a}$$

$$\phi_L = 1 - h(f(x,y,z,t)) \quad \text{(liquid phase)} \tag{12b}$$

where suffixes G and L denote gas and liquid phase respectively. ϕ_k is the local instant void fraction of each phase and takes the value of unity when phase k exists and takes the value of zero when phase k doesn't exist. Here, $h(w)$ is Heaviside function which is defined by

$$\begin{aligned} h(w) &= 1 \ (w>0) \\ &= 0 \ (w<0) \end{aligned} \tag{13}$$

Heaviside function and the delta function are related by

$$\delta(w) = \frac{dh(w)}{dw} \tag{14}$$

Using above equations, the derivatives of characteristic function are related to interfacial area concentration as follows.

$$\text{grad } \phi_k = -n_{ki} a_i \quad (k = G, L) \tag{15}$$

$$\frac{\partial \phi_k}{\partial t} = v_i \bullet n_{ki} a_i \quad (k = G, L) \tag{16}$$

Here, $\mathbf{n}_{ki}$ is unit normal outward vector of phase k as shown in Fig.5 and $\mathbf{v}_i$ is the velocity of interface.

Using above-mentioned relations, it is shown that local instant interfacial area concentration is given in term of correlation function of characteristic function (Kataoka (2008)). As shown in

Eq.(15), local instant interfacial concentration is related to the derivative of characteristic function of each phase. Here, directional differentiation of characteristic function is considered. Spatial coordinate (x,y,z) is denoted by vector **x** and displacement vector of **r** is defined.

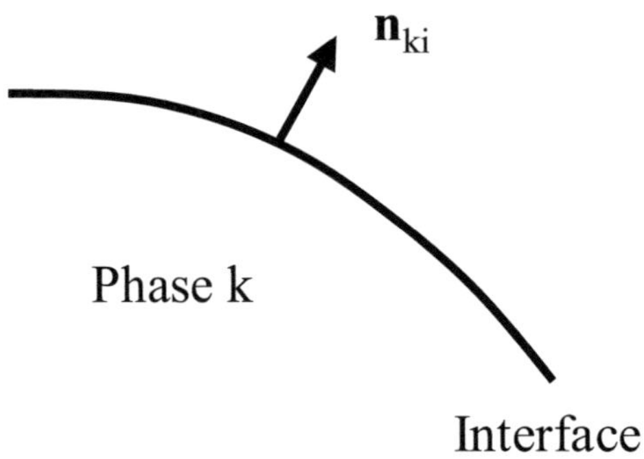

Fig. 5. Unit normal outward vector of phase k

$$x=(x,y,z) \tag{17}$$

$$r=(r_x,r_y,r_z) \tag{18}$$

At position **x**, directional differentiation of characteristic function $\phi_k(\mathbf{x})$ in **r** direction (denoted by $\frac{\partial}{\partial r}$) is defined by

$$\begin{aligned} \frac{\partial}{\partial r}\phi_k(x) &= n_r \bullet \text{grad}\phi_k(x) \\ &= -n_r \bullet n_{ki} a_i \\ &= -\cos\theta \;\; a_i \end{aligned} \tag{19}$$

Here, $\mathbf{n}_r$ is unit vector of **r** direction and θ is the angle between $\mathbf{n}_r$ and $\mathbf{n}_{ki}$ as shown in Fig. 6.

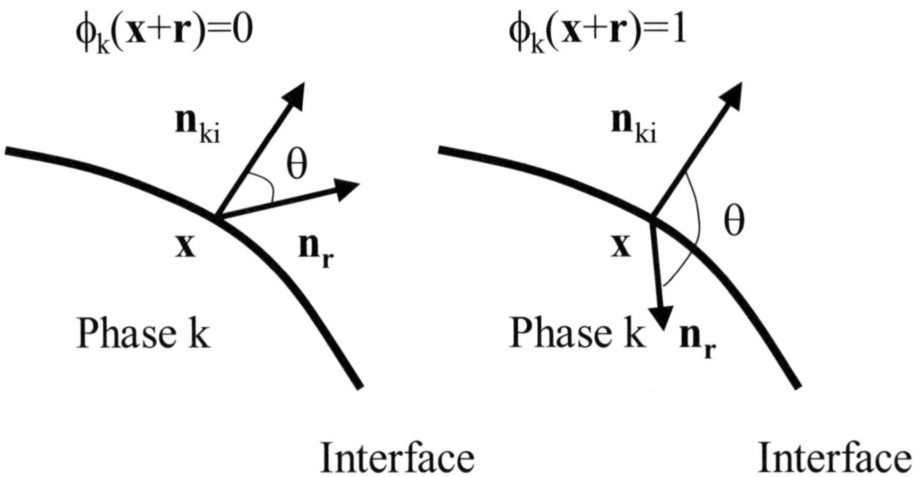

Fig. 6. Configuration of $\mathbf{n}_r$ and $\mathbf{n}_{ki}$

In view of Fig.6, the product of $\phi_k(\mathbf{x+r})$ and Eq(19) is given by

$$\begin{aligned}\phi_k(x+r)\frac{\partial}{\partial r}\phi_k(x) &= 0 \qquad (0 \le \theta \le \pi/2)\\ &= -\cos\theta \; a_i \quad (\pi/2 \le \theta \le \pi)\end{aligned} \tag{20}$$

Equation (19) is rewritten by

$$\phi_k(x+r)\frac{\partial}{\partial r}\phi_k(x) = \frac{1}{2}(-\cos\theta + |\cos\theta|)a_i \tag{21}$$

From Eqs.(19) and (21), one obtains

$$\frac{\partial}{\partial r}\phi_k(x) - 2\phi_k(x+r)\frac{\partial}{\partial r}\phi_k(x) = -|\cos\theta| \; a_i \tag{22}$$

Integrating Eq.(22) for all **r** directions, one obtains

$$\int_0^{2\pi}\int_0^{\pi}\{\frac{\partial}{\partial r}\phi_k(x) - 2\phi_k(x+r)\frac{\partial}{\partial r}\phi_k(x)\}\sin\theta d\theta d\eta = \int_0^{2\pi}\int_0^{\pi} -|\cos\theta| \; a_i \sin\theta d\theta d\eta = -2\pi a_i \tag{23}$$

Rearranging Eq(23), one obtains

$$a_i = -\frac{1}{2\pi}\int_0^{2\pi}\int_0^{\pi}\{\frac{\partial}{\partial r}\phi_k(x) - 2\phi_k(x+r)\frac{\partial}{\partial r}\phi_k(x)\}\sin\theta d\theta d\eta \tag{24}$$

As stated above, $\frac{\partial}{\partial r}$ is directional differentiation of characteristic function $\phi_k(\mathbf{x})$ in **r** direction.

When one approximates the directional differentiation of characteristic function in Eq.(24) in the interval of $|\mathbf{r}|$, one obtains,

$$\frac{\partial}{\partial r}\phi_k(x) \approx \frac{\phi_k(x+r) - \phi_k(x)}{|r|} \tag{25}$$

$$\phi_k(x+r)\frac{\partial}{\partial r}\phi_k(x) \approx \frac{\phi_k(x+r)\phi_k(x+r) - \phi_k(x+r)\phi_k(x)}{|r|} = \frac{\phi_k(x+r) - \phi_k(x+r)\phi_k(x)}{|r|} \tag{26}$$

Then, the integrated function in Eq.(24) can be given by

$$\frac{\partial}{\partial r}\phi_k(x) - 2\phi_k(x+r)\frac{\partial}{\partial r}\phi_k(x) \approx \frac{\phi_k(x)\{\phi_k(x+r) - 1\} + \phi_k(x+r)\{\phi_k(x) - 1\}}{|r|} \tag{27}$$

$$\left(a_i|\cos\theta| \approx \frac{\phi_k(\mathbf{x})\{1 - \phi_k(\mathbf{x+r})\} + \phi_k(\mathbf{x+r})\{1 - \phi_k(\mathbf{x})\}}{|\mathbf{r}|} \right)$$

Using this relation , Eq.(24) can be rewritten by

$$a_i = \frac{1}{2\pi}\int_0^{2\pi}\int_0^{\pi} \lim_{|r|\to 0}\{\frac{\phi_k(x)\{1-\phi_k(x+r)\}+\phi_k(x+r)\{1-\phi_k(x)\}}{|r|}\}\sin\theta d\theta d\eta \tag{28}$$

Averaging Eq.(28), one obtains,

$$\overline{a_i} = \frac{1}{2\pi}\int_0^{2\pi}\int_0^{\pi} \lim_{|r|\to 0}\{\frac{\overline{\phi_k(x)\{1-\phi_k(x+r)\}}+\overline{\phi_k(x+r)\{1-\phi_k(x)\}}}{|r|}\}\sin\theta d\theta d\eta \tag{29}$$

In the right hand side of Eq.(29), the term

$$\overline{\phi_k(x)\{1-\phi_k(x+r)\}}+\overline{\phi_k(x+r)\{1-\phi_k(x)\}}$$

represents the probability where gas-liquid interface exists between x and x+r as shown in Fig.7.

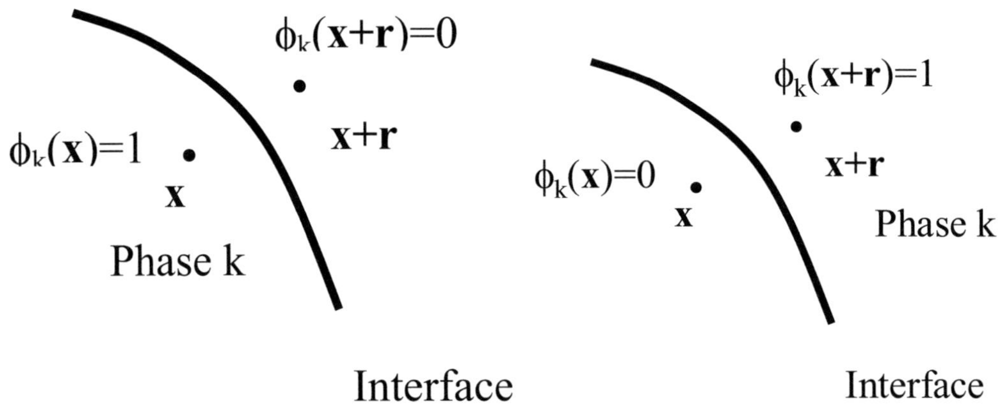

Fig. 7. The case where interface exists between **x** and **x+r**.

3. Basic transport equations of interfacial area concentration

Based on the rigorous formulation of interfacial area concentration, one can derive transport equation of interfacial area concentration. The transport equations of interfacial area concentration consist of two equations. One is the conservation equation of interfacial area concentration and the other is the conservation equation of interfacial velocity (velocity of interface), $\mathbf{V}_i$.

Kataoka (2008) derived the local instant conservation equation of interfacial area concentration based on the formulation given by Eq.(24). In order to obtain the local instant conservation equation of interfacial area concentration, characteristic function of each phase (denoted by ϕ_k) given by Eqs.(11) or (12) is needed. Kataoka(1986) also derived local instant formulation of two-phase flow which gives the local instant conservation equations of mass momentum and energy in each phase. The conservation equation of characteristic function of each phase (denoted by ϕ_k) given by

$$\frac{\partial}{\partial t}(\phi_k \rho_k) + \mathrm{div}(\phi_k \rho_k \mathbf{v}_k) = -\rho_{ki}(\mathbf{v}_{ki} - \mathbf{V}_i) \bullet \mathbf{n}_{ki} a_i \quad (k = G, L) \tag{30}$$

Here, ρ_k, $\mathbf{v}_k$ are density, velocity of each phase. Suffix ki is value of phase k at interface. Using Eqs.(24) and (30), Local instant conservation equation of interfacial area concentration is given by

$$\frac{\partial}{\partial t}(a_i) + \mathrm{grad}(a_i \mathbf{V}_i) = \frac{1}{2\pi}\int_0^{2\pi}\int_0^{\pi}\{\frac{\partial v_k}{\partial r}\mathrm{grad}(\varphi_k) - 2\phi_k(\mathbf{x}+\mathbf{r})\frac{\partial v_k}{\partial r}\mathrm{grad}(\varphi_k)\}\sin\theta d\theta d\eta \tag{31}$$

Averaging Eq.(31), one obtains the conservation equation of averaged interfacial area concentration by

$$\frac{\partial}{\partial t}(\overline{a_i}) + \mathrm{grad}(\overline{a_i \mathbf{V}_i}) = \frac{1}{2\pi}\int_0^{2\pi}\int_0^{\pi}\overline{\{\frac{\partial v_k}{\partial r}\mathrm{grad}(\phi_k) - 2\phi_k(\mathbf{x}+\mathbf{r})\frac{\partial v_k}{\partial r}\mathrm{grad}(\phi_k)\}}\sin\theta d\theta d\eta \tag{32}$$

where averaged interfacial velocity $\overline{\mathbf{V}_i}$ is defined by

$$\overline{\mathbf{V}_i} = \overline{(a_i \mathbf{V}_i)} / \overline{a_i} \tag{33}$$

The right hand side of Eqs.(31) and(32) represent the source term of interfacial area concentration due to the deformation of interface. In the dispersed flows such as bubbly flow and droplet flow, this term correspond breakup or coalescence of bubbles and droplets.

Morel (2007) derived the conservation equation of averaged interfacial area concentration based on the detailed geometrical consideration of interface.

$$\frac{\partial \overline{a_i}}{\partial t} + \nabla \bullet \overline{a_i \mathbf{V}_i} = \overline{a_i (\mathbf{V}_i \bullet \mathbf{n}_{Gi}) \nabla \bullet \mathbf{n}_{Gi}} \tag{34}$$

Here, $\overline{}$ denotes time averaging and $\overline{\mathbf{V}_i}$ is the time averaged velocity of interface which is given by

$$\overline{\mathbf{V}_i} = \overline{a_i (\mathbf{V}_i \bullet \mathbf{n}_{Gi}) \mathbf{n}_{Gi}} / \overline{a_i} \tag{35}$$

The research group directed by Prof. Ishii in Purdue university derived the transport equation of interfacial area concentration of time averaged interfacial area concentration based on the transport equation of number density function of bubbles (Kocamustafaogullari and Ishii (1995), Hibiki and Ishii (2000a)) . It is given by

$$\frac{\partial \overline{a_i}}{\partial t} + \nabla \bullet \overline{a_i \mathbf{V}_i} = \sum_{j=1}^{4}\phi_j + \phi_{ph} \tag{36}$$

Here, the first term in right hand side of Eq.(36) represent the source and sink terms due to bubble coalescence and break up. Interfacial area decreases when bubbles coalescence and increases when bubbles break up. This term is quite important in interfacial area transport. Therefore, the constitutive equations of this term are given by Hibiki and Ishii (2000a, 2000b) and Ishii and Kim (2004) based on detailed mechanistic modeling. The second term in right hand side of Eq.(36) represent the source and sink terms due to phase change. Equation (36) is practical transport equation of interfacial area concentration.

As for the conservation equation of interfacial velocity, Kataoka et al. (2010,2011a) have derived rigorous formulation based on the local instant formulation of interfacial area concentration and interfacial velocity, which is shown below. Since interface has no mass, momentum equation of interface cannot be formulated. Therefore, the conservation equation of interfacial velocity (or governing equation of interfacial velocity) has to be derived in collaboration with the momentum equation of each phase. Since interfacial velocity is only defined at interface, local instant formulation of interfacial velocity must be expressed in the form of

$$a_i V_i$$

Using Eq.(22), interfacial velocity is expressed by

$$a_i V_i \left|\cos\theta\right| = -V_i \frac{\partial}{\partial r} \phi_k(x) + 2V_i \phi_k(x+r) \frac{\partial}{\partial r} \phi_k(x) \tag{37}$$

Considering Fig.7, interfacial velocity is approximated by velocity of each phase without phase change.

$$V_i \approx \phi_k(x)\{1-\phi_k(x+r)\} v_k(x) + \phi_k(x+r)\{1-\phi_k(x)\} v_k(x+r) \tag{38}$$

From Eqs.(27) and (38) with some rearrangements, one obtains following approximate expression.

$$a_i V_i \left|\cos\theta\right| \approx \frac{\phi_k(x)\{1-\phi_k(x+r)\} v_k(x) + \phi_k(x+r)\{1-\phi_k(x)\} v_k(x+r)}{|r|} \tag{39}$$

When one takes the limit of $|r| \to 0$, one obtains

$$a_i V_i \left|\cos\theta\right| = \lim_{|r|\to 0} \frac{\phi_k(x)\{1-\phi_k(x+r)\} v_k(x) + \phi_k(x+r)\{1-\phi_k(x)\} v_k(x+r)}{|r|} \tag{40}$$

Integrating Eq.(40) in all direction, one finally obtains

$$a_i V_i = \frac{1}{2\pi} \int_0^{2\pi} \int_0^{\pi} \lim_{|r|\to 0} \frac{\phi_k(x)\{1-\phi_k(x+r)\} v_k(x) + \{1-\phi_k(x)\}\phi_k(x+r) v_k(x+r)}{|r|} \sin\theta d\theta d\phi \tag{41}$$

Averaging Eq.(41), averaged interfacial velocity is given by

$$\overline{V_i a_i} = \frac{1}{2\pi} \int_0^{2\pi} \int_0^{\pi} \lim_{|r|\to 0} \frac{\overline{\phi_k(x)\{1-\phi_k(x+r)\} v_k(x)} + \overline{\{1-\phi_k(x)\}\phi_k(x+r) v_k(x+r)}}{|r|} \sin\theta d\theta d\phi \tag{42}$$

On the other hand, using Eq.(29), following relation can be obtained for averaged velocity of each phase, $\overline{\overline{v_k}}$ by

$$\overline{\overline{v_k}}\,\overline{a_i} = \frac{1}{2\pi}\int_0^{2\pi}\int_0^{\pi} \lim_{|r|\to 0} \frac{\overline{\phi_k(x)\{1-\phi_k(x+r)\}\overline{\overline{v_k(x)}}} + \overline{\{1-\phi_k(x)\}\phi_k(x+r)\overline{\overline{v_k(x)}}}}{|r|} \sin\theta d\theta d\phi \quad (43)$$

where $\overline{\overline{v_k}}$ is defined by

$$\overline{\overline{v_k}} = (\overline{\phi_k v_k}) / \overline{\phi_k} \quad (44)$$

Using, Eqs.(42) and (43), the difference between time averaged interfacial velocity, $\overline{V_i}$ and time average velocity of phase k, $\overline{\overline{v_k}}$ is given by

$$(\overline{V_i} - \overline{\overline{v_k}})\overline{a_i}$$
$$= \frac{1}{2\pi}\int_0^{2\pi}\int_0^{\pi} \lim_{|r|\to 0} \frac{\overline{\phi_k(x)\{1-\phi_k(x+r)\}\{v_k(x) - \overline{\overline{v_k(x)}}\}} + \overline{\{1-\phi_k(x)\}\phi_k(x+r)\{v_k(x+r) - \overline{\overline{v_k(x)}}\}}}{|r|} \sin\theta d\theta d\phi \quad (45)$$

Rearranging the term in integration in the right hand side of Eq.(45) one obtains

$$\lim_{|r|\to 0} \frac{\overline{\phi_k(x)\{1-\phi_k(x+r)\}\{v_k(x) - \overline{\overline{v_k(x)}}\}} + \overline{\{1-\phi_k(x)\}\phi_k(x+r)\{v_k(x+r) - \overline{\overline{v_k(x)}}\}}}{|r|}$$
$$= -\lim_{|r|\to 0} \frac{\overline{\phi_k'(x)\phi_k(x+r)v_k'(x+r)} + \overline{\phi_k(x)\phi_k'(x+r)v_k'(x)}}{|r|} \quad (46)$$

Here, ϕ_k' and v_k' are fluctuating terms of local instant volume fraction and velocity of phase k which are given by

$$v_k' = v_k - \overline{\overline{v_k}} \quad (47)$$

$$\phi_k' = \phi_k - \overline{\phi_k} \quad (48)$$

Equations (45) and (46) indicate that the difference between time averaged interfacial velocity, $\overline{V_i}$ and time averaged velocity of phase k, $\overline{\overline{v_k}}$ is given in terms of correlations between fluctuating terms of local instant volume fraction and velocity of phase k which are related to turbulence terms of phase k.

Then, it is important to derive the governing equation of the correlation term given by Eq.(46). In what follows, one derives the governing equation based on the local instant basic equations of mass conservation and momentum conservation of phase k which are given below (Kataoka (1986)). In these conservation equations, tensor representation is used. Einstein abbreviation rule is also applied. When the same suffix appear, summation for that suffix is carried out except for the suffix k denoting gas and liquid phases.

(Mass conservation)

$$\phi_k \frac{\partial v_{k\beta}}{\partial x_\beta} = 0 \tag{49}$$

(Momentum conservation)

$$\phi_k \frac{\partial v_{k\alpha}}{\partial t} + \phi_k \frac{\partial}{\partial x_\beta}(v_{k\alpha} v_{k\beta}) = -\phi_k \frac{1}{\rho_k}\frac{\partial P_k}{\partial x_\alpha} + \phi_k \frac{1}{\rho_k}\frac{\partial \tau_{k\alpha\beta}}{\partial x_\beta} + \phi_k F_{k\alpha} \tag{50}$$

Averaging Eqs.(49) and (50) , one obtains time averaged conservation equation of mass and momentum conservation of phase k.

(Time averaged mass conservation)

$$\frac{\partial \overline{\overline{v_{k\beta}}}}{\partial x_\beta} = -\frac{1}{\overline{\overline{\phi_k}}}\overline{v'_{k\beta i} n_{k\beta i} a_i} \tag{51}$$

(Time averaged momentum conservation)

$$\begin{aligned}\frac{\partial \overline{\overline{v_{k\alpha}}}}{\partial t} + \frac{\partial}{\partial x_\beta}\left(\overline{\overline{v_{k\alpha} v_{k\beta}}}\right) = -\frac{1}{\rho_k}\frac{\partial \overline{\overline{P_k}}}{\partial x_\alpha} + \frac{1}{\rho_k}\frac{\partial}{\partial x_\beta}\left(\overline{\overline{\tau_{k\alpha\beta}}} - \rho_k \overline{\overline{v'_{k\alpha} v'_{k\beta}}}\right) + \overline{\overline{F_{k\alpha}}} - \frac{\overline{\overline{v_{k\alpha}}}}{\overline{\overline{\phi_k}}}\overline{v'_{k\beta i} n_{k\beta i} a_i} \\ -\frac{1}{\overline{\overline{\phi_k}}}\frac{1}{\rho_k}\overline{P'_{ki} n_{k\alpha i} a_i} + \frac{1}{\overline{\overline{\phi_k}}}\frac{1}{\rho_k}\overline{\tau'_{k\alpha\beta i} n_{k\beta i} a_i} + \frac{1}{\overline{\overline{\phi_k}}}\overline{\overline{v'_{k\alpha} v'_{k\beta}}}\,\overline{n_{k\beta i} a_i}\end{aligned} \tag{52}$$

Subtracting Eqs(51) and (52) from Eqs.(49) and (50), the conservation equations of fluctuating terms are obtained.

(Conservation equation of mass fluctuation)

$$\phi_k \frac{\partial v'_{k\beta}}{\partial x_\beta} = \frac{\phi_k}{\overline{\overline{\phi_k}}}\overline{v'_{k\beta i} n_{k\beta i} a_i} \tag{53}$$

(Conservation equation of momentum fluctuation)

$$\begin{aligned}\phi_k \frac{\partial v'_{k\alpha}}{\partial t} + \phi_k \frac{\partial}{\partial x_\beta}\left(v'_{k\alpha} v'_{k\beta} + v'_{k\alpha}\overline{\overline{v_{k\beta}}} + v'_{k\beta}\overline{\overline{v_{k\alpha}}}\right) = -\phi_k \frac{1}{\rho_k}\frac{\partial P'_k}{\partial x_\alpha} + \phi_k \frac{1}{\rho_k}\frac{\partial}{\partial x_\beta}\left(\tau'_{k\alpha\beta} + \rho_k \overline{\overline{v'_{k\alpha} v'_{k\beta}}}\right) \\ + \phi_k F_{k\alpha}{}' + \frac{\phi_k}{\overline{\overline{\phi_k}}}\overline{\overline{v_{k\alpha}}}\,\overline{v'_{k\beta i} n_{k\beta i} a_i} + \frac{\phi_k}{\overline{\overline{\phi_k}}}\frac{1}{\rho_k}\overline{P'_{ki} n_{k\alpha i} a_i} - \frac{\phi_k}{\overline{\overline{\phi_k}}}\frac{1}{\rho_k}\overline{\tau'_{k\alpha\beta i} n_{k\beta i} a_i} - \frac{\phi_k}{\overline{\overline{\phi_k}}}\overline{\overline{v'_{k\alpha} v'_{k\beta}}}\,\overline{n_{k\beta i} a_i}\end{aligned} \tag{54}$$

Using Eqs(53) and (54), one can derive conservation equation of

$$\overline{\phi_k{}'(x)\phi_k(x+r)v_k{}'(x+r)} + \overline{\phi_k(x)\phi_k{}'(x+r)v_k{}'(x)}$$

Then, conservation equation of the difference between interfacial velocity and averaged velocity of each phase is derived. The result is given by

$$\frac{\partial}{\partial t}\{\left(\overline{V_i}-\overline{\overline{v_k}}\right)\overline{a_i}\}+\frac{\partial}{\partial x_\beta}(\left(\overline{V_i}-\overline{\overline{v_k}}\right)\overline{a_i}\overline{\overline{v_{k\beta}}})$$

$$=\frac{1}{\rho_k}\frac{1}{2\pi}\int_0^{2\pi}\int_0^{\pi}\lim_{|r|\to 0}\frac{1}{|r|}(\overline{\phi_k'\phi_{kr}\frac{\partial P_{kr}'}{\partial x_\alpha}}+\overline{\varphi_k\varphi_{kr}'\frac{\partial P_k'}{\partial x_\alpha}})\sin\theta d\theta d\phi$$

$$-\frac{1}{\rho_k}\frac{1}{2\pi}\int_0^{2\pi}\int_0^{\pi}\lim_{|r|\to 0}\frac{1}{|r|}\{\overline{\phi_k'\phi_{kr}\frac{\partial}{\partial x_\beta}(\tau_{k\alpha\beta r}'+\rho_k\overline{\overline{v_{k\alpha r}'v_{k\beta r}'}})}+$$

$$+\overline{\phi_k\phi_{kr}'\frac{\partial}{\partial x_\beta}(\tau_{k\alpha\beta}'+\rho_k\overline{\overline{v_{k\alpha}'v_{k\beta}'}})}\}\sin\theta d\theta d\phi$$

$$-\frac{1}{2\pi}\int_0^{2\pi}\int_0^{\pi}\lim_{|r|\to 0}\frac{1}{|r|}(\overline{\phi_k'\phi_{kr}F'_{k\alpha r}}+\overline{\phi_k\phi_{kr}'F'_{k\alpha}})\sin\theta d\theta d\phi$$

$$-\frac{1}{2\pi}\int_0^{\pi}\int_0^{2\rho}\lim_{|r|\to 0}\frac{1}{|r|}(\frac{\overline{\phi_k'\phi_{kr}}}{\overline{\overline{\varphi_{kr}}}}\frac{1}{\rho_k}\overline{P_{ki}'n_{k\alpha i}a_{ir}}+\frac{\overline{\phi_k\phi_{kr}'}}{\overline{\overline{\varphi_k}}}\frac{1}{\rho_k}\overline{P_{ki}'n_{k\alpha i}a_i})\sin\theta d\theta d\phi$$

$$+\frac{1}{2\pi}\int_0^{\pi}\int_0^{2\rho}\lim_{|r|\to 0}\frac{1}{|r|}(\frac{\overline{\phi_k'\phi_{kr}}}{\overline{\overline{\varphi_{kr}}}}\frac{1}{\rho_k}\overline{\tau_{ki}'n_{k\alpha i}a_{ir}}+\frac{\overline{\phi_k\phi_{kr}'}}{\overline{\overline{\varphi_k}}}\frac{1}{\rho_k}\overline{\tau_{ki}'n_{k\alpha i}a_i})\sin\theta d\theta d\phi$$

$$+\frac{1}{2\pi}\int_0^{\pi}\int_0^{2\rho}\lim_{|r|\to 0}\frac{1}{|r|}(\frac{\overline{\phi_k'\phi_{kr}}}{\overline{\overline{\varphi_{kr}}}}\overline{\overline{v_{k\alpha r}'v_{k\beta r}'}\,n_{k\alpha i}a_{ir}}+\frac{\overline{\phi_k\phi_{kr}'}}{\overline{\overline{\varphi_k}}}\overline{\overline{v_{k\alpha}'v_{k\beta}'}\,n_{k\alpha i}a_i})\sin\theta d\theta d\phi$$

$$+\frac{1}{2\pi}\int_0^{\pi}\int_0^{2\rho}\lim_{|r|\to 0}\frac{1}{|r|}\{\frac{\partial}{\partial x_\beta}(\overline{\phi_k'\phi_{kr}v_{k\alpha r}'v_{k\beta r}'})+\frac{\partial}{\partial x_\beta}(\overline{\phi_k\phi_{kr}'v_{kr}'v_{k\beta}'})\}\sin\theta d\theta d\phi$$

$$+\frac{1}{2\pi}\int_0^{\pi}\int_0^{2\rho}\lim_{|r|\to 0}\frac{1}{|r|}(\overline{\phi_k'\phi_{kr}v_{k\beta r}'}\frac{\partial\overline{\overline{v_{k\beta r}}}}{\partial x_\beta}+\overline{\phi_k\phi_{kr}'v_{k\beta}'}\frac{\partial\overline{\overline{v_{k\beta}}}}{\partial x_\beta})\sin\theta d\theta d\phi$$

$$-\frac{1}{2\pi}\int_0^{\pi}\int_0^{2\rho}\lim_{|r|\to 0}\frac{1}{|r|}\{\overline{v_{k\alpha r}'(\overline{\overline{v_{k\beta r}}}+v_{k\beta r}')\phi_{kr}\frac{\partial}{\partial x_\beta}\phi_k'}+$$

$$+\{\overline{v_{k\alpha}'(\overline{\overline{v_{k\beta}}}+v_{k\beta}')\phi_k\frac{\partial}{\partial x_\beta}\phi_{kr}'}\}\sin\theta d\theta d\phi \tag{55}$$

$$+\frac{1}{2\pi}\int_0^{\pi}\int_0^{2\rho}\lim_{|r|\to 0}\frac{1}{|r|}\{\overline{v_{k\alpha r}'\phi_{kr}\frac{\partial}{\partial x_\beta}(\phi_r v_{k\beta}'+\phi_k'\overline{\overline{v_{k\beta r}}})}+$$

$$+\overline{v_{k\alpha}'\phi_k\frac{\partial}{\partial x_\beta}(\phi_{kr}v_{k\beta r}'+\phi_{kr}'\overline{\overline{v_{k\beta}}})}\}\sin\theta d\theta d\phi$$

As shown above, the formulation of governing equation of interfacial velocity is derived. Then, the most strict formulation of transport equations of interfacial area concentration is given by conservation equation of interfacial area concentration (Eq.(32) , Eq(34), or Eq.(36)) and conservation equation of interfacial velocity (Eq.(55)). As shown in Eq.(55), the conservation equation of interfacial velocity consists of various correlation terms of fluctuating terms of velocity and local instant volume fractions. These correlation terms represent the turbulent transport of interfacial area, which reflects the interactions between gas liquid interface and turbulence of gas and liquid phases. Equation (55) represents such turbulence transport terms of interfacial area concentration. Accurate predictions of interfacial area transport can be possible by solving the transport equations derived here. However, Eq.(55)

consists of complicated correlation terms of fluctuating terms of local instant volume fraction, velocity, pressure and shear stress. The detailed knowledge of these correlation terms is not available. Therefore, solving Eq.(55) together with basic equations of two-fluid model is difficult at present. More detailed analytical and experimental works on turbulence transport terms of interfacial area concentration are necessary for solving practically Eq.(55).

4. Constitutive equations of transport equations of interfacial area concentration. Source and sink terms, diffusion term, turbulence transport term

As shown in the previous section, the rigorous formulation of transport equation of interfacial area concentration are given by conservation equation of interfacial area concentration (Eq.(32), Eq.(34) or Eq(36)) and conservation equation of interfacial velocity (Eq.(55)). However, Eq.(55) consists of complicated correlation terms of fluctuating terms of local instant volume fraction, velocity, pressure and shear stress. The detailed knowledge of these correlation terms is not available. Therefore, solving Eq.(55) together with basic equations of two-fluid model is difficult at present. More detailed analytical and experimental works on turbulence transport terms of interfacial area concentration are necessary for solving practically Eq.(55). From Eqs.(45) and (46), interfacial velocity is related to averaged velocity of phase k (gas phase or liquid phase)by following equation.

$$\overline{\overline{V_i}\,\overline{a_i}} = \overline{\overline{v_k}}\,\overline{a_i} - \frac{1}{2\pi}\int_0^{2\pi}\int_0^{\pi} \lim_{|r|\to 0}\frac{1}{|r|}\left(\overline{\phi'_k \varphi_{kr} v'_{kr}} + \overline{\phi_k \phi'_{kr} v'_k}\right)\sin\theta d\theta d\phi \tag{56}$$

When one considers bubbly flow and phase k is gas phase, Eq.(56) can be rewritten by

$$\overline{\overline{V_i}\,\overline{a_i}} = \overline{\overline{v_G}}\,\overline{a_i} - \frac{1}{2\pi}\int_0^{2\pi}\int_0^{\pi} \lim_{|r|\to 0}\frac{1}{|r|}\left(\overline{\phi'_G \phi_{Gr} v'_{Gr}} + \overline{\phi_G \phi'_{Gr} v'_G}\right)\sin\theta d\theta d\phi \tag{57}$$

From Eqs.(28) and (29) , following relation is derived.

$$a'_i = a_i - \overline{a_i} = -\frac{1}{2\pi}\int_0^{2\pi}\int_0^{\pi} \lim_{|r|\to 0}\frac{1}{|r|}\left\{\phi'_k(1-2\overline{\phi_{kr}}) + \phi'_{kr}(1-2\overline{\phi_k}) - 2(\phi'_k\phi'_{kr} - \overline{\phi'_k\phi'_{kr}})\right\}\sin\theta d\theta d\phi \tag{58}$$

In Eq.(57), the terms, $\phi'_G/|r|$ and $\phi'_{Gr}/|r|$ are related to the fluctuating term of interfacial area concentration. On the other hand, the terms, $\phi_{Gr}v'_{Gr}$ and $\phi_G v'_G$ are the fluctuating term of gas phase velocity at the location, **x+r.** and **x**. Therefore, the second term of right hand side of Eq.(57) is considered to correspond to turbulent transport term due to the turbulent velocity fluctuation. In analogous to the turbulent transport of momentum, energy (temperature) and mass, the correlation term described above is assumed to be proportional to the gradient of interfacial area concentration which is transported by turbulence (diffusion model). Then, one can assume following relation.

$$-\frac{1}{2\pi}\int_0^{2\pi}\int_0^{\pi} \lim_{|r|\to 0}\frac{1}{|r|}\left(\overline{\phi'_G \phi_{Gr} v'_{Gr}} + \overline{\phi_G \phi'_{Gr} v'_G}\right)\sin\theta d\theta d\phi = -D_{ai}\,\mathrm{grad}\,\overline{a_i} \tag{59}$$

Here, the coefficient, D_{ai} is considered to correspond to turbulent diffusion coefficient of interfacial area concentration. In analogy to the turbulent transport of momentum, energy (temperature) and mass, this coefficient is assumed to be given by

$$D_{ai} \propto |v'_G| L \tag{60}$$

Here, L is the length scale of turbulent mixing of gas liquid interface and $|v'_G|$ is the turbulent velocity of gas phase. In bubbly flow, it is considered that turbulent mixing of gas liquid interface is proportional to bubble diameter, d_B and the turbulent velocity of gas phase is proportional to the turbulent velocity of liquid phase. These assumptions were confirmed by experiment and analysis of turbulent diffusion of bubbles in bubbly flow (Kataoka and Serizawa (1991a)). Therefore, turbulent diffusion coefficient of interfacial area concentration is assumed by following equation.

$$D_{ai} = K_1 |v'_L| d_B = 6K_1 \frac{\alpha}{a_i} |v'_L| \tag{61}$$

Here, α is the averaged void fraction and $|v'_L|$ is the turbulent velocity of liquid phase. K_1 is empirical coefficient. For the case of turbulent diffusion of bubble, experimental data were well predicted assuming $K_1=1/3$ For the case of turbulent diffusion of interfacial area concentration, there are no direct experimental data of turbulent diffusion. However, the diffusion of bubble is closely related to the diffusion of interfacial area (surface area of bubble). Therefore, as first approximation, the value of K_1 for bubble diffusion can be applied to diffusion of interfacial area concentration in bubbly flow.

Equations (61) is based on the model of turbulent diffusion of interfacial area concentration. In this model, it is assumed that turbulence is isotropic. However, in the practical two-phase flow in the flow passages turbulence is not isotropic and averaged velocities and turbulent velocity have distribution in the radial direction of flow passage. In such non-isotropic turbulence, the correlation terms of turbulent fluctuation of velocity and interfacial area concentration given by Eq.(57) is largely dependent on anisotropy of turbulence field. Such non-isotropic turbulence is related to the various terms consisting of turbulent stress which appear in the right hand side of Eq.(55). Assuming that turbulent stress of gas phase is proportional to that of liquid phase and turbulence model in single phase flow, turbulent stress is given by

$$\overline{\overline{v'_L v'_L}} = -\varepsilon_{LTP}\{\nabla \overline{\overline{v_L}} + {}^t(\nabla \overline{\overline{v_L}})\} + \frac{2}{3}k\delta_{ij} \tag{62}$$

Here, ε_{LTP} is the turbulent diffusivity of momentum in gas-liquid two-phase flow. For bubbly flow, this turbulent diffusivity is given by various researchers (Kataoka and Serizawa (1991b,1993)).

$$\varepsilon_{LTP} = \frac{1}{3}\alpha d_B |v'_L| \tag{63}$$

Based on the model of turbulent stress in gas-liquid two-phase flow and Eq.(55), it is assumed that turbulent diffusion of interfacial area concentration due to non-isotropic turbulence is proportional to the velocity gradient of liquid phase. For the diffusion of bubble due to non-isotropic turbulence in bubbly flow in pipe, Kataoka and Serizawa (1991b,1993) proposed the following correlation based on the analysis of radial distributions of void fraction and bubble number density.

$$J_B = K_2 \alpha d_B n_B \frac{\partial \overline{\overline{v_L}}}{\partial y} \tag{64}$$

Here, J_B is the bubble flux in radial direction and n_B is the number density of bubble. y is radial distance from wall of flow passage. K_2 is empirical coefficient and experimental data were well predicted assuming K_2=10. In analogous to Eq.(64), it is assumed that turbulent diffusion of interfacial area concentration due to non-isotropic turbulence is given by following equation.

$$J_{ai} = K_2 \alpha d_B a_i \frac{\partial \overline{\overline{v_L}}}{\partial y} \tag{65}$$

Here, J_{ai} is the flux of interfacial area concentration in radial direction. Equation (64) can be interpreted as equation of bubble flux due to the lift force due to liquid velocity gradient.

As shown above, turbulent diffusion of interfacial area concentration due to non-isotropic turbulence is related to the gradient of averaged velocity of liquid phase and using analogy to the lift force of bubble, Eq.(58) can be rewritten in three dimensional form by

$$\begin{aligned} &-\frac{1}{2\pi}\int_0^{2\pi}\int_0^{\pi} \lim_{|r|\to 0}\frac{1}{|r|}\left(\overline{\phi'_G \phi_{Gr} v'_{Gr}} + \overline{\phi_G \phi'_{Gr} v'_G}\right)\sin\theta\, d\theta\, d\phi = \\ &-D_{ai}\,\mathrm{grad}\,\overline{a_i} + C\alpha a_i(\overline{\overline{v_G}} - \overline{\overline{v_L}}) \times \mathrm{rot}(\overline{\overline{v_L}}) \end{aligned} \tag{66}$$

Empirical coefficient C in the right hand side of Eq.(66) should be determined based on the experimental data of spatial distribution of interfacial area concentration and averaged velocity of each phase. However, at present, there are not sufficient experimental data. Therefore, as first approximation, the value of coefficient C can be given by Eq.(67)

$$C = K_2 d_B / u_R \ \ \text{(based on Eq.(65))} \tag{67}$$

Using Eqs(55) and (66), transport equation of interfacial area concentration (Eq.(32),(34) or (36)) can be given by following equation for gas-liquid two-phase flow where gas phase is dispersed in liquid phase for bubbly flow.

$$\begin{aligned} &\frac{\partial \overline{a_i}}{\partial t} + \mathrm{div}(\overline{a_i}\,\overline{\overline{v_G}}) = \mathrm{div}(D_{ai}\,\mathrm{grad}\,\overline{a_i}) - \mathrm{div}\{C\alpha a_i(\overline{\overline{v_G}} - \overline{\overline{v_L}}) \times \mathrm{rot}(\overline{\overline{v_L}})\} + \\ &+\frac{2}{3}\frac{\overline{a_i}}{\alpha \rho_G}\left(\Gamma_G - \alpha \frac{D\rho_G}{Dt}\right) + \phi_{CO} + \phi_{BK} \end{aligned} \tag{68}$$

Here, D/Dt denotes material derivative following the gas phase motion and turbulent diffusion coefficient of interfacial area concentration, D_{ai} is given by Eq.(61). Coefficient of turbulent diffusion of interfacial area concentration due to non-isotropic turbulence, C is given by Eq.(67). The third term in the right hand side of Eq.(68) is source term of interfacial area concentration due to phase change and density change of gas phase due to pressure change. Γ_G is the mass generation rate of gas phase per unit volume of two-phase flow due to evaporation. ϕ_{CO} and ϕ_{Bk} are sink and source term due to bubble coalescence and break up

Similarly, the transport equation of interfacial area concentration for droplet flow is given by

$$\frac{\partial \overline{a_i}}{\partial t} + \mathrm{div}(\overline{a_i}\,\overline{\overline{v_L}}) = \mathrm{div}(D_{ai}\,\mathrm{garda}_i) - \mathrm{div}\{C(1-\alpha)\overline{a_i}(\overline{\overline{v_L}} - \overline{\overline{v_G}}) \times \mathrm{rot}(\overline{\overline{v_G}})\} + \frac{2}{3}\frac{\overline{a_i}}{(1-\alpha)\rho_L}\left(\Gamma_L - (1-\alpha)\frac{D\rho_L}{Dt}\right) + \phi_{CO} + \phi_{BK} \tag{69}$$

Here, D/Dt denotes material derivative following the liquid phase motion and ϕ_{CO} and ϕ_{Bk} are sink and source terms due to droplet coalescence and break up and Γ_L is the mass generation rate of liquid phase per unit volume of two-phase flow due to condensation. Here, turbulent diffusion coefficient of interfacial area concentration is approximated by turbulent diffusion coefficient of droplet (Cousins and Hewitt (1968)) as first approximation. The coefficient C for turbulent diffusion of interfacial area concentration due to non-isotropic turbulence (or lift force term) can be approximated by lift force coefficient of solid sphere as first approximation.

The research and development of source and sink terms in transport equation of interfacial area concentration have been carried out mainly for the bubbly flow based on detailed analysis and experiment of interfacial area concentration which is shown below.

Hibiki and Ishii(2000a,2002) developed the transport equation of interfacial area mentioned above and carried out detailed modeling of source and sink terms of interfacial area concentration. They assumed that the sink term of interfacial area concentration is mainly due to the coalescence of bubble. On the other hand, they assumed the source term is mainly contributed by the break up of bubble due to liquid phase turbulence. Based on detailed mechanistic modeling of bubble liquid interactions, they finally obtained the constitutive equations for sink and source terms of interfacial area transport.

The sink term of interfacial area concentration due to the coalescence of bubbles, ϕ_{CO} is composed of number of collisions of bubbles per unit volume and the probability of coalescence at collision and given by

$$\phi_{CO} = -\left(\frac{\alpha}{\overline{\overline{a_i}}}\right)^2 \frac{\Gamma_C \alpha^2 \varepsilon^{1/3}}{d_b^{11/3}(\alpha_{max} - \alpha)} \exp\left(-K_C \sqrt[6]{\frac{d_b^5 \rho_L^3 \varepsilon^2}{\sigma^3}}\right) \tag{70}$$

where the term $\frac{\Gamma_C \alpha^2 \varepsilon^{1/3}}{d_b^{11/3}(\alpha_{max} - \alpha)}$ represents the number of collisions of bubbles per unit volume and the term $\exp\left(-K_C \sqrt[6]{\frac{d_b^5 \rho_L^3 \varepsilon^2}{\sigma^3}}\right)$ represents the probability of coalescence at collision. d_b, ε and σ are bubble diameter, turbulent dissipation and surface tension. α_{max} is maximum permissible void fraction in bubbly flow and assumed to be 0.52. Γ_C and K_C are empirical constants and following values are given

$$\Gamma_C = 0.188,\ K_C = 1.29 \tag{71}$$

As for the source term due to the break up of bubble, it is assumed that bubble break up mainly occurs due to the collision between bubble and turbulence eddy of liquid phase. The constitutive equation is given based on the detailed mechanistic modeling of this phenomenon as

$$\phi_{BK} = \left(\frac{\alpha}{a_i}\right)^2 \frac{\Gamma_B \alpha(1-\alpha)\varepsilon^{1/3}}{d_b^{11/3}(\alpha_{max}-\alpha)} \exp\left(-\frac{K_B \sigma}{\rho_L d_b^{5/3} \varepsilon^{2/3}}\right) \tag{72}$$

where the term $\frac{\Gamma_B \alpha(1-\alpha)\varepsilon^{1/3}}{d_b^{11/3}(\alpha_{max}-\alpha)}$ represents the number of collisions of bubble and turbulence eddy per unit volume and the term $\exp\left(-\frac{K_B \sigma}{\rho_L d_b^{5/3} \varepsilon^{2/3}}\right)$ represents the probability of break up at collision. Γ_B and K_B are empirical constants and following values are given.

$$\Gamma_B = 0.264,\ K_B = 1.37 \tag{73}$$

The validity of transport equation of interfacial area concentration (Eq.(68)) and constitutive equations for sink term due to bubble coalescence (Eq.(70)) and source term due to bubble break up (Eq.(72)) are confirmed by experimental data as will be described later in details.

Hibiki and Ishii (2000b) further modified their model of interfacial area transport and applied to bubbly-to-slug flow transition. In bubbly-to-slug flow transition, bubbles are classified into two groups that are small spherical/distorted bubble (group I) and large cap/slug bubble (group II). They derived transport equations of interfacial area concentration and constitutive equations for sink and source terms for group I and group II bubbles based on the transport equation and constitutive equations for bubbly flow mentioned above.

Yao and More (2004) developed more practical transport equation of interfacial area concentration and constitutive correlations of source terms. They derived these equations based on the basic transport equation developed in CEA and models of source terms developed at Purdue University (Ishii 's group). They also developed sink term due to coalescence of bubbles which is given by

$$\phi_{CO} = -107.8\left(\frac{\alpha}{a_i}\right)^2 \frac{\alpha^2 \varepsilon^{1/3}}{d_b^{11/3}} \frac{1}{g(\alpha) + 1.922\alpha\sqrt{We/1.24}} \exp\left(-1.017\sqrt{\frac{We}{1.24}}\right) \tag{74}$$

where $g(\alpha)$ and We is given by

$$g(\alpha) = \frac{(\alpha_{max}^{1/3} - \alpha^{1/3})}{\alpha_{max}^{1/3}} \qquad (\alpha_{max} = 0.52) \tag{75}$$

$$We = \frac{2\rho_L(\varepsilon d_b)^{2/3} d_b}{\sigma} \tag{76}$$

On the other hand, source term due to break up of bubble by liquid phase turbulence is given by

$$\phi_{BK} = 60.3\left(\frac{\alpha}{a_i}\right)^2 \frac{\varepsilon^{1/3}\alpha(1-\alpha)}{d_b^{11/3}} \frac{1}{1+0.42(1-\alpha)\sqrt{We/1.24}} \exp\left(-\sqrt{\frac{1.24}{We}}\right) \tag{77}$$

The transport equation of interfacial area concentration and constitutive equations of source terms described above are implemented to CATHARE code which is developed at CEA using three-dimensional two-fluid model and k-ε turbulence model. Predictions were carried out for thermal hydrodynamic structure of boiling and non-boiling (air-water) two-phase flow including of interfacial area concentration. Comparisons were made with experimental data of DEBORA experiment which is boiling experiment using R-12 carried out at CEA and DEDALE experiment which is air-water experiment carried out at EDF, Electricite de France. The predictions reasonably agreed with experimental data of boiling and non-boiling two-phase flow for distribution of void fraction, velocities of gas and liquid phase, turbulent velocity and interfacial area concentration and the validity of transport equation and constitutive equations described above was confirmed.

5. Experimental researches on interfacial area concentration

The measurements of interfacial area have been carried out earlier in the field of chemical engineering using chemical reaction and/or chemical absorption at gas-liquid interface (Sharma and Danckwerts (1970)). A lot of experimental studies have been reported and reviewed (Ishii et al.,(1982), Kocamustafaogullari and Ishii (1983)). However, in this method, measured quantity is the product of interfacial area concentration and mass transfer coefficient. Light attenuation method and photographic method were also developed and measurement of interfacial area concentration was carried out. However, the measured interfacial area concentration using these methods is volumetric averaged value and measurement of local interfacial area concentration is impossible. In the detailed analysis of multidimensional two-phase flow, measurements of distribution of local interfacial area concentration are indispensable for the validation of interfacial area transport model. Therefore, the establishment of the measurement method of local interfacial area concentration was strongly required. Ishii (1975) and Delhaye (1968) derived following relation among time averaged interfacial area concentration, number of interfaces and velocity of interface. They pointed out local interfacial area concentration can be measured using two or three sensor probe based on this relation.

$$\overline{a_i} = \frac{1}{T}\sum_{j=1}^{N}\frac{1}{\left|\mathbf{n}_{ij}\bullet\mathbf{v}_{ij}\right|} \tag{78}$$

Here, T and N are time interval of measurement and number of interfaces passing a measuring point during time interval T. **n**ij and **v**ij are unit normal vector and interfacial velocity of j-th interface. For bubbly flow, assuming that shape of bubble is spherical and sensor of probe passes any part of bubble with equal probability, Eq.(78) can be simplified to

$$\overline{a_i} = 4\frac{N}{T}\overline{\frac{1}{|v_{sz}|}} \tag{79}$$

Here, v_{sz} is the z directional (flow directional) component of velocity of interface measured by double sensor probe as shown in Fig.8. v_{sz} is obtained by

$$v_{sz} = \frac{\Delta s}{\Delta t} \tag{80}$$

where Δs is spacing of two sensors (Fig.8) and Δt is the time interval where interface passes upstream sensor and downstream sensor.

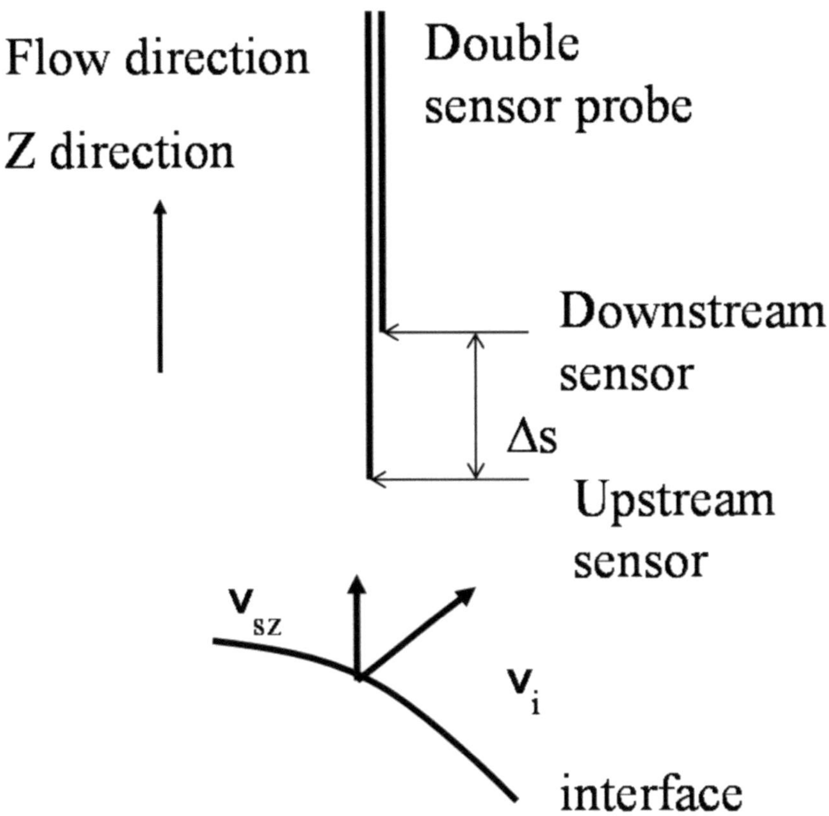

Fig. 8. Double sensor probe and velocity of interface

Later, based on local instant formulation of interfacial area concentration, Kataoka et al. (1986) proposed three double sensor probe method (four sensor probe method) as shown in Fig.9. Using this method, time averaged interfacial area is measured without assuming spherical bubble and statistical behavior of bubbles. The passing velocities measured by each double sensor probe are denoted by v_{sk} which are given by

$$v_{sk} = \frac{\Delta s_k}{\Delta t_k} \quad (k=1,2,3) \tag{81}$$

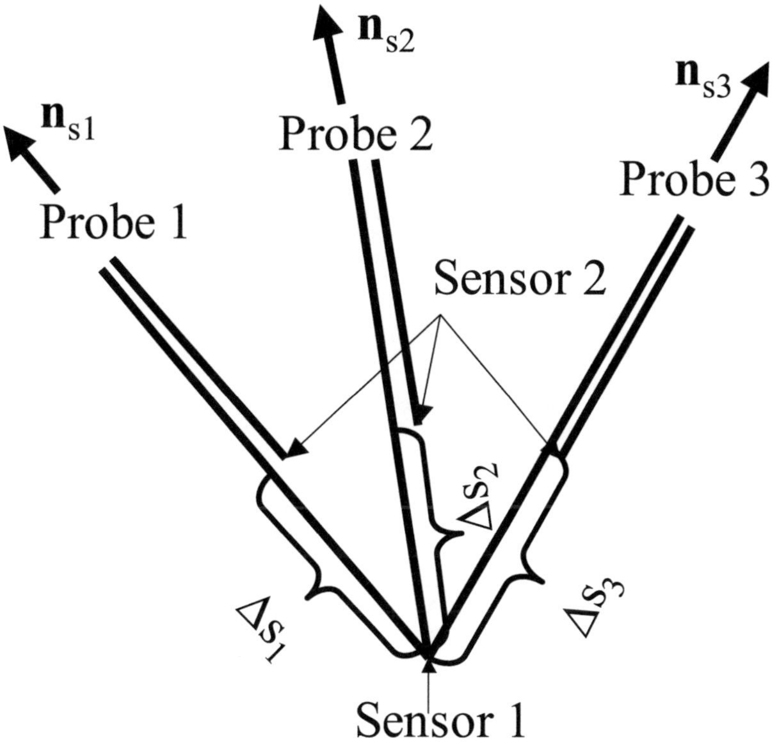

Fig. 9. Three double sensor probe (four sensor probe)

The direction cosines of unit vector of each double sensor probe ($\mathbf{n}_{sk}$, as shown in Fig.9) are denoted by $\cos\eta_{xk}$, $\cos\eta_{yk}$, $\cos\eta_{zk}$. Then, the inverse of product of interfacial velocity and unit normal vector of interface which appears in Eq.(78) is given by

$$\frac{1}{|n_i \bullet v_i|} = \frac{\sqrt{|A_1|^2 + |A_2|^2 + |A_3|^2}}{\sqrt{|A_0|^2}} \tag{82}$$

Here, $|A_0|$, $|A_1|$, $|A_2|$ and $|A_3|$ are given by

$$|A_0| = \begin{vmatrix} \cos\eta_{x1} & \cos\eta_{y1} & \cos\eta_{z1} \\ \cos\eta_{x2} & \cos\eta_{y2} & \cos\eta_{z2} \\ \cos\eta_{x3} & \cos\eta_{y3} & \cos\eta_{z3} \end{vmatrix} \tag{83}$$

$$|A_1| = \begin{vmatrix} 1/v_{s1} & \cos\eta_{y1} & \cos\eta_{z1} \\ 1/v_{s2} & \cos\eta_{y2} & \cos\eta_{z2} \\ 1/v_{s3} & \cos\eta_{y3} & \cos\eta_{z3} \end{vmatrix} \tag{84}$$

$$|A_2| = \begin{vmatrix} \cos\eta_{x1} & 1/v_{s1} & \cos\eta_{z1} \\ \cos\eta_{x2} & 1/v_{s2} & \cos\eta_{z2} \\ \cos\eta_{x3} & 1/v_{s3} & \cos\eta_{z3} \end{vmatrix} \tag{85}$$

$$|A_3| = \begin{vmatrix} \cos\eta_{x1} & \cos\eta_{y1} & 1/v_{s1} \\ \cos\eta_{x2} & \cos\eta_{y2} & 1/v_{s2} \\ \cos\eta_{x3} & \cos\eta_{y3} & 1/v_{s3} \end{vmatrix} \tag{86}$$

When three double sensor probes as shown in Fig.9 are orthogonal (perpendicular to each other), Eq.(82) is simply given by

$$\frac{1}{|n_i \bullet v_i|} = \sqrt{\left(\frac{1}{v_{s1}}\right)^2 + \left(\frac{1}{v_{s2}}\right)^2 + \left(\frac{1}{v_{s3}}\right)^2} \tag{87}$$

Then time averaged interfacial area concentration is given by

$$\overline{a_i} = \frac{1}{T}\sum_{j=1}^{N}\sqrt{\left(\frac{1}{v_{s1j}}\right)^2 + \left(\frac{1}{v_{s2j}}\right)^2 + \left(\frac{1}{v_{s3j}}\right)^2} \tag{88}$$

Most of recent experimental works of local interfacial area measurement are carried out by double sensor probe or three double sensor probe (four sensor probe) using electrical resistivity probe or optical probe.

For practical application, Kataoka et al.(1986) further proposed a simplified expression of Eq.(88) for double sensor probe which is given by

$$\overline{a_i} = 4\frac{1}{T}\sum_{j=1}^{N}\sqrt{\left(\frac{1}{v_{szj}}\right)^2}\frac{1}{1-\cot\frac{1}{2}\alpha_0\ln(\cos\frac{1}{2}\alpha_0) - \tan\frac{1}{2}\alpha_0\ln(\sin\frac{1}{2}\alpha_0)} \tag{89}$$

where α_0 is given by

$$\frac{\sin 2\alpha_0}{2\alpha_0} = \frac{1-(\sigma_z^2/|\overline{v_{iz}}|^2)}{1+3(\sigma_z^2/|\overline{v_{iz}}|^2)} \tag{90}$$

Here, $|\overline{v_{iz}}|$ and σ_z are the mean value and fluctuation of the z component interfacial velocity.

Hibiki, Hognet and Ishii (1998) carried out more detailed analysis of configuration of gas-liquid interface and double sensor probe and proposed more accurate formulation of interfacial area concentration measurement using double sensor probe. It is given by

$$\overline{a_i} = 2\frac{1}{T}\sum_{j=1}^{N}\sqrt{\left(\frac{1}{v_{szj}}\right)^2}\, I(\omega_0)\frac{\omega_0^{\ 3}}{3(\omega_0 - \sin\omega_0)} \tag{91}$$

Here ω_0 is given by

$$\frac{3}{2\omega_0^{\ 2}}\left(1-\frac{\sin 2\omega_0}{2\omega_0}\right) = \frac{1-(\sigma_z^{\ 2}/\left|\overline{v_{iz}}\right|^2)}{1+3(\sigma_z^{\ 2}/\left|\overline{v_{iz}}\right|^2)} \tag{92}$$

Double sensor probe or three double sensor probe (four sensor probe) has finite spacing between sensors. In relation to sensor spacing and size of bubble, some measurement errors are inevitable. In order to evaluate such measurement errors, a numerical simulation method using Monte Carlo approach is proposed (Kataoka et al., (1994), Wu and Ishii (1999)) for sensitivity analysis of measurement errors of double sensor probe or three double sensor probe. Using this method, Wu and Ishii (1999) carried out comprehensive analysis of accuracy of interfacial area measurement using double sensor probe including the probability of missing bubbles. They obtained formulation of interfacial area concentration measurement similar to Eqs.(91) and (92). The method using Eqs.(89) and (90) underestimated the interfacial area concentration up to 50%.

For adiabatic two-phase flow, many research groups all over the world, carried out measurements of interfacial area concentration mainly using double sensor or four sensor electrical resistivity probes. Most of experiments were carried out for vertical upward air-water two-phase flow in pipe. Some data were reported in annulus or downward flow. Flow regime covers bubbly flow to bubbly-to-slug transition. Some data are reported for annular flow. The experimental database of interfacial area concentration for non-boiling system described above is summarized in Table 1 (Kataoka (2010)).

Measurement of interfacial area concentration in boiling two-phase flow is quite important in view of practical application to nuclear reactor technology. However, in boiling two-phase flow, measurement of interfacial area is much more difficult compared with the measurement in non-boiling two-phase flow because of the durability of electrical resistivity and optical probes in high temperature liquid. Therefore, the accumulation of experimental data in boiling system was not sufficient compared with those in non-boiling system. However, recently, based on the establishment of measurement method of interfacial area as described above and improvement of electrical resistivity and optical probes, detailed measurements of interfacial area concentration become possible and experimental works have been carried out by various research groups. Most of experiments are carried out in annulus test section where inner pipe is heated. However, recently, some experimental studies are reported in rod bundle geometry. The experimental database of interfacial area concentration for boiling system described above is summarized in Table 2 (Kataoka (2010)).

Serizawa et al.1975,1992			
Air-Water	Pipe, D=60mm, 30mm	j_L=0.5 – 5.0 m/s	Double sensor,
Vertical up	L=1800mm,2500mm	j_G=0.047 –0.54 m/s	electrical resistivity
Grossetete 1995			
Air-Water	Pipe, D=38.1mm	j_L=0.526 – 0.877 m/s	Double sensor optical fiber
Vertical up	L=5906 mm	j_G=0.0588 –0.322 m/s	
Hibiki et al 1998, 1999,2001.			
Air-Water	Pipe, D=25.4mm, 50.8mm	j_L=0.292 – 5.0m/s	Double sensor,
Vertical up	L=3150 mm	j_G=0.0162 –3.9 m/s	electrical resistivity
Hibiki et al. 2003a,20003b			
Air-Water	Annulus, Di=19.2mm		
	Do=38.1mm	j_L=0.272 – 2.08 m/s	Double & four sensor
Vertical up	L=4730 mm	j_G=0.0313 –3.8 m/s	electrical resistivity
Hibiki et a l2004,2005.			
Air-Water	Pipe, D=25.4mm, 50.8mm	j_L=-0.62 – -3.11 m/s	Four sensor
Vertical down	L=3400 mm	j_G=-0.00427 – -0.486 m/s	electrical resistivity
Takamasa et al.2003a,2003b			
Air-Water	Pipe, D=9mm	j_L=0.138 – 1.0m/s	Stereo image processing
Nitrogen-Water	L=819 mm	j_G=0.0084 – 0.052m/s	Micro & normal gravity
Vertical up			
Hazuku et al. 2007			
Air-Water	Pipe, D=11mm	j_L=0.0878 – 0.790m/s	Laser focus
Vertical up	L=2750 mm	j_G=39.5 – 73.0 m/s	displacement meter
Shen et al . 2005			
Air-Water	Pipe, D=200mm	j_L=0.035 – 0.277 m/s	Double sensor,
Vertical up	L=22600mm	j_G=0.186 – 0.372 m/s	electrical resistivity
Ohuki and Akimoto 2000			
Air-Water	Pipe, D=200mm	j_L==0.06 – 1.06 m/s	Double sensor,
Vertical up	L=12000mm	j_G=0.03 – 4.7 m/s	electrical resistivity
			Optical fiber probe
Shawkat et al. 2008			
Air-Water	Pipe, D=200mm	j_L==0.2 – 0.68 m/s	X type anemometer
Vertical up	L=8400mm	j_G=0.005 – 0.18 m/s	
Prasser 2007			
Air-Water	Pipe, D=51.2, 195.3 mm	j_L==1.02 m/s	Wire mesh sensor
Vertical up	L=3072, 7812mm	j_G=0.0094 – 0.53 m/s	

Table 1. Summary of Experimental Database of Interfacial Area Concentration for Non-Boiling System

Roy and et al. 1994			
R-113	Annulus, Di=19.2mm Do=38.1mm	G=579 - 801Kg/m^2/s	Double sensor optical fiber
Vertical up	L=2750mm	q= 0.79 – 116kW/m^2	0.269 MPa
T_L=43.0 – 50.3 C, ΔTsub=37.1 – 29.8 C			
Zeitoun et al. 1994, 1996			
Water	Annulus, Di=12.7mm Do=25.4mm	G=151.4 – 411.7Kg/m^2/s	High speed video camera
Vertical up	L=306mm	q= 287 – 796kW/m^2	0.117 – 0.166 MPa
T_L =11.6 – 31.1 C			
Situ et al. 2004a,2004b			
Water	Annulus, Di=19.0mm Do=38.1mm	j_L==0.5 – 2.02m/s	Double sensor electrical resistivity
Vertical up	L=1700mm	q= 5 – 200kW/m^2	0.1292 – 0.1481 MPa
T_L =95.0– 99.0 C			
Bae et al. 2008			
Water	Annulus, Di=21.0mm Do=40.0mm	G=340 – 674Kg/m^2/s	Double sensor electrical resistivity
Vertical up	L=1870mm	q= 97 – 359kW/m^2	0.121 –0.142 MPa
T_L =95.0– 99.0 C			
Yun et al. 2008			
Water	3x3 rod bundle pitch: 16.6mm, diameter 8.2 mm		
	G=250 – 552Kg/m^2/s	Double sensor electrical resistivity	
Vertical up	L=1700mm	q= 25 – 185kW/m^2	0.12 MPa
T_L =96.0– 104.9 C			
Lee et al. 2002,2008			
Water	Annulus, Di=19mm Do=37.5mm	G=478 – 1049.5Kg/m^2/s	Double sensor electrical resistivity
Vertical up	L=1670mm	q= 88 – 359kW/m^2	0.01147 – 0.1698 MPa
T_L =84.3– 100.4 C			

Table 2. Summary of Experimental Database of Interfacial Area Concentration for Boiling System

6. Validation of interfacial area transport models by experimental data

In order to confirm the validity of transport equation of interfacial area, comparisons with experimental data were carried out mainly for bubbly flow and churn flow. The transport equation for bubbly flow is given by Eq.(68). This equation includes turbulent diffusion term of interfacial area, turbulent diffusion term due to non-isotropic turbulence, sink term due to bubble coalescence and source term due to bubble break up. Each term is separately validated by experimental data.

Kataoka et al. (2011b, 2011c) carried out the validation of turbulent diffusion term of interfacial area, turbulent diffusion term due to non-isotropic turbulence using experimental data of radial distributions in air-water two-phase flow in round pipe under developed region. Under steady state and developed region without phase change, coalescence and break up of bubbles are negligible. Under such assumptions, transport equation of interfacial area concentration based on turbulent transport model, Eq.(68) can be simplified and given by following equation.

$$K_1 d_B |v'_L| \frac{1}{R-y}\frac{\partial}{\partial y}\left((R-y)\frac{\partial \overline{a_i}}{\partial y}\right) + K_2 \alpha d_B \overline{a_i} \frac{1}{R-y}\frac{\partial}{\partial y}\left((R-y)\frac{\partial \overline{\overline{V_L}}}{\partial y}\right) = 0 \qquad (93)$$

Here, R is pipe radius and y is distance from pipe wall. Kataoka's model for turbulent diffusion of interfacial area concentration, (Eqs.(61) , (65) and (67)) was used.

Kataoka et al. (2011c) further developed the model of turbulent diffusion term due to non-isotropic turbulence for churn flow. In the churn flow, additional turbulence void transport terms appear due to the wake of large babble as schematically shown in Fig.10.

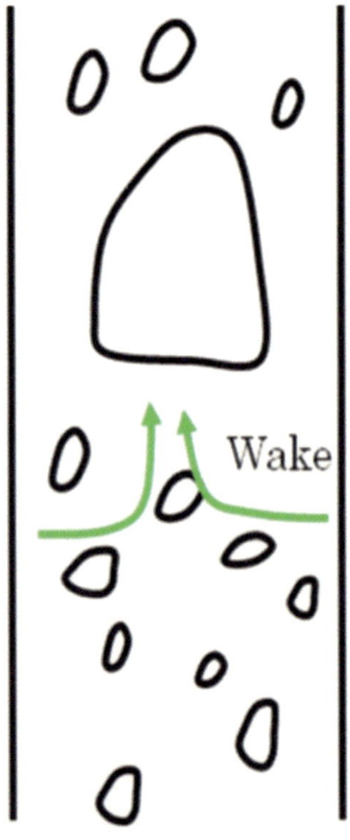

Fig. 10. Wake in Churn Flow Regime

For interfacial area transport due to wake of churn bubble, interfacial area is transported toward the center of pipe. The flux of interfacial area concentration in radial direction J_{ai}, due to churn bubble is related to the terminal velocity of churn bubble. The flux of interfacial area concentration toward the center of pipe is large at near wall and small at the center of pipe. Then, it is simply assumed to be proportional to the distance from pipe center. Finally, the flux of interfacial area concentration in radial direction, J_{ai} due to churn bubble is assumed to be given by

$$J_{ai} = K_{Cai}\frac{R-y}{R}\{0.35\sqrt{gD}\}\overline{a_i} \tag{94}$$

Then, transport equation of interfacial area concentration based on turbulent transport model in churn flow is given by

$$\frac{1}{R-y}\frac{\partial}{\partial y}\left((R-y)K_1 d_B |v_L'|\frac{\partial \overline{a_i}}{\partial y}\right) + \frac{1}{R-y}\frac{\partial}{\partial y}\left((R-y)^2 K_{Cai}\frac{0.35\sqrt{gD}}{R}\overline{a_i}\right) = 0 \tag{95}$$

In order to predict radial distribution of interfacial area concentration using Eq.(93) or Eq.(95), radial distributions of void fraction, averaged liquid velocity and turbulent liquid velocity are needed. These distributions were already predicted based on the turbulence model of two-phase flow for bubbly flow and churn flow (Kataoka et al. (2011d)).

Using transport equation of interfacial area concentration for bubbly flow (Eq.(93) and churn flow (Eq.(95)), the radial distributions of interfacial area concentration are predicted and compared with experimental data. Serizawa et al. (1975, 1992) measured distributions of void fraction, interfacial area concentration, averaged liquid velocity and turbulent liquid velocity for vertical upward air-water two-phase flow in bubbly and churn flow regimes in round tube of 60mm diameter. Void fraction and interfacial area were measured by electrical resistivity probe and averaged liquid velocity and turbulent liquid velocity were measured by anemometer using conical type film probe with quartz coating. Their experimental conditions are

Liquid flux, J_L: 0.44 - 1.03 m/s

Gas flux, J_G: 0 - 0.403 m/s

For empirical coefficient, K_{cai} is assumed to be 0.01 based on experimental data. The condition of flow regime transition from bubbly to churn flow is given in terms of area averaged void fraction, $\overline{\alpha}$ based on experimental results which is given by

$$\overline{\alpha} = 0.2 \tag{96}$$

Figures 11 and 12 show some examples of the comparison between experimental data and prediction of radial distributions of interfacial area concentration in bubbly flow and churn flow. In bubbly flow regime, distributions of interfacial area concentration show wall peak of which magnitude is larger for larger liquid flux whereas distributions interfacial area concentration in churn flow show core peak. The prediction based on the present model well reproduces the experimental data.

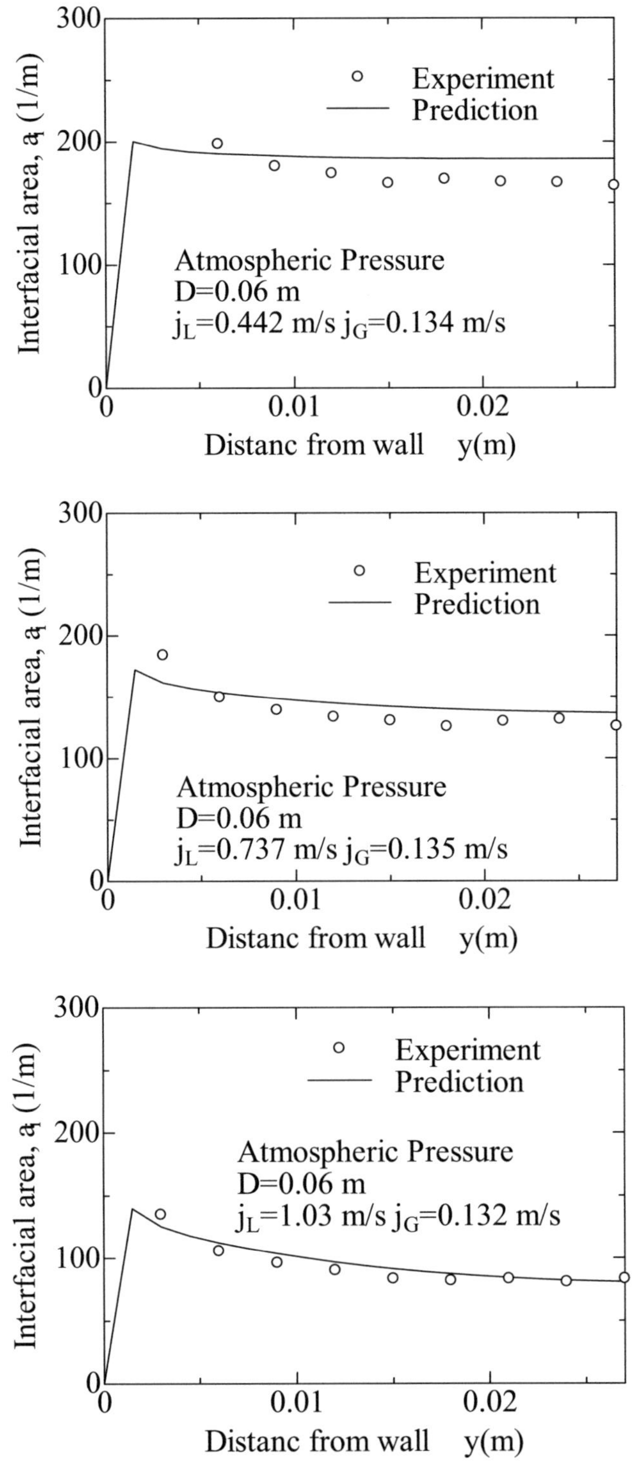

Fig. 11. Distributions of Interfacial Area Concentration for Bubbly Flow

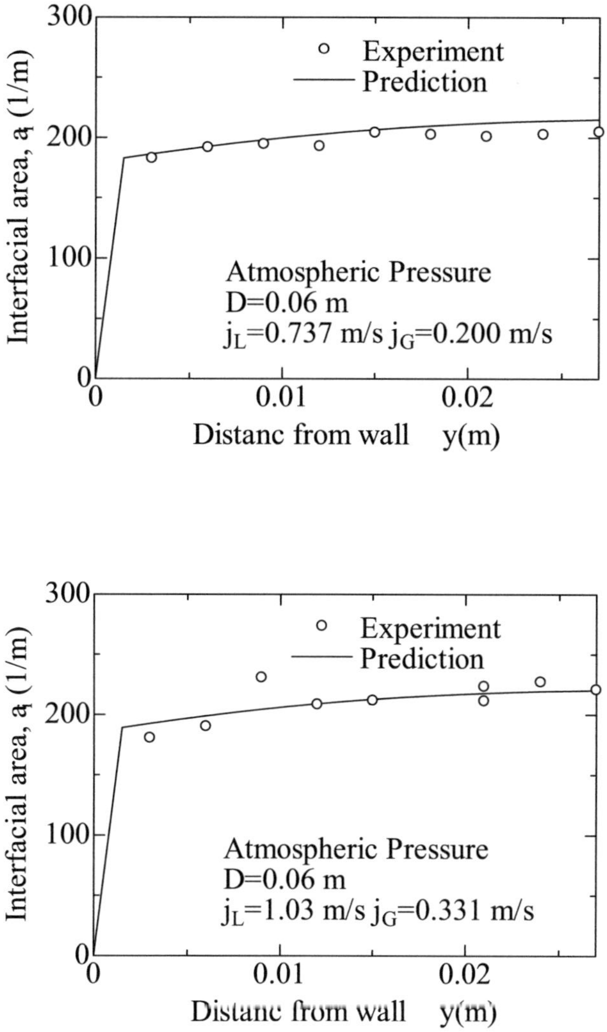

Fig. 12. Distributions of Interfacial Area Concentration for Churn Flow

Hibiki and Ishii (2000a) carried out the validation of their own correlations of sink term due to bubble coalescence (Eq.(72)) and source term due to bubble break up (Eq.(70)) using experimental data. They carried out experiments in vertical upward air water two-phase flow in pipe under atmospheric pressure. In order to validate their interfacial transport model, evolutions of radial distributions of interfacial area concentration in the flow direction were systematically measured. Experimental conditions are as follows.

Condition I

Pipe diameter D: 25.4mm, Measuring positions z from inlet: (z/D=12, 65,125),

Liquid flux j_L=0.292 - 3.49 m/s, Gas flux j_G=0.05098 -0.0931 m/s

Condition II

Pipe diameter D: 50.8mm, Measuring positions z from inlet: (z/D=6,30.3, 53.5)

Liquid flux j_L=0.491 - 5.0 m/s, Gas flux j_G=0.0556 -3.9 m/s

Figures 13 and 14 show the result of comparison between experimental data and prediction using transport equation of interfacial area with sink term due to bubble coalescence (Eq.(70)) and source term due to bubble break up (Eq.(72)). Predictions agree with experimental data within 10% accuracy.

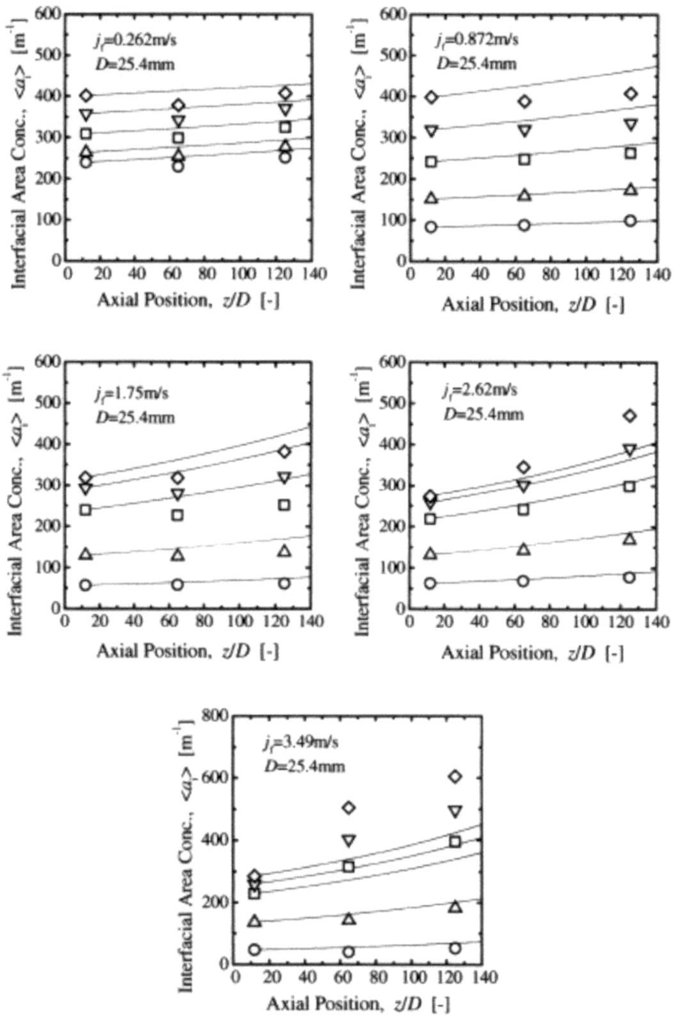

Fig. 13. Comparison between Experimental data and prediction for the variation of interfacial area concentration along flow direction for 25.4mm diameter pipe (Hibiki, T. and Ishii, M. 2000a One-Group Interfacial Area Transport of Bubbly Flows in Vertical Round Tubes, International Journal of Heat and Mass Transfer, 43, 2711-2726.Fig.8)

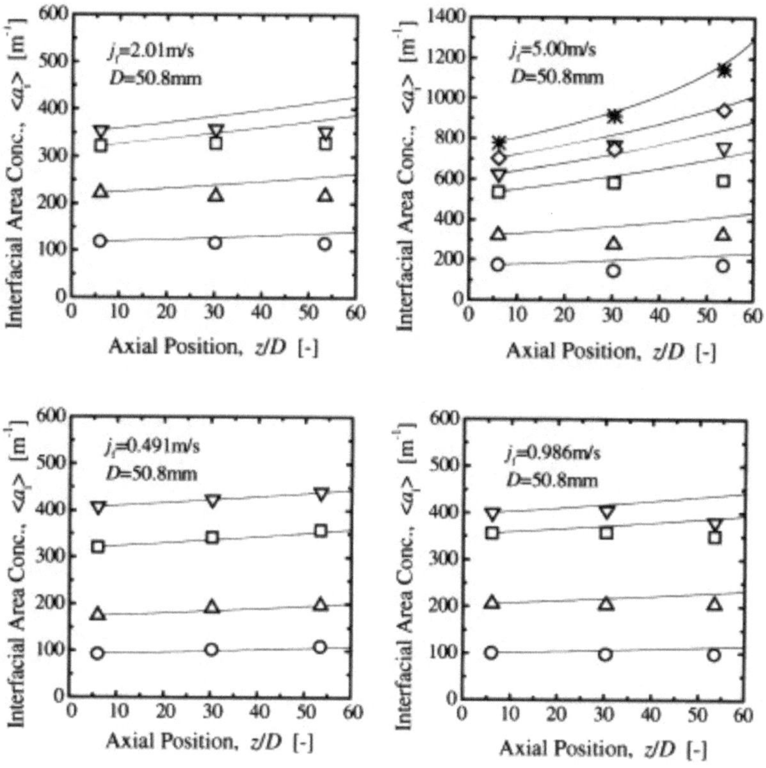

Fig. 14. Comparison between Experimental data and prediction for the variation of interfacial area concentration along flow direction for 50.8mm diameter pipe (Hibiki, T. and Ishii, M. 2000a One-Group Interfacial Area Transport of Bubbly Flows in Vertical Round Tubes, International Journal of Heat and Mass Transfer, 43, 2711-2726.Fig.9)

7. Conclusion

In this chapter, intensive review on recent developments and present status of interfacial area concentration and its transport model was carried out. Definition of interfacial area and rigorous formulation of local instant interfacial area concentration was introduced. Using this formulation, transport equations of interfacial area concentration were derived in details. Transport equations of interfacial area concentration consist of conservation equation of interfacial area concentration and conservation equation of interfacial velocity. For practical application, simplified transport equation of interfacial area concentration was derived with appropriate constitutive correlations. For bubbly flow, constitutive correlations of turbulent diffusion, turbulent diffusion due to non-isotropic turbulence, sink term due to bubble coalescence and source term due to bubble break up were developed. Measurement methods on interfacial area concentration were reviewed and experiments of interfacial area concentration for non-boiling system and boiling system were reviewed. Validation of transport equations of interfacial area concentration was carried out for bubbly and churn flow with satisfactory agreement with experimental data. At present, transport equations of interfacial area concentration can be applied to analysis of two-phase flow with considerable accuracy. However, the developments of constitutive correlations are limited to bubbly and churn flow regimes. Much more researches are needed for more systematic developments of transport equations of interfacial area concentration.

8. References

Bae, B.U., Yoon, H.Y., Euh, D.J., Song, C.H., and Park, G.C., 2008 Computational Analysis of a Subcooled Boiling Flow with a One-group Interfacial Area Transport Equation, " Journal of Nuclear Science and Technology, 45[4], 341-351.

Cousins L.B. and Hewitt, G.F. 1968 Liquid Phase Mass Transfer in Annular Two-Phase Flow: Radial Liquid Mixing, AERE-R 5693.

Delhaye, J.M. 1968 Equations Fondamentales des Ecoulments Diphasiques, Part 1 and 2, CEA-R-3429, Centre d'Etudes Nucleaires de Grenoble, France.

C. Grossetete, C., 1995 Experimental Investigation and Preliminary Numerical Simulations of Void Profile Development in a Vertical Cylindrical Pipe, Proceedings of The 2nd International Conference on Multiphase Flow '95-Kyoto, , Kyoto Japan , April 3-7, 1995, paper IF-1.

Hazuku, T., Takamasa, T., Hibiki, T., and Ishii, M., 2007 Interfacial area concentration in annular two-phase flow, International Journal of Heat and Mass Transfer, 50, 2986-2995.

Hibiki, T., Hogsett, S., and Ishii, M., 1998 Local measurement of interfacial area, interfacial velocity and liquid turbulence in two-phase flow, Nuclear Engineering and Design, 184, 287–304.

Hibiki, T., Ishii, M., and Xiao, Z., 1998 Local flow measurements of vertical upward air-water flow in a round tube, Proceedings of Third International Conference on Multiphase Flow, ICMF'98, Lyon, France, June 8-12, 1998, paper 210.

Hibiki, T., and M. Ishii, M., 1999 Experimental study on interfacial area transport in bubbly two-phase Flows, International Journal of Heat and Mass Transfer, 42 3019-3035.

Hibiki, T. and Ishii, M. 2000a One-Group Interfacial Area Transport of Bubbly Flows in Vertical Round Tubes, International Journal of Heat and Mass Transfer, 43, 2711-2726.

Hibiki, T. and Ishii, M. 2000b Two-group interfacial area transport equations at bubbly-to-slug flow transition, " Nuclear Engineering and Design, 202[1], 39-76.

Hibiki, T., Ishii, M., and Z. Xiao, Z., 2001 Axial interfacial area transport of vertical bubbly flows, " International Journal of Heat and Mass Transfer, 44, 1869-1888.

Hibiki, T. and Ishii, M. 2002 Development of one-group interfacial area transport equation in bubbly flow systems, International Journal of Heat and Mass Transfer, Volume 45[11], 2351-2372.

Hibiki, T., Situ, R., Mi, Y., and Ishii, M., 2003a Local flow measurements of vertical upward bubbly flow in an annulus, International Journal of Heat and Mass Transfer, 46, 1479–1496.

Hibiki, T., Mi, Y., Situ, R., and Ishii, M., 2003b Interfacial area transport of vertical upward bubbly two-phase flow in an annulus, International Journal of Heat and Mass Transfer, 46, 4949–4962.

Hibiki, T., Goda, H., Kim, S., Ishii, M., and Uhle, J., 2004 Structure of vertical downward bubbly flow, International Journal of Heat and Mass Transfer, 47, 1847–1862.

Hibiki, T., Goda, H., Kim, S., Ishii, M., and Uhle, J., 2005 Axial development of interfacial structure of vertical downward bubbly flow, International Journal of Heat and Mass Transfer, 48, 749–764.

Ishii, M. 1975 Thermo-Fluid Dynamic Theory of Two-Phase Flow, Eyrolles, Paris.

Ishii, M. and Kim, S. 2004 Development of One-Group and Two-Group Interfacial Area Transport Equations, Nucl. Sci. Eng., 146, 257-273.

Ishii, M., Mishima, K., Kataoka, I., Kocamustafaogullari, G. 1982 Two-Fluid Model and Importance of the Interfacial Area in Two-Phase Flow Analysis, Proceedings of the 9th US National Congress of Applied Mechanics, Ithaca, USA, June 21-25 1982, pp.73-80

Kataoka, I. 1986 Local Instant Formulation of Two-Phase Flow, Int. J. Multiphase Flow, 12, 745-758.

Kataoka, I. Ishii, M. and Serizawa, A. 1986 Local formulation and measurements of interfacial area concentration, Int. J. Multiphase Flow, 12, 505-527.

Kataoka, I. and Serizawa, A. 1990 Interfacial Area Concentration in Bubbly Flow, Nuclear Engineering and Design, 120, 163-180.

Kataoka, I. and Serizawa, A. 1991a Bubble Dispersion Coefficient and Turbulent Diffusivity in Bubbly Two-Phase Flow, Turbulence Modification in Multiphase Flows -1991-, ASME Publication FED-Vol.110, pp.59-66.

Kataoka, I. and Serizawa, A. 1991b Statistical Behaviors of Bubbles and Its Application to Prediction of Phase Distribution in Bubbly Two-Phase Flow, Proceedings of The International Conference on Multiphase Flow '91-Tsukuba, Vol.1, pp.459-462, Tsukuba, Japan, September 24-27.

Kataoka, I.and Serizawa, A. 1993 Analyses of the Radial Distributions of Average Velocity and Turbulent Velocity of the Liquid Phase in Bubbly Two-Phase Flow, JSME International Journal, Series B, 36-3, 404-411

Kataoka, I..Ishii, M., and Serizawa, A., 1994 Sensitivity analysis of bubble size and probe geometry on the measurements of interfacial area concentration in gas-liquid two-phase flow, Nuclear Engineering & Design, 146, 53-70.

Kataoka, I. et al. 2008 Basic Transport Equation of Interfacial Area Concentration In Two-Phase Flow, Proc. NTHAS6: Sixth Japan-Korea Symposium on Nuclear Thermal Hydraulics and Safety, N6P1126, Okinawa, Japan, Nov. 24- 27.

Kataoka, I. 2010 Development of Researches on Interfacial Area Transport, Journal of Nuclear Science and Technology, 47. 1-19.

Kataoka, et al., 2010 Modeling of Turbulent Transport Term of Interfacial Area Concentration in Gas-Liquid Two-Phase Flow, The Third CFD4NRS (CFD for Nuclear Reactor Safety Applications) workshop, September 14-16, 2010, Bethesda, MD, USA

Kataoka, et al., 2011a Modeling of Turbulent Transport Term of Interfacial Area Concentration in Gas-Liquid Two-Phase Flow, To be published in Nuclear Engineering & Design

Kataoka, I. 2011b, Modeling and Verification of Turbulent Transport of Interfacial Area Concentration in Gas-Liquid Two-Phase Flow, ICONE19-43077, Proceedings of ICONE19, 19th International Conference on Nuclear Engineering, May 16-19, 2011, Chiba, Japan

Kataoka, I., et al., 2011c Basic Equations of Interfacial Area Transport in Gas-Liquid Two-Phase Flow, The 14th International Topical Meeting on Nuclear Reactor Thermal Hydraulics (NURETH-14) paper Log Number: 166, Hilton Toronto Hotel, Toronto, Ontario, Canada, September 25-29, 2011.

Kataoka, I., et al., 2011d Analysis Of Turbulence Structure And Void Fraction Distribution In Gas-Liquid Two-Phase Flow Under Bubbly And Churn Flow Regime, Proceedings of ASME-JSME-KSME Joint Fluids Engineering Conference , AJK2011-10003, Hamamatsu.

Kocamustafaogullari, G. and Ishii, M, 1983 Interfacial area and nucleation site density in boiling systems, International Journal of Heat and Mass Transfer, 26, 1377-1389

Kocamustafaogullari, G and Ishii, M. 1995 Foundation of the Interfacial Area Transport Equation and Its Closure Relations, International Journal of Heat and Mass Transfer, 38, 481-493.

Lee, T.H., Park, G.C., Lee, D.J., 2002 Local flow characteristics of subcooled boiling flow of water in a vertical concentric annulus, International Journal of Multiphase Flow, 28, 1351-1368.

Lee, T.H., Yun, B.J., Park, G.C., Kim, S.O., and Hibiki, T., 2008 Local interfacial structure of subcooled boiling flow in a heated annulus, Journal of Nuclear Science and Technology, 45 [7], 683-697.

Liles, D. et al. 1984 TRAC-PF-1: An Advanced Best Estimate Computer Program for Pressurized Water Reactor Analysis, NUREG/CR-3567, LA-10157-MS.

Morel, C. 2007 On the Surface Equations in Two-Phase Flows and Reacting Single-Phase Flows, International Journal of Multiphase Flow, 33, 1045-1073.

Ohnuki, A., and Akimoto, H., 2000 Experimental study on transition of flow pattern and phase distribution in upward air-water two-phase flow along a large vertical pipe, International Journal of Multiphase Flow, 26, 367-386.

Prasser, H.-M., 2007 Evolution of interfacial area concentration in a vertical air–water flow measured by wire–mesh sensors, " Nuclear Engineering and Design, 237, 1608–1617.

Ransom. V.H. et al. 1985 RELAP/MOD2 Code Manual, Volume 1; Code Structure, System Models and Solution Methods, NUREG/CR-4312, EGG-2796.

Roy, R.P., Velidandla, V., Kalra, S.P., and Peturaud, P., 1994Local measurements in the two-phase region of turbulent subcooled boiling flow, Transactions of the ASME, Journal of Heat Transfer, 116, 660-669.

Shawkat, M.E., Ching, C.Y., and Shoukri, M., 2008 Bubble and liquid turbulence characteristics of bubbly flow in a large diameter vertical pipe, " International Journal of Multiphase Flow, 34, 767-785.

Shen, X., Mishima, K., and Nakamura, H., 2005 Two-phase phase distribution in a vertical large diameter pipe, International Journal of Heat and Mass Transfer, 48, 211–225.

Serizawa, A., Kataoka, I., and Michiyoshi, I., 1975 Turbulence Structure of Air-Water Bubbly flow-I –III, Int. J. Multiphase Flow, 2, 221-259.

Serizawa, A., Kataoka, I., and Michiyoshi, I., 1992 Phase distribution in bubbly flow, " Multiphase Science and Technology, ed. by G.F.Hewitt, J.M.Delhaye and N. Zuber, pp.257-302, Hemisphere, N.Y.

Sharma, M.M., Danckwerts, P.V. 1970 Chemical method of measuring interfacial area and mass transfer coefficients in two-fluid systems, Br. Chem. Engng., 15[4] 522-528.

Situ, R., Hibiki, T., Sun, X., Mi, Y., and Ishii, M., 2004a Axial development of subcooled boiling flow in an internally heated annulus, Experiments in Fluids, 37, 589–603.

Situ, R., Hibiki, T., Sun, X., Mi, Y., and Ishii, M., 2004b Flow structure of subcooled boiling flow in an internally heated annulus, International Journal of Heat and Mass Transfer, 47, 5351–5364.

Takamasa, T., Goto, T., Hibiki, T., and Ishii, M., 2003a Experimental study of interfacial area transport of bubbly flow in small-diameter tube, International Journal of Multiphase Flow, 29, 395–409.

Takamasa, T., Iguchi, T., Hazuku, T., Hibiki, T., and Ishii, M., 2003b Interfacial area transport of bubbly flow under micro gravity environment, International Journal of Multiphase Flow, 29, 291–304.

Wu, Q., and Ishii, M., 1999 Sensitivity study on double sensor conductivity probe for the measurement of interfacial area concentration in bubbly flow, International Journal of Multiphase Flow,.25, 155-173.

Yao, W. and Morel, C. 2004 Volumetric Interfacial Area Prediction in Upward Bubbly Two-Phase Flow, International Journal of Heat and Mass Transfer, 47[2], 307-328.

Yun, B.J., Park, G.C., Julia, J.E., and Hibiki, Y., 2008 Flow Structure of Subcooled Boiling Water Flow in a Subchannel of 3x3 Rod Bundles, Journal of Nuclear Science and Technology, 45[5], 402-422.

Zeitoun, O., Shoukri, M., and Chatoorgoon, V., 1994 Measurement of interfacial area concentration in subcooled liquid vapor flow, Nuclear Engineering and Design, 152, 243-255.

Zeitoun, O., and Shoukri, M., 1996 Bubble behavior and mean diameter in subcooled flow boiling, Transactions of the ASME, Journal of Heat Transfer, 118, 110-116.

7

Thermal Aspects of Conventional and Alternative Fuels in SuperCritical Water-Cooled Reactor (SCWR) Applications

Wargha Peiman, Igor Pioro and Kamiel Gabriel
University of Ontario Institute of Technology
Canada

1. Introduction

The demand for clean, non-fossil based electricity is growing; therefore, the world needs to develop new nuclear reactors with higher thermal efficiency in order to increase electricity generation and decrease the detrimental effects on the environment. The current fleet of nuclear power plants is classified as Generation III or less. However, these models are not as energy efficient as they should be because the operating temperatures are relatively low. Currently, a group of countries have initiated an international collaboration to develop the next generation of nuclear reactors called Generation IV. The ultimate goal of developing such reactors is to increase the thermal efficiency from what currently is in the range of 30 - 35% to 45 - 50%. This increase in thermal efficiency would result in a higher production of electricity compared to current Pressurized Water Reactor (PWR) or Boiling Water Reactor (BWR) technologies.

The Generation IV International Forum (GIF) Program has narrowed design options of the nuclear reactors to six concepts. These concepts are Gas-cooled Fast Reactor (GFR), Very High Temperature Reactor (VHTR), Sodium-cooled Fast Reactor (SFR), Lead-cooled Fast Reactor (LFR), Molten Salt Reactor (MSR), and SuperCritical Water-cooled Reactor (SCWR). These nuclear-reactor concepts differ in their design in aspects such as the neutron spectrum, coolant, moderator, and operating temperature and pressure.

A SuperCritical Water-cooled Reactor can be designed as a thermal-neutron-spectrum or fast-neutron-spectrum system. SCWR operates above the critical point of water which is at a temperature of 374°C and a pressure of 22.1 MPa. The operating pressure of SCWR is 25 MPa and the outlet temperature of the coolant is 550 - 625°C depending on the design chosen by the respective country that is developing it. The primary choice of fuel for SCWR is an oxide fuel while a metallic fuel has been considered as the secondary choice for the fast-neutron-spectrum SCWRs. A supercritical-water Rankine cycle has been chosen as the power cycle (US DOE, 2002). The thermal efficiency of SCWR is in the range of 45 - 50 %. Figure 1 shows a schematic diagram of a SCWR.

Some of the advantages of SCW Nuclear Power Plants (NPPs) over the conventional NPPs include higher thermal efficiency within a range of 45–50% (Pioro and Duffey, 2007) compared to 30 - 35% for the current NPPs, lower capital costs per kWh of electricity, and the possibility for co-generation of hydrogen. For instance, the copper-chlorine cycle requires steam at temperatures between 500 and 530°C (Naterer et al., 2009, 2010), which is within the operating range of some SCWR designs. These systems work when supercritical water from a reactor flows through a heat exchanger and transfers heat to a low-pressure steam, which becomes a superheated steam. This superheated steam is transferred at the outlet of the heat exchanger to an adjacent hydrogen plant at a lower pressure (Naterer et al., 2009, 2010).

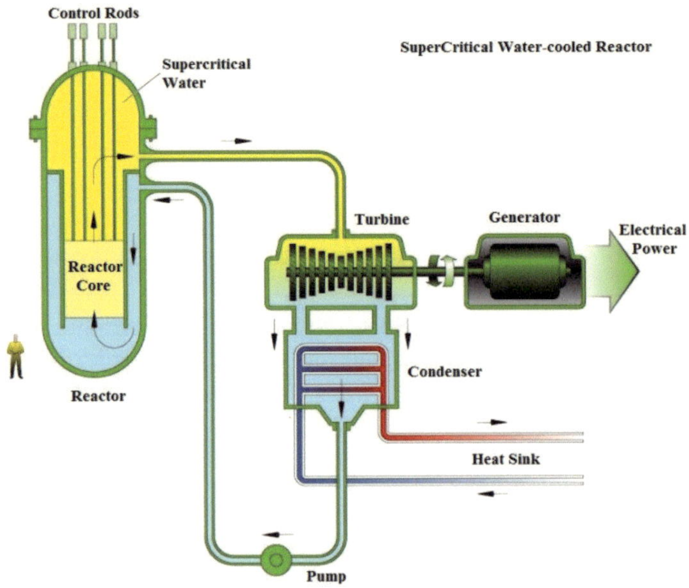

Fig. 1. Schematic diagram of PV SCWR (US DOE, 2002).

In general, SCWRs can be classified based on the neutron spectrum, moderator, or pressure boundary. In terms of the pressure boundary, SCWRs are classified into two categories, a) Pressure Vessel (PV) SCWRs, and b) Pressure Tube (PT) or Pressure Channel (PCh) SCWRs (Oka et al., 2010; Pioro and Duffey, 2007). The PV SCWR requires a thick pressure vessel with a thickness of about 50 cm (Pioro and Duffey, 2007) in order to withstand high pressures. The vast majority of conventional PWRs and BWRs are examples of PV reactors. Figure 1 shows a schematic diagram of a PV SCWR. On the other hand, the core of a PT SCWR consists of distributed pressure channels, with a thickness of 10 - 15 mm, which might be oriented vertically or horizontally, analogous to RBMK and CANDU reactors, respectively. For instance, SCW CANDU (CANada Deuterium Uranium) reactor consists of 300 horizontal fuel channels with coolant inlet and outlet temperatures of 350 and 625°C at a pressure of 25 MPa (Pioro and Duffey, 2007). It should be noted that a vertical core option has not been ruled out; both horizontal and vertical cores are being studied by the Atomic Energy of Canada Limited (AECL) (Diamond, 2010). Nevertheless, PT SCWRs provide a better control of flow and density variations. On the other hand, in PV SCWRs, there is a non-uniform temperature variation of coolant at the outlet of the pressure vessel.

In terms of the neutron spectrum, most SCWR designs are thermal-spectrum; however, fast-spectrum SCWR designs are studied. Recently, Liu et al. (2010) have proposed a mixed spectrum SCWR core, which consists of fast and thermal regions. In general, various solid or liquid moderator options can be utilized in thermal-spectrum SCWRs. These options include light-water, heavy-water, graphite, beryllium oxide, and zirconium hydride (Kirillov et al., 2007). This liquid moderator concept can be used in both PV and PT SCWRs. The only difference is that in a PV SCWR, the moderator and the coolant are the same fluid. Thus, light-water is a practical choice for the moderator. In contrast, in PT SCWRs the moderator and the coolant are separated. As a result, there are a variety of options in PT SCWRs, mostly due to the separation of the coolant and the moderator.

One of these options is to use a liquid moderator such as light-water or heavy-water. One of the advantages of using a liquid moderator in PT SCWRs is that the moderator acts as a passive heat sink in the event of a Loss Of Coolant Accident (LOCA). A liquid moderator provides an additional safety feature[1], which enhances the safety of operation. On the other hand, one disadvantage of liquid moderators is an increased heat loss from the fuel channels to the liquid moderator, especially at SCWR conditions.

The second option is to use a solid moderator. Currently, in RBMK reactors and some other types of reactors such as AGR and HTR, graphite is used as the moderator. However, graphite may catch fire at high temperatures under some conditions when exposed to water or oxygen. Other materials such as beryllium oxide and zirconium hydride may be used as solid moderators (Kirillov et al., 2007). In this case, heat losses are reduced significantly. On the contrary, the solid moderators do not provide a passive-safety feature.

High operating temperatures of SCWRs leads to high fuel centerline temperatures. Currently, UO_2 has been used in Light Water Reactors (LWRs) and Pressurized Heavy Water Reactors (PHWRs); however, it has a low thermal conductivity which may result in high fuel centerline temperatures. Previous studies (Grande et al., 2010; Pioro et al., 2010; Villamere et al., 2009) have shown that the fuel centerline temperatures could exceed the industry limit of 1850°C (Reisch, 2009) when UO_2 is used at SCWR conditions. These studies have been conducted based on an average thermal power per channel and have not taken into account the effects of fuel-sheath gap on the sheath and fuel centreline temperatures. Additionally, the possibility of using enhanced thermal-conductivity fuels in SCWRs has not been examined by previous studies. Moreover, previous studies have focused on the fuel without any emphasis on the fuel channel. Therefore, there is a need to investigate the potential use of conventional and alternative fuels for future use in SCWRs.

2. Heat transfer at supercritical conditions

Heat transfer at supercritical conditions is characterized by changes in the thermophysical properties of the fluid specifically at pseudocritical points. A pseudocritical point exists at a pressure above the critical pressure of a fluid and at a temperature corresponding to the maximum value of the specific heat for this particular pressure (Pioro and Duffey, 2007). The increase in the specific heat reaches its maximum at the critical point and then decreases as the pressure increases. Furthermore, the pseudocritical temperature increases as the pressure increases. For instance, the corresponding pseudocritical temperatures of light-water at 23 and 25 MPa are approximately 377.5 and 384.9°C, respectively. Nevertheless, as the temperature passes through the pseudocritical temperature, the specific heat increases. This increase in the specific heat of the fluid allows for the deposition of a significant

[1]Currently, such option is used in CANDU-6 reactors.

amount of heat into the fluid. Eventually, this deposited heat can be converted into mechanical energy in steam turbines.

In addition to the specific heat, other thermophysical properties of a fluid undergo significant changes at the pseudocritical point. These changes affect the heat transfer capabilities of the fluid. Therefore, it is important to ensure that the thermophysical properties of a supercritical fluid are determined with accuracy. Figure 2 shows density and specific heat of water at 22.064 and 25 MPa. These thermophysical properties of water have been determined using the NIST REFPROP software.

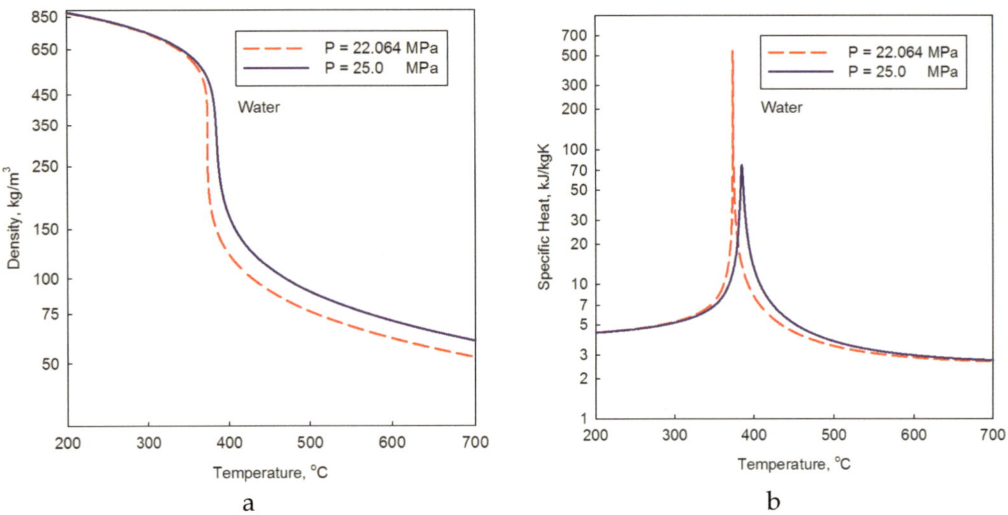

Fig. 2. a) Density and b) specific heat of water at 22.064 and 25 MPa.

In general, all thermophysical properties experience considerable changes near the critical and pseudocritical points. These changes are the greatest near the critical point; whereas, they become more gradual in the vicinity of the pseudocritical point. This gradual change in the thermophysical peroperties of fluids results in asingle-phase flow at supercritical conditions. In contrast, at subcritical conditions, a two-phase flow exists as the temperature of the fluid reaches the saturation temperature corresponding to the operating pressure. At the saturation tempertaure, the fluid undergoes a phase change from liquid to vapor when heat is added to the fluid. As a result of this phase change, there is a discontinuity in the variation of the thermophysical properties of the fluid. Figure 3a shows the density of water at 7, 11, and 15 MPa pressures, which correspond to the operating pressures of BWRs, CANDU reactors, and PWRs. As shown in Fig. 3a, there is a sharp drop in the density of water as the saturation temperatures of the corresponding pressures are reached.

The thermal efficiency of a Nuclear Power Plant (NPP) to a large extent depends on the pressure and temperature of the steam at the inlet to the turbine when the Rankine cycle is considered. In the case of either a direct cycle or an indirect cycle, the physical properties of the steam at the inlet of the turbine depend on the operating temperature and pressure of the reactor coolant. Figure 3b shows the operating pressures and temperatures of BWRs, PWRs, and PHWRs (e.g., CANDU reactors), which comprise the vast majority of the currently operating NPPs.

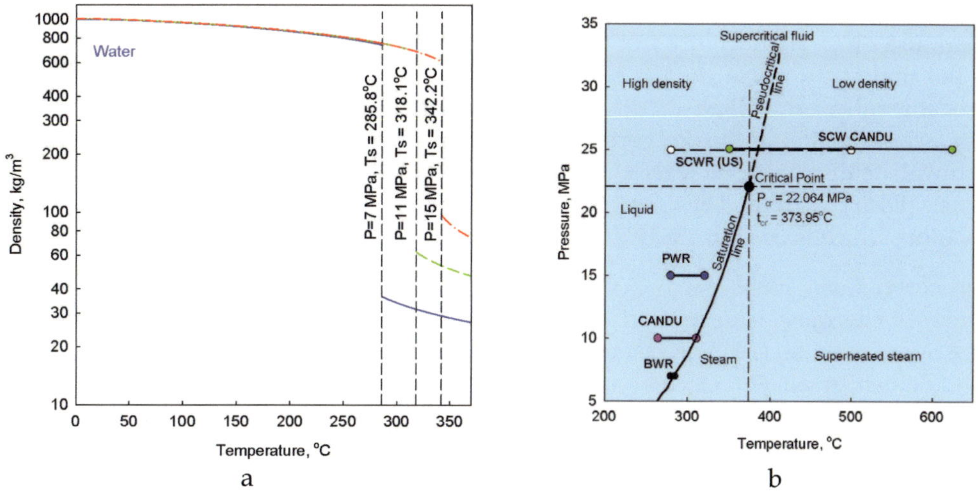

Fig. 3. a) Density of water at 7, 11, and 15 MPa and b) Operating parameters of several reactors (Pioro and Duffey, 2007).

In terms of the operating conditions of the coolant, these reactors are all categorized as subcritical. PWRs have the highest operating pressure approximately at 15 MPa followed by CANDU reactors and BWRs, which operate at a pressure of 11 and 7 MPa, respectively. The outlet temperature of the coolant depends on the operating pressure of the reactor. In PWRs and CANDU reactors, the outlet temperature of the coolant is slightly below the saturation temperature of their corresponding operating pressures in order to avoid boiling of the coolant inside the reactor and achieve a high enthalpy rise across the reactor core. In addition, it is necessary to maintain the pressure within an operational margin due to pressure fluctuation during operation. As a result, the thermal efficiency of NPPs is limited by operating at subcritical pressures. Consequently, the operating pressure must be increased to pressures above the critical pressure in order to achieve higher thermal efficiencies compared to those of the current NPPs.

As shown in Fig. 3b, SCWRs operate at pressures and temperatures above the critical pressure and temperature of water. These high temperatures and pressures make it possible to use supercritical "steam" turbines, which have led to high thermal efficiencies when used in coal-fired power plants. As a result, SCWRs will use a proven technology, which has been examined over 50 years of operation in coal-fired power plants. The use of such technology minimizes the technological barriers for the development of suitable turbines for use in the SCW NPPs.

2.1 Heat-transfer correlations

The development of SCWRs requires an intensive study of convective heat transfer at supercritical pressures. Heat transfer at a supercritical pressure is different from that of a subcritical pressure because the thermophysical properties of a light-water coolant undergo significant variations as the temperature of the coolant passes through the pseudocritical point. Therefore, the traditional Nusselt number and other related non-dimensional parameters developed at subcritical pressures based on the bulk-fluid temperature cannot be used (Bae and Kim, 2009).

At a supercritical pressure, the thermophysical properties of a coolant at the sheath-wall temperature differ significantly from those at the bulk-fluid temperature. Although, a fluid does not undergo a phase change at a supercritical pressure, a low-density fluid separates the sheath-wall from a high-density fluid at high heat fluxes and low mass fluxes. This results in a reduction in the convective Heat Transfer Coefficient (HTC). Consequently, the sheath-wall temperature increases. This phenomenon is known as the Deteriorated Heat Transfer (DHT) regime. Therefore, the sheath-wall temperature must be reflected in a correlation, which is used to study the heat transfer at supercritical conditions.

Many correlations have been developed for the calculation of HTC at supercritical conditions. The most widely used correlations include those developed by Bishop et al. (1964); Swenson et al. (1965); Krasnoscheckov et al. (1967); Jackson (2002); and Mokry et al. (2009). Zahlan et al. (2011) compared the prediction capabilities of sixteen correlations including the aforementioned correlations. The conclusion of the Zahlan et al. (2011) comparison study showed that the Mokry et al. (2009) correlation resulted in the lowest Root-Mean-Square (RMS) error within the supercritical region compared to all other examined correlations.

3. Specifications of generic 1200-MW$_{el}$ PT SCWR

The core of a generic 1200-MW$_{el}$ PT SCWR consists of 300 fuel channels that are located inside a cylindrical tank called the calandria vessel. There are 220 SuperCritical-Water (SCW) fuel channels and 80 Steam Re-Heat (SRH) fuel channels. SRH and SCW fuel channels are located on the periphery and at the center of the core, respectively. In terms of neutron spectrum, the studied PT SCWR is a thermal-spectrum reactor. In this thermal-spectrum PT SCWR, light-water and heavy-water have been chosen as the coolant and the moderator, respectively. The coolant enters the supercritical fuel channels at an inlet temperature of 350°C and reaches an outlet temperature of 625°C at a pressure of 25 MPa. The inlet temperature of the SuperHeated Steam (SHS), which is used as the coolant, in the SRH fuel channels, is 400°C and the corresponding outlet temperature is 625°C at an operating pressure of 5.7 MPa. Table 1 lists the operating parameters of the generic 1200-MW$_{el}$ PT SCWR (Naidin et al., 2009).

Parameters	Unit	Generic PT SCWR	
Electric Power	MW	1143-1270	
Thermal Power	MW	2540	
Thermal Efficiency	%	45 - 50	
Coolant/Moderator	-	H_2O/D_2O	
Pressure of SCW at Inlet | Outlet	MPa	25.8	25
Pressure of SHS at Inlet | Outlet	MPa	6.1	5.7
T_{in} | T_{out} Coolant (SCW)	°C	350	625
T_{in} | T_{out} Coolant (SHS)	°C	400	625
Mass Flow Rate per SCW | SRH Channel	kg/s	4.4	9.8
Thermal Power per SCW | SRH Channel	MW	8.5	5.5
# of SCW | SRH Channels	-	220	80

Table 1. Operating parameters of generic PT SCWR (Naidin et al., 2009).

3.1 Thermal cycles

The use of supercritical "steam" turbines in NPPs leads to higher thermal efficiencies compared to those of the current NPPs. There are several design options of Rankin cycles in order to convert the thermal energy of the supercritical "steam" into mechanical energy in a supercritical turbine. These design options include direct, indirect, and dual cycles. In a direct cycle, supercritical "steam" from the reactor passes directly through a supercritical turbine eliminating the need for the steam generators. This elimination reduces the costs and leads to higher thermal efficiencies compared to those produced in indirect cycles. In an indirect cycle, the supercritical coolant passes through the heat exchangers or steam generators to transfer heat to a secondary fluid, which passes through the turbine(s). The advantage of an indirect cycle is that potential radioactive particles would be contained inside the steam generators. On the other hand, the temperature of the secondary loop fluid is lower than that of the primary loop (e.g., reactor heat transport system loop). As a result, the thermal efficiency of an indirect cycle is lower than that of a direct cycle (Pioro et al., 2010). Figure 4 shows a single-reheat cycle for SCW NPPs.

With direct cycles, the thermal efficiency can be increased further through a combination of reheat and regeneration options. As shown in Fig. 4, in a single-reheat cycle, supercritical "steam" from the reactor passes through a high pressure turbine where its temperature and pressure drop. Then, the steam from the outlet of the high pressure turbine is sent through the SRH fuel channels inside the reactor core, but at a lower pressure. As the steam passes through the SRH fuel channels its temperature increases to an outlet temperature of 625°C at a pressure between 3 and 7 MPa (Pioro et al., 2010). At the outlet of the SRH channels, SHS passes through the intermediate pressure turbines. When a regenerative option is

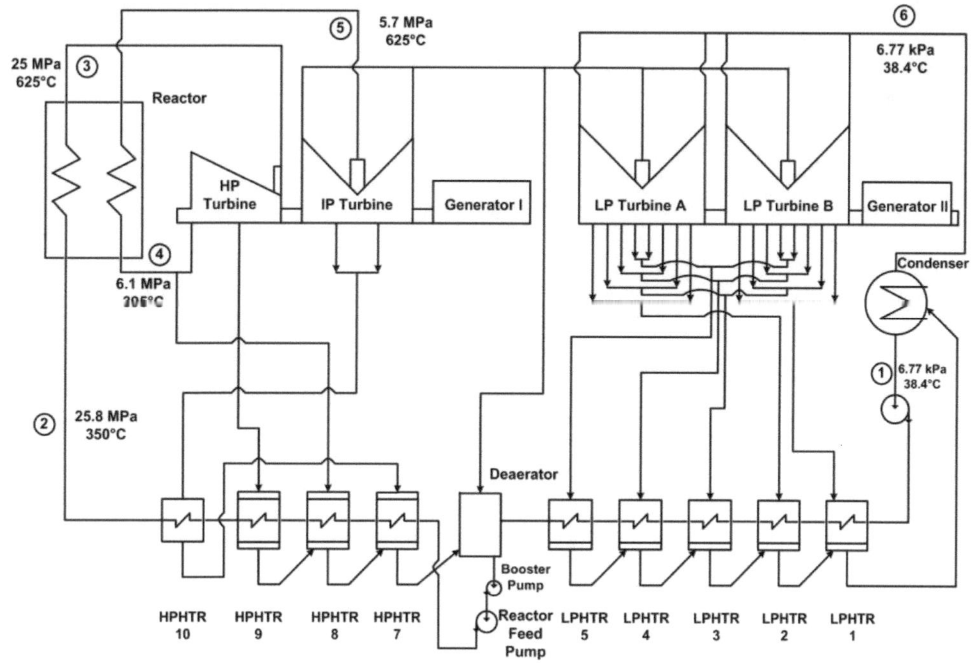

Fig. 4. Single-reheat cycle for SCW NPPs (Naidin et al., 2009).

considered, steam from high and intermediate turbines are extracted and sent to a series of open and closed feed-water heat exchangers. The steam is used to increase the temperature of the feed-water.

4. Fuel channel designs

The design of a fuel channel for SCWRs is an arduous undertaking due to high operating temperatures, which require materials that withstand temperatures as high as 625°C under normal operating conditions. In contrast, current materials, which withstand such design temperatures, have high absorption cross-sections for thermal neutrons. Consequently, a fuel-channel design must address the limitations due to material options to allow for maximum performance using available materials. AECL has proposed several fuel-channel designs for SCWRs. These fuel-channel designs can be classified into two categories: direct-flow and re-entrant channel concepts, which will be described in Sections 4.1 and 4.2. It should be noted that a re-entrant fuel-channel concept was developed by Russian scientists and was utilized at Unit 1 of the Beloyarskaya NPP in the 1960s (Saltanov et al., 2009).

4.1 High-Efficiency fuel Channel

The High Efficiency fuel Channel (HEC) consists of a pressure tube, a ceramic insulator, a liner tube, and fuel bundles. Figure 5 shows a 3-D view of HEC. The outer surface of the pressure tube is exposed to a moderator. The moderator could be a liquid moderator such as heavy-water or a solid moderator. The purpose of using an insulator is to reduce the operating temperature of the pressure tube and heat losses from the coolant to the moderator. Low operating temperatures of the pressure tube would allow for the use of available materials such as Zr-2.5%Nb, which has low absorption cross-sections for thermal neutrons (Chow and Khartabil, 2008).

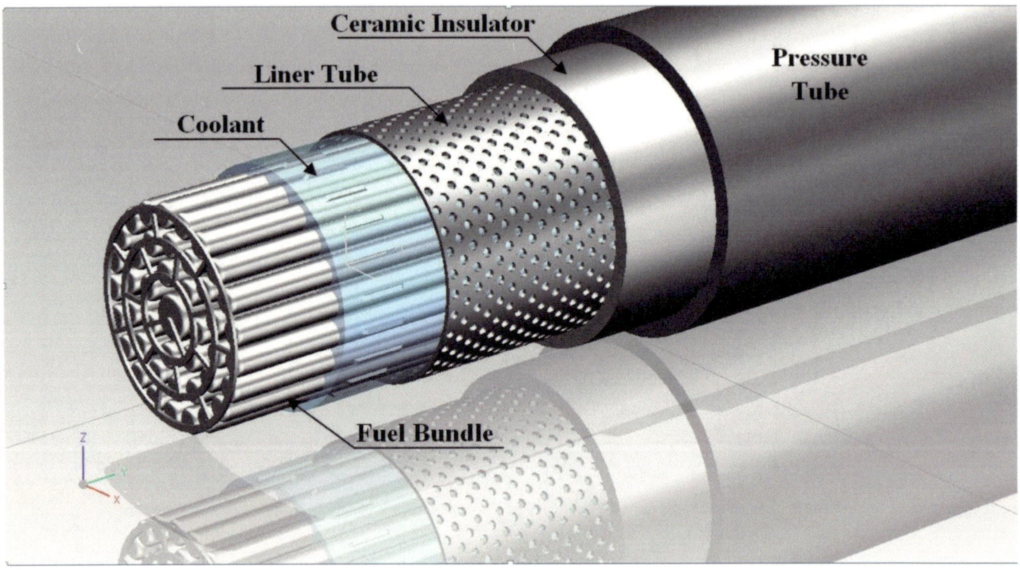

Fig. 5. High efficiency fuel channel (based on Chow and Khartabil, 2008).

The proposed material for the ceramic insulator is Yttria Stabilized Zirconia (YSZ) (Chow and Khartabil, 2008). YSZ has a low neutron absorption cross-section, low thermal-conductivity and high corrosion resistance in exposure to water at supercritical conditions (Chow and Khartabil, 2008). These properties make YSZ a good candidate as an insulator. The liner, which is a perforated tube and made of stainless steel, intends to protect the ceramic insulator from being damaged during operation or possible refuelling due to stresses introduced by fuel bundles and from erosion by the coolant flow.

4.2 Re-Entrant fuel Channels

There are several Re-Entrant fuel Channel (REC) designs. As shown in Fig. 6, the first design consists of a pressure tube and a flow tube which are separated by a gap. The coolant flows along the gap between the pressure tube and the flow tube. Then, at the end of the fuel channel, the coolant flows inside the flow tube where a bundle string is placed. The outer surface of the pressure tube is in contact with the moderator. The use of this fuel-channel design is possible only if the liquid moderator is pressurized to reduce heat loss.

Since the heat loss from the aforementioned fuel channel is significantly high, this design has been modified in the form of the fuel channels shown in Figs. 7 and 8. The second design (see Fig. 7) consists of a calandria tube, a pressure tube, and a flow tube. The gap between the pressure tube and the calandria tube is filled with an inert gas, which provides thermal insulation, reducing the heat losses from the 'hot' pressure tube to the moderator. As shown in Fig. 7, the outer surface of the calandria tube is exposed to a liquid moderator.

Unlike the HEC design, forces due to fuelling/refuelling are not exerted directly on the ceramic in the third design shown in Fig. 8, ensuring that the mechanical integrity of the ceramic insulator is maintained. In addition, the ceramic insulator acts as a thermal barrier, which in turn results in relatively lower operating temperatures of the pressure tube while reducing the heat loss from the coolant to the moderator. Such low operating temperatures allow for the use of Zr-2.5%Nb, which has low absorption cross-sections for thermal neutrons, as the material of the pressure tube. Therefore, lower heat losses, a better protection of the ceramic insulator, and the possibility of using Zr-2.5%Nb as the material of the pressure tube are several advantages of this fuel channel.

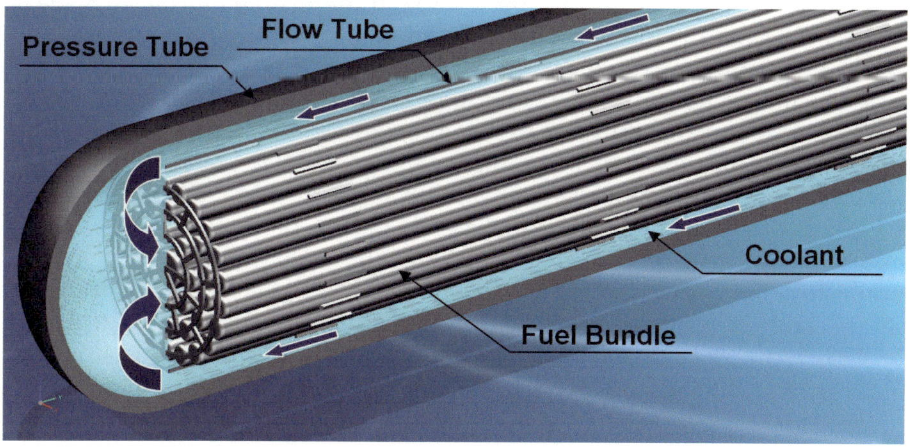

Fig. 6. Re-entrant fuel channel (based on Chow and Khartabil, 2008).

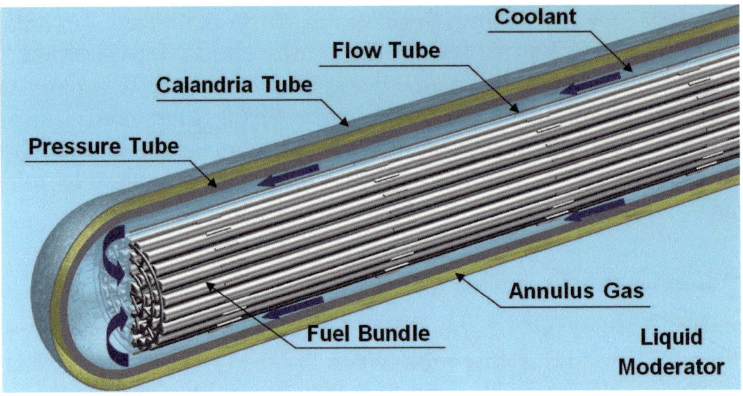

Fig. 7. Re-entrant fuel channel with gaseous insulator.

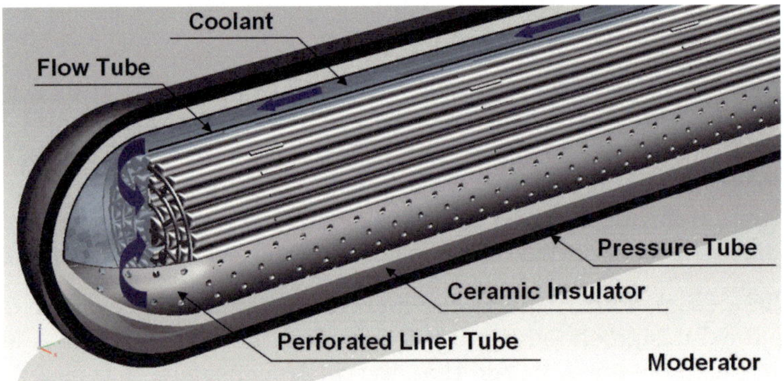

Fig. 8. Re-entrant fuel channel with ceramic insulator.

5. Nuclear fuels

Nuclear fuels can be classified into two main categories; metallic fuels and ceramic fuels. The most common metallic fuels include uranium, plutonium, and thorium (Kirillov et al., 2007). The advantage of metallic fuels is their high thermal conductivity; however, they suffer from low melting points and also that the fuel undergoes phase change. The three phases in a metallic uranium fuel includes α-, β-, and γ-phase. A phase changes to another phase as a function of temperature, resulting in a volume change in the fuel. In addition, metallic fuels undergo oxidation when exposed to air or water. For use in high-temperature applications, a potential fuel must have a high melting point, high thermal conductivity, and good irradiation and mechanical stability (Ma, 1983). These requirements eliminate various nuclear fuels categorized under the metallic fuels mainly due to their low melting points and high irradiation creep and swelling rates (Ma, 1983). On the other hand, ceramic fuels have promising properties, which make these fuels suitable candidates for SCWR applications. Table 2 provides basic properties of selected fuels at 0.1 MPa and 25°C (Chirkin, 1968; IAEA, 2008; Frost, 1963; Cox and Cronenberg, 1977; Leitnaker and Godfrey, 1967; Lundberg and Hobbins, 1992).

In general, ceramic fuels have good dimensional and radiation stability and are chemically compatible with most coolants and sheath materials. Consequently, this section focuses only on ceramic fuels. The ceramic fuels examined in this chapter are UO_2, MOX, ThO_2, UC, UN, UO_2–SiC, UO_2–C, and UO_2–BeO. Further, these ceramic fuels can be classified into three categories: 1) low thermal-conductivity fuels, 2) enhanced thermal-conductivity fuels, and 3) high thermal-conductivity fuels. Low thermal-conductivity fuels are UO_2, MOX, and ThO_2. Enhanced thermal-conductivity fuels are UO_2-SiC, UO_2–C, and UO_2–BeO; and high thermal-conductivity fuels are UC and UN.

Property	Unit	UO_2	MOX	ThO_2	UC	UN
Molecular Mass	amu	270.3	271.2	264	250.04	252.03
Theoretical density	kg/m^3	10960	11,074	10,000	13630[2]	14420
Melting Point	°C	2847±30	2750	3227±150	2507[3] 2520 2532[4]	2850±30[5]
Heat Capacity	J/kgK	235	240	235	203[6]	190
Heat of Vaporization	kJ/kg	1530	1498	-	2120	1144[7] 3325[8]
Thermal Conductivity	W/mK	8.7	7.8	9.7	21.2	14.6
Linear Expansion Coefficient	1/K	9.75×10^{-6}	9.43×10^{-6}	8.9[9]$\times10^{-6}$	10.1×10^{-6}	7.52×10^{-6}
Crystal Structure	-	FCC[10]	FCC	FCC	FCC	FCC

Table 2. Basic properties of selected fuels at 0.1 MPa and 25°C.

In addition to the melting point of a fuel, the thermal conductivity of the fuel is a critical property that affects the operating temperature of the fuel under specific conditions. UO_2 has been used as the fuel of choice in BWRs, PWRs, and CANDU reactors. The thermal conductivity of UO_2 is between 2 and 3 W/m K within the operating temperature range of SCWRs. On the other hand, fuels such as UC and UN have significantly higher thermal conductivities compared to that of UO_2 as shown in Fig. 9 (Cox and Cronenberg, 1977; Frost et al., 1963; IAEA, 2008; Ishimoto et al., 1995; Leitnaker and Godfrey, 1967; Khan et al., 2010, Kirillov et al., 2007; Lundberg and Hobbins, 1992; Solomon et al., 2005). Thus, under the same operating conditions, the fuel centerline temperature of high thermal conductivity fuels should be lower than that of UO_2 fuel.

[2] Frost(1963)
[3] Cox and Cronenberg (1977)
[4] Lundberg and Hobbins (1992)
[5] at nitrogen pressure ≥ 0.25 MPa
[6] Leitnaker & Godfrey (1967)
[7] $UN(s)=U(l)+0.5N_2(g)$, Gingerich (1969)
[8] $UN(s)=U(g)+0.5N_2(g)$, Gingerich (1969)
[9] at 1000°C, Bowman et al.(1965;1966)
[10] Faced-Centered Cubic (FCC)

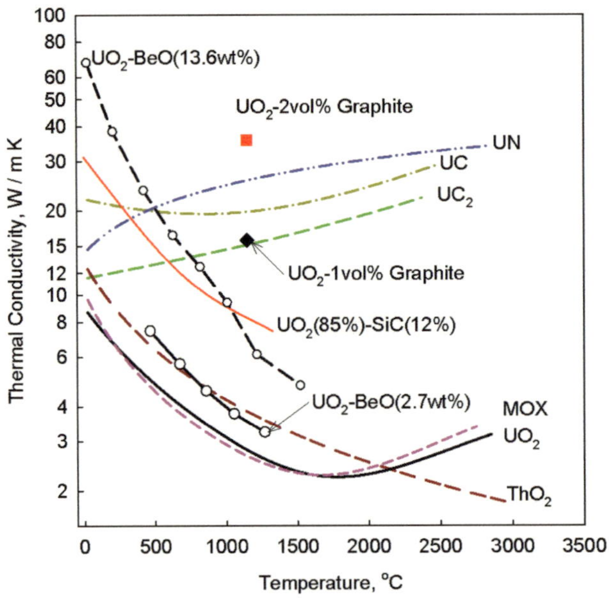

Fig. 9. Thermal conductivities of several fuels.

5.1 Low Thermal-Conductivity Fuels: UO_2, MOX, and ThO_2

5.1.1 UO_2 and MOX

As a ceramic fuel, Uranium Dioxide (UO_2) is a hard and brittle material due to its ionic or covalent interatomic bonding. In spite of that, the uranium dioxide fuel is currently used in PWRs, BWRs, and CANDU reactors because of its properties. Firstly, oxygen has a very low thermal-neutron absorption cross-section, which does not result in a serious loss of neutrons. Secondly, UO_2 is chemically stable and does not react with water within the operating temperatures of these reactors. Thirdly, UO_2 is structurally very stable. Additionally, the crystal structure of the UO_2 fuel retains most of fission products even at high burn-up (Cochran and Tsoulfanidis, 1999). Moreover, UO_2 has a high melting point; however, its thermal conductivity is very low, minimizing the possibility of using UO_2 as a fuel of choice for SCWRs. The thermal conductivity of 95% Theoretical Density (TD) UO_2 can be calculated using the Frank correlation, shown as Eq. (1) (Carbajo et al., 2001). This correlation is valid for temperatures in the range of 25 to 2847°C.

$$k_{uo_2}(T) = \frac{100}{7.5408 + 17.692\,(10^{-3}T) + 3.6142\,(10^{-3}T)^2} + \frac{6400}{(10^{-3}T)^{5/2}} \exp^{-16.35/(10^{-3}T)} \quad (1)$$

Mixed Oxide (MOX) fuel refers to nuclear fuels consisting of UO_2 and plutonium dioxide (PuO_2). MOX fuel was initially designed for use in Liquid-Metal Fast Breeder Reactors (LMFBRs) and in LWRs when reprocessing and recycling of the used fuel is adopted (Cochran and Tsoulfanidis, 1999). The uranium dioxide content of MOX may be natural, enriched, or depleted uranium, depending on the application of MOX fuel. In general, MOX fuel contains between 3 and 5% PuO_2 blended with 95 – 97 % natural or depleted uranium dioxide (Carbajo et al., 2001). The small fraction of PuO_2 slightly changes the thermophysical properties of MOX fuel compared with those of UO_2 fuel. Nonetheless, the thermophysical properties of MOX fuel should be selected when a study of the fuel is undertaken.

Most thermophysical properties of UO_2 and MOX (3 – 5 % PuO_2) have similar trends. For instance, thermal conductivities of UO_2 and MOX fuels decrease as the temperature increases up to 1700°C (see Fig. 9). The most significant differences between these two fuels have been summarized in Table 2. Firstly, MOX fuel has a lower melting temperature, lower heat of fusion, and lower thermal conductivity than UO_2 fuel. For the same power, MOX fuel has a higher stored energy which results in a higher fuel centerline temperature compared with UO_2 fuel. Secondly, the density of MOX fuel is slightly higher than that of UO_2 fuel.

The thermal conductivity of the fuel is of importance in the calculation of the fuel centerline temperature. The thermal conductivities of MOX and UO_2 decrease as functions of temperature up to temperatures around 1527 – 1727°C, and then it increases as the temperature increases (see Fig. 9). In general, the thermal conductivity of MOX fuel is slightly lower than that of UO_2. In other words, addition of small amounts of PuO_2 decreases the thermal conductivity of the mixed oxide fuel. However, the thermal conductivity of MOX does not decrease significantly when the PuO_2 content of the fuel is between 3 and 15%. But, the thermal conductivity of MOX fuel decreases as the concentration of PuO_2 increases beyond 15%. As a result, the concentration of PuO_2 in commercial MOX fuels is kept below 5% (Carbajo et al., 2001). Carbajo et al. (2001) recommended the following correlation shown as Eq. (2) for the calculation of the thermal conductivity of 95% TD MOX fuel. This correlation is valid for temperatures between 427 and 2827°C, x less than 0.05, and PuO_2 concentrations between 3 and 15%. In Eq. (2), T indicates temperature in Kelvin.

$$k(T,x) = \frac{1}{A + C(10^{-3}T)} + \frac{6400}{(10^{-3}T)^{5/2}} \exp^{-16.35/(10^{-3}T)}, \quad x = 2 - O/M \qquad (2)$$

Where x is a function of oxygen to heavy metal ration and

$$A(x) = 2.58x + 0.035 \quad (\mathrm{mK/W}), \quad C(x) = -0.715x + 0.286 \quad (\mathrm{m/K})$$

5.1.2 ThO_2

Currently, there is an interest in using thorium based fuels in nuclear reactors. Thorium is widely distributed in nature and is approximately three times as abundant as uranium. However, ThO_2 does not have any fissile elements to fission with thermal neutrons. Consequently, ThO_2 must be used in combination with a "driver" fuel (e.g., UO_2 or UC), which has ^{235}U as its initial fissile elements. The presence of a "driver" fuel such as UO_2 in a nuclear-reactor core results in the production of enough neutrons, which in turn start the thorium cycle. In this cycle, ^{232}Th is converted into ^{233}Th, which decays to ^{233}Pa. The latter element eventually results in the formation of ^{233}U, which is a fissile element (Cochran and Tsoulfanidis, 1999).

In regards to PT reactors, there are two possibilities when ThO_2 is used. One option is to place ThO_2 and a "driver" fuel in different fuel channels. The separation between ThO_2 fuel and the "driver" fuel allows ThO_2 fuel to stay longer inside the core. The second option is to enclose ThO_2 and the "driver" in same fuel bundles, which are placed inside the fuel channels throughout the reactor core. This option requires the enrichment of the "driver" fuel since it has to be irradiated as long as ThO_2 fuel stays inside the core (IAEA, 2005). Nevertheless, the current study considers the thermal aspects of one single fuel channel, which consists of ThO_2 fuel bundles (i.e., first Option). However, this assumption does not suggest that the whole core is composed of fuel channels containing ThO_2.

The use of thorium based fuels in nuclear reactors requires information on the thermophysical properties of these fuels, especially thermal conductivity. Jain et al. (2006) conducted experiments on thorium dioxide (ThO_2). In their analysis, the thermal conductivity values were calculated based on Eq. (3), which requires the measured values of the density, thermal diffusivity, and specific heat of ThO_2. These properties were measured for temperatures between 100 and 1500°C (Jain et al., 2006). In the current study, the correlation developed by Jain et al. (2006), which is shown as Eq. (4), has been used.

$$k = a\rho c_p \tag{3}$$

$$k_{ThO_2} = \frac{1}{0.0327 + 1.603 \times 10^{-4}\, T} \tag{4}$$

5.2 High Thermal-Conductivity Fuels: UC and UN

5.2.1 UC

From a heat transfer point of view, there is an interest on carbides of uranium as nuclear fuels due to their high thermal conductivities and high melting points. Carbides of uranium usable for nuclear fuels are Uranium Carbide (UC) and Uranium Dicarbide (UC_2). For instance, UC has been proposed as the fuel of choice for a SCWR concept in Russia (Pioro and Duffey, 2007). Uranium sesquicarbide (U_2C_3) is another carbide of uranium; however, it cannot be manufactured through casting or compaction of a powder. However, UC_2 may transform to U_2C_3 at high temperatures and under stress (Frost, 1963).

UC, which has a Faced-Centered Cubic (FCC) crystal structure similar to those of UN and NaCl, has a high melting point approximately 2507°C and a high thermal conductivity, above 19 W/m K at all temperatures up to the melting point. UC has a density of 13630 kg/m^3, which is lower than that of UN but higher than those of UO_2. It should be noted that the density of hypo-stoichiometric UC is slightly higher than that of stoichiometric UC, which is listed in Table 2. Coninck et al. (1975) reported densities between 13730 and 13820 kg/m^3 at 25°C for hypo-stoichiometric UC. Moreover, UC has a higher uranium atom density compared to UO_2 but lower than that of UN. The uranium atom densities of UC and UN are 1.34 and 1.4 times that of UO_2, respectively.

For hypo-stoichiometric UC, the thermal diffusivity a, in m^2/s, and thermal conductivity k, in W/m K, correlations are valid for a temperature range of 570 and 2000°C. In Eqs. (5) and (6), T is in degrees Kelvin (Coninck et al., 1975). For stoichiometric UC, Coninck et al. (1975) provided two correlations, shown as Eqs. (7) and (8), which can be used to determine the mean values of the thermal diffusivity and thermal conductivity of stoichiometric UC for a temperature range between 850 and 2250°C, in m^2/s and W/m K, respectively.

$$\alpha = 10^{-4} \cdot \left[5.75 \cdot 10^{-2} + 1.25 \cdot 10^{-6}(T\text{-}273.15)\right] \tag{5}$$

$$k = 100 \cdot \left[2.04 \cdot 10^{-1} + 2.836 \cdot 10^{-8}(T - 843.15)^2\right] \tag{6}$$

$$\alpha = 10^{-4} \cdot \left[5.7 \cdot 10^{-2} + 1.82 \cdot 10^{-12}(T\text{-}1123.15)^3\right] \tag{7}$$

$$k = 100 \cdot \left[1.95 \cdot 10^{-1} + 3.57 \cdot 10^{-8}(T\text{-}1123.15)^2\right] \tag{8}$$

In addition to Eqs. (6) and (8), Kirillov et al. (2007) have recommended another correlation, shown as Eqs. (9) and (10), for the calculation of the thermal conductivity of UC in W/m K. In the current study, Eq. (21) have been used to determine the thermal conductivity of UC for the calculation of the UC fuel centerline temperature at SCWR conditions, because this equation provides the lowest thermal conductivity values for a wide temperature range, leading to a conservative calculation of the fuel centerline temperature. In Eqs. (9) and (10), T is in degrees Kelvin.

$$k = 21.7 - 3.04 \cdot 10^{-3}(T\text{-}273.15) + 3.61 \cdot 10^{-6}(T\text{-}273.15)^2, \quad 323<T<973\,\text{K} \tag{9}$$

$$k = 20.2 + 1.48 \times 10^{-3}(T\text{-}273.15), \quad 973<T<2573\,\text{K} \tag{10}$$

Frost (1963) developed a correlation shown as Eq. (11), which can be used to determine the diametric increase of UC fuel as a function of time-averaged fuel centerline temperature. According to Eq. (11), UC fuel undergoes significant swelling for temperatures above 1000°C. In Eq. (11), R_D and T are percent diametric increase per atom % burn-up and time-averaged fuel centerline temperature in K, respectively. In addition, as shown in Fig. 10, Harrison (1969) provided the volumetric swelling of UC as a function of burn-up for various temperatures.

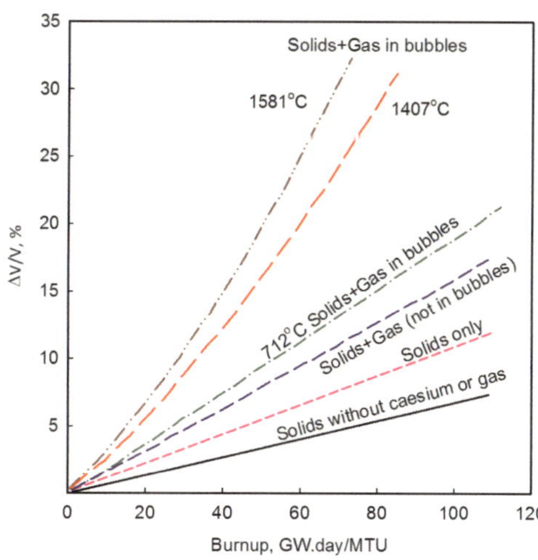

Fig. 10. Volumetric swelling of UC as function of temperature and burn-up.

$$R_D = 0.6 + 0.77\left(\frac{9 \cdot T}{5000} - 1\right) \tag{11}$$

5.2.2 UN

Uranium mononitride or uranium nitride (UN), which is a ceramic fuel, can be produced by the carbothermic reduction of uranium dioxide plus carbon in nitrogen. This process produces UN with densities in the range of 65 to 90% of TD (Shoup and Grace, 1977). UN has a high melting point, high thermal conductivity, and high radiation stability. These properties enhance the safety of operation and allow the fuel to achieve high burn-ups (IAEA, 2008). In addition, UN has the highest fissile atom density, which is approximately 1.4 times that of UO_2 and greater than those of other examined fuels. In other words, when UN is used as a fuel, a smaller volume of fuel is required, which leads to a smaller core. In contrast, one disadvantage of the UN fuel is that under some conditions it decomposes to liquid uranium and gaseous nitrogen (IAEA, 2008), which in turn results in the formation of cracks in the fuel. These cracks increase the chance of the release of gaseous fission products. In addition, the formation of cracks in nuclear fuels has adverse effects on their mechanical and thermophysical properties.

Hayes et al. (1990a) developed a correlation shown as Eq. (12), which calculates the thermal conductivity of UN, in W/m K. This correlation, which is a function of both temperature and percent porosity, can be applied when porosity changes between 0 and 20% for temperatures in the range of 25°C and 1650°C (Hayes et al., 1990a). The standard deviation of the Hayes et al. correlation is ±2.3%.

$$k = 1.864\exp(-2.14\,P)\,T^{0.361} \tag{12}$$

Irradiation swelling, growth, and creep are the primary effects of irradiation on a nuclear fuel. Irradiation swelling results in volumetric instability of the fuel at high temperatures while irradiation growth causes dimensional instability of the fuel at temperatures lower than 2/3 of the melting point of the fuel (Ma, 1983). In addition to dimensional and volumetric instability, a continuous and plastic deformation of the fuel due to creep may adversely affect its mechanical properties. Thus, it is required to study the behaviour of the fuel under irradiation specifically the irradiation-induced swelling, irradiation-induced growth and irradiation-induced creep of the fuel.

Ross et al. (1990) developed a correlation for the prediction of percent volumetric swelling of UN fuel. This correlation is shown as Eq. (13), where T_{avg} is the volume average fuel temperature in K, B is the fuel burn-up in MW day/M g(U), and $\rho_{\%TD}$ is the percent theoretical density of the fuel (e.g., $\rho_{\%TD}$ equals to 0.95 for a fuel with 5% porosity). In addition to this correlation, the volumetric swelling of UN can be calculated based on fuel centerline temperature using Eq. (14) (Ross et al., 1990). The uncertainty associated with Eq. (14) is ±25% for burn-ups above 10,000 MW day/Mg (U) while at lower burn-ups the uncertainty increases to ±60% (Ross et al., 1990). Figure 11 shows the volume expansion of 95% TD UN based on Eq. (14).

$$\Delta V/V(\%) = 4.7 \cdot 10^{-11}\, T_{avg}^{3.12}\left(\frac{B}{9008.1}\right)^{0.83} \rho_{\%TD}^{0.5} \tag{13}$$

$$\Delta V/V(\%) = 1.16 \cdot 10^{-8}\, T_{CLT}^{2.36} \left(\frac{B}{9008.1} \right)^{0.82} \rho_{\%TD}^{0.5} \tag{14}$$

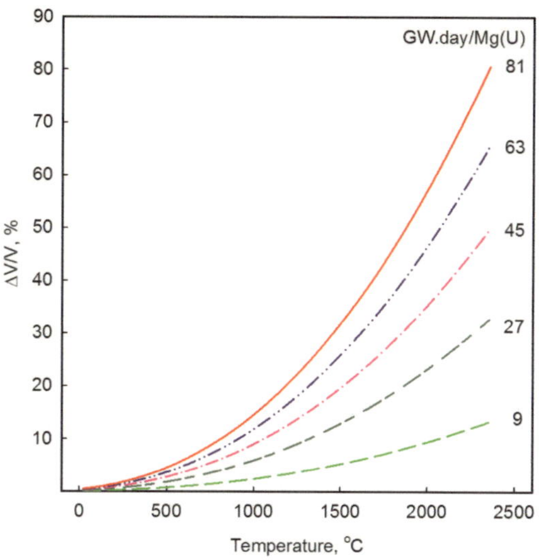

Fig. 11. Percent volumetric swelling of UN as function of burn-up and temperature.

5.3 Composite fuels with enhanced thermal-conductivity

Currently, there is a high interest in developing high thermal-conductivity fuels, and improving the thermal conductivity of low thermal-conductivity fuels such as UO_2. High thermal conductivities result in lower fuel centerline temperatures and limit the release of gaseous fission products (Hollenbach and Ott, 2010). As shown previously, UO_2 has a very low thermal conductivity at high temperatures compared to other fuels such as UC and UN. However, there is a possibility to increase the thermal conductivity of UO_2. This increase in the thermal conductivity of UO_2 can be performed either by adding a continuous solid phase or long, thin fibbers of a high thermal-conductivity material (Hollenbach and Ott, 2010; Solomon et al., 2005).

A high thermal-conductivity material must have a low thermal-neutron absorption cross-section, assuming that the fuel will be used in a thermal-spectrum nuclear reactor (Hollenbach and Ott, 2010). In addition, it must have a high melting point and be chemically compatible with the fuel, the cladding, and the coolant. The need to meet these requirements narrows the potential materials to silicon carbide (SiC), beryllium oxide (BeO), and graphite (C). The following sections provide some information about UO_2 fuel composed of the aforementioned high thermal-conductivity materials.

5.3.1 UO_2 - SiC

The thermal conductivity of UO_2 fuel can be improved by incorporating silicon carbide (SiC) into the matrix of the fuel. SiC has a high melting point approximately at 2800°C, high thermal conductivity (78 W/m K at 727°C), high corrosion resistance even at high temperatures, low thermal neutron absorption, and dimensional stability (Khan et al., 2010). Therefore, when used with UO_2, SiC can address the problem of low thermal conductivity of UO_2 fuel.

Calculation of the thermal conductivity of UO_2 plus SiC the fuel falls under the theories of composites. Generally, theories contemplating the thermal conductivity of composites are classified into two categories. One category assumes that inclusions are randomly distributed in a homogeneous mixture. The effective thermal conductivities of the composites, based on the aforementioned principle, are formulated by Maxwell. The other category, which is based on the work performed by Rayleigh, assumes that particles are distributed in a regular manner within the matrix.

Khan et al. (2010) provided the thermal conductivity of UO_2–SiC fuel as a function of temperature and weight percent of SiC. Khan et al. (2010) assumed that the thin coat of SiC covered UO_2 particles and determined the thermal conductivity of the composite fuel for three cases. The results of the study conducted by Khan et al. (2010) indicate that the continuity of SiC layer leads to a relatively significant increase in thermal conductivity. However, the discontinuity of SiC resulted in little improvement in the ETC of the fuel. Thus, the addition of a continuous solid phase of SiC to UO_2 fuel increases the effective thermal conductivity of the fuel. In the present study, UO_2–SiC fuel with 12wt% SiC with an overall 97 percent TD has been examined and its thermal conductivity has been calculated using Eq. (15).

$$k_{\mathrm{eff}} = -9.59 \cdot 10^{-9}\, T^3 + 4.29 \cdot 10^{-5}\, T^2 - 6.87 \cdot 10^{-2}\, T + 4.68 \cdot 10 \qquad (15)$$

5.3.2 UO_2-C

Hollenbach and Ott (2010) studied the effects of the addition of graphite fibbers on thermal conductivity of UO_2 fuel. Theoretically, the thermal conductivity of graphite varies along different crystallographic planes. For instance, the thermal conductivity of perfect graphite along basal planes is more than 2000 W/m K (Hollenbach and Ott, 2010). On the other hand, it is less than 10 W/m K in the direction perpendicular to the basal planes. Hollenbach and Ott (2010) performed computer analyses in order to determine the effectiveness of adding long, thin fibbers of high thermal-conductivity materials to low thermal-conductivity materials to determine the effective thermal conductivity. In their studies, the high thermal-conductivity material had a thermal conductivity of 2000 W/m K along the axis, and a thermal conductivity of 10 W/m K radially, similar to perfect graphite. The low thermal-conductivity material had properties similar to UO_2 (e.g., with 95% TD at ~1100°C) with a thermal conductivity of 3 W/m K.

Hollenbach and Ott (2010) examined the effective thermal conductivity of the composite for various volume percentages of the high thermal-conductivity material, varying from 0 to 3%. The results show if the amount of the high thermal-conductivity material increases to 2 % by volume, the effective thermal conductivity of the composite reaches the range of high thermal-conductivity fuels, such as UC and UN.

5.3.3 UO_2–BeO

Beryllium Oxide (BeO) is a metallic oxide with a very high thermal conductivity. BeO is chemically compatible with water, UO_2, and most sheath materials including zirconium alloys. In addition to its chemical compatibility, BeO is insoluble with UO_2 at temperatures up to 2160°C. As a result, BeO remains as a continuous second solid phase in the UO_2 fuel matrix while being in good contact with UO_2 molecules at the grain boundaries. BeO has desirable thermochemical and neutronic properties, which have resulted in the use of BeO in aerospace, electrical and nuclear applications. For example, BeO has been used as the moderator and the reflector in some nuclear reactors. However, the major concern with beryllium is its toxicity. But, the requirements for safe handling of BeO are similar to those of UO_2. Therefore, the toxicity of BeO is not a limiting factor in the use of this material with UO_2 (Solomon et al., 2005).

Similar to other enhanced thermal-conductivity fuels, the thermal conductivity of UO_2 can be increased by introducing a continuous phase of BeO at the grain boundaries. The effects of the present of such second solid phase on the thermal conductivity of UO_2 is significant such that only 10% by volume of BeO would improve the thermal conductivity of the composite fuel by 50% compared to that of UO_2 with 95% TD. For the purpose of this study, UO_2–BeO fuel with 13.6 wt% of BeO has been examined.

6. Fuel centerline temperature calculations

In order to calculate the fuel centerline temperature, steady-state one-dimensional heat-transfer analysis was conducted. The MATLAB and NIST REFPROP software were used for programming and retrieving thermophysical properties of a light-water coolant, respectively. First, the heated length of the fuel channel was divided into small segments of one-millimeter lengths. Second, the temperature profile of the coolant was calculated. Third, sheath-outer and inner surface temperatures were calculated. Fourth, the heat transfer through the gap between the sheath and the fuel was determined and used to calculate the outer surface temperature of the fuel. Finally, the temperature of the fuel in the radial and axial directions was calculated. It should be noted that the radius of the fuel pellet was divided into 20 segments. The results will be presented for fuel-sheath gap widths of zero, 20 μm and 36 μm. Moreover, the fuel centerline temperature profiles have been calculated based on a no-gap condition in order to determine the effect of gap conductance on the fuel centerline temperature. Figure 12 illustrates the methodology based on which fuel centerline temperature was calculated. The following section provides more information about each step shown in Fig. 12.

As shown in Fig. 12, the convective heat transfer between the sheath and the coolant is the only heat transfer mode which has been taken directly into consideration. In radiative heat transfer, energy is transferred in the form of electromagnetic waves. Unlike convection and conduction heat transfer modes in which the rate of heat transfer is linearly proportional to temperature differences, a radiative heat transfer depends on the difference between absolute temperatures to the fourth power. The sheath temperature is high[11] at SCWR conditions; therefore, it is necessary to take into account the radiative heat transfer.

In the case of the sheath and the coolant, the radiative heat transfer has been taken into consideration in the Nusselt number correlation, which has been used to calculate the HTC. In general, the Nusselt number correlations are empirical equations, which are developed

[11] It might be as high as 850°C.

based on experiments conducted in water using either bare tubes or tubes containing electrically heated elements simulating the fuel bundles. To develop a correlation, surface temperatures of the bare tube and/or simulating rods are measured along the heated length of the test section by the use of thermocouples or Resistance Temperature Detectors (RTDs). These measured surface temperatures already include the effect of the radiative heat transfer; therefore, the developed Nusselt number correlations represent both radiative and convection heat transfer modes. Consequently, the radiative heat transfer has been taken indirectly into consideration in the calculations.

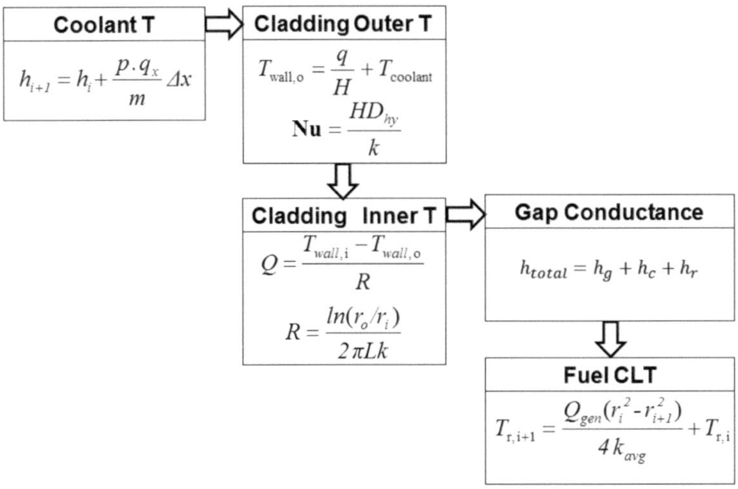

Fig. 12. Fuel centerline temperature calculations.

6.1 Bulk-fluid temperature profile

The temperature profile of the coolant along the heated length of the fuel channel can be calculated based on the heat balance. Equation (16) was used to calculate the temperature profile of the coolant. The NIST REPFROP software Version 8.0 was used to determine the thermophysical properties at a bulk-fluid temperature corresponding to each one-millimeter interval.

$$h_{i+1} = h_i + \frac{p \cdot q_x}{\dot{m}} \cdot \Delta x \quad (16)$$

In Eq. (16), q_x is the axial heat flux value, which is variable along the heated length of the fuel channel if a non-uniform Axial Heat Flux Profile (AHFP) is used. In the present chapter, four AHFPs have been applied in order to calculate the fuel centerline temperature in fuel channels at the maximum channel thermal power. These AHFPs are cosine, upstream-skewed cosine, downstream-skewed cosine, and uniform. The aforementioned AHFPs were calculated based on power profiles listed in Leung (2008) while the downstream-skewed AHFP was determined as the mirror image of the upstream-skewed AHFP. A local heat flux can be calculated by multiplying the average heat flux by the corresponding power ratio from Fig. 13.

It should be noted that there are many power profiles in a reactor core. In other words, the axial heat flux profile in each fuel channel differs from those of the other fuel channels. This variation in power profiles is due to the radial and axial power distribution, fuel burn-up, presence of reactivity control mechanisms, and refuelling scheme. Thus, a detailed design requires the maximum thermal power in the core, which can be determined based on neutronic analysis of the core which is beyond the scope of this chapter. However, the four examined AHFPs envelope a wide range of power profiles.

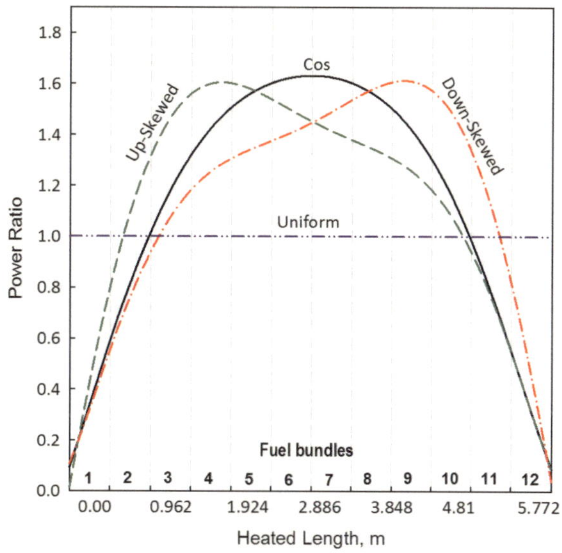

Fig. 13. Power ratios along heated length of fuel channel (based on Leung (2008)).

6.2 Sheath temperature

The calculation of the sheath temperature requires HTC values along the heated length of the fuel channel. In this study, the Mokry et al. correlation, shown as Eq. (17), has been used to determine HTC. The average Prandtl number in the Mokry correlation is calculated based on the average specific heat using Eq. (18). In Eq. (18) μ and k are the dynamic viscosity and thermal conductivity of the coolant at bulk temperature. The experimental data, based on which this correlation was developed, was obtained within conditions similar to those of proposed SCWR concepts. The experimental dataset was obtained for supercritical water flowing upward in a 4-m-long vertical bare tube. The data was collected at a pressure of approximately 24 MPa for several combinations of wall and bulk fluid temperatures. The temperatures were below, at, or above the pseudocritical temperature. The mass flux ranged from 200-1500 kg/m^{2}s; coolant inlet temperature varied from 320 to 350°C, for heat flux up to 1250 kW/m^2 (Mokry et al., 2009). The Mokry correlation requires iterations to be solved, because it contains two unknowns, which are HTC and sheath wall temperature. To solve this problem through iterations, Newton's law of cooling should be used.

From a safety point of view, it is necessary to know the uncertainty of a correlation in calculating the HTC and sheath wall temperature. As shown in Fig. 14, the uncertainty associated in the prediction of the HTC using the Mokry et al. correlation is ±25%. In other words, the HTC values calculated by the Mokry correlation are within ±25% deviation from the corresponding experimental values. However, the uncertainty associated with wall temperature is smaller and lies within ±15%. Figure 15 shows the uncertainty in the prediction of the wall temperature associated with the Mokry et al. correlation.

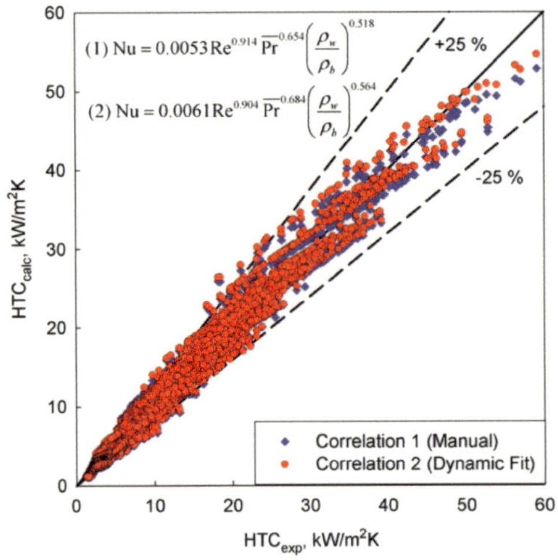

Fig. 14. Uncertainty in predicting HTC based on the Mokry et al. correlation (Mokry et al., 2011).

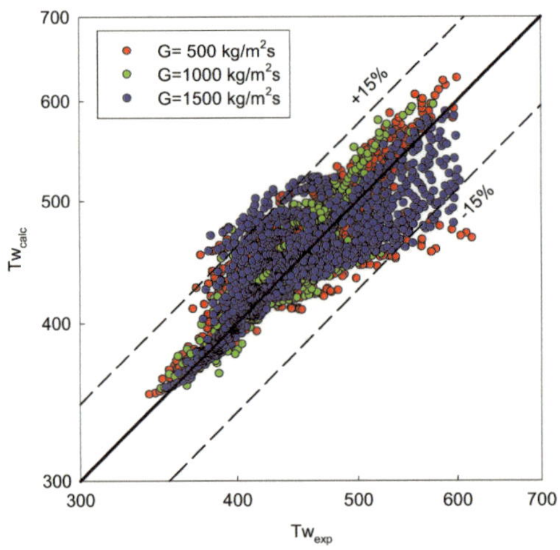

Fig. 15. Uncertainty in predicting wall temperature using the Mokry et al. correlation (Mokry et al., 2011).

6.2.1 Outer-surface temperature of sheath

The following sequence of equations can be used in order to calculate the outer surface temperature of the sheath along the heated length of the fuel channel.

Assumption to start the iteration: $T_{sheath,wall\ o} = T_{bulk} + 50^{\circ}\mathrm{C}$

$$\mathbf{Nu}_{\mathrm{b}} = 0.0061\,\mathbf{Re}_{\mathrm{b}}^{0.904}\,\overline{\mathbf{Pr}}_{\mathrm{b}}^{0.684}\left(\frac{\rho_w}{\rho_b}\right)^{0.564} \tag{17}$$

$$\overline{\mathbf{Pr}} = \frac{\mu\bar{c}_p}{k},\ \bar{c}_p = \frac{H_{sheath,T} - H_{bulk,T}}{T_{sheath} - T_{bulk}} \tag{18}$$

$$q = h\left(T_{sheath,wall\ o} - T_{bulk}\right) \tag{19}$$

The developed MATLAB code uses an iterative technique to determine the sheath-wall temperature. Initially, the sheath-wall temperature is unknown. Therefore, an initial guess is needed for the sheath-wall temperature (i.e., 50°C above the bulk-fluid temperature). Then, the code calculates the HTC using Eq. (17), which requires the thermophysical properties of the light-water coolant at bulk-fluid and sheath-wall temperatures. Next, the code calculates a "new" sheath-wall temperature using the Newton's law of cooling shown as Eq. (19). In the next iteration, the code uses an average temperature between the two consecutive temperatures. The iterations continue until the difference between the two consecutive temperatures is less than 0.1 K. It should be noted that the initial guessed sheath-wall temperature could have any value, because regardless of the value the temperature converges. The only difference caused by different guessed sheath-wall temperatures is in the number of iterations and required time to complete the execution of the code.

As mentioned previously, the thermophysical properties of the coolant undergo significant changes as the temperature passes through the pseudocritical point. Since the operating pressure of the coolant is 25 MPa, the pseudocritical point is reached at 384.9°C. As shown in Fig. 16, the changes in the thermophysical properties of the coolant were captured by the Nusselt number correlation, Eq. (16). The Prandtl number in Eq. (16) is responsible for taking into account the thermophysical properties of the coolant. Figure 16 shows the thermophysical properties of the light-water coolant along the length of the fuel channel. The use of these thermophysical properties in the Nusselt number correlation indicates that the correlation takes into account the effect of the pseudocritical point on the HTC between the sheath and the coolant.

6.2.2 Inner-sheath temperature

The inner surface temperature of the sheath can be calculated using Eq. (20). In Eq. (20), *k* is the thermal conductivity of the sheath, which is calculated based on the average temperature of the outer and inner wall surface temperatures. This inner-sheath temperature calculation is conducted through the use of an iteration, which requires an initial guess for the inner surface temperature of the sheath.

$$Q = \frac{T_{sheath,wall\ i} - T_{sheath,wall\ o}}{\dfrac{ln(r_o / r_i)}{2\pi Lk}} \tag{20}$$

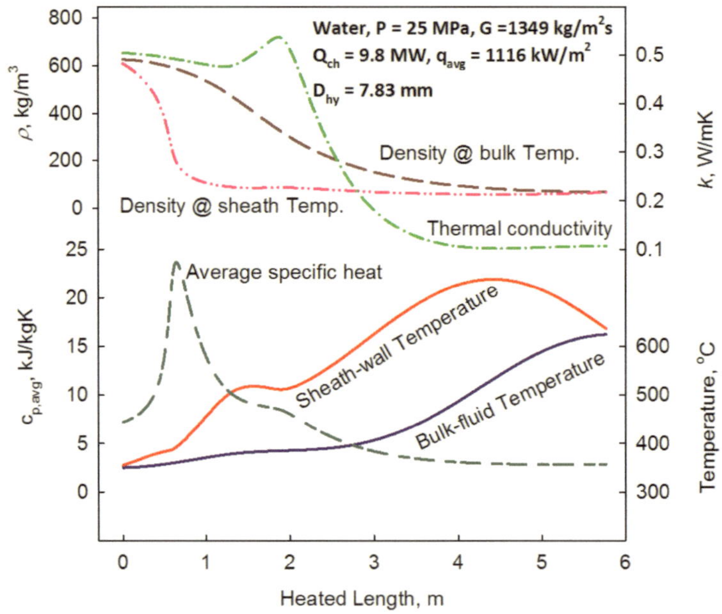

Fig. 16. Thermophysical properties of light-water coolant as function of temperature.

6.3 Gap conductance

Heat transfer through the fuel-sheath gap is governed by three primary mechanisms (Lee et al., 1995). These mechanisms are 1) conduction through the gas, 2) conduction due to fuel-sheath contacts, and 3) radiation. Furthermore, there are several models for the calculation of heat transfer rate through the fuel-sheath gap. These models include the offset gap conductance model, relocated gap conductance model, Ross and Stoute model, and modified Ross and Stoute model.

In the present study, the modified Ross and Stoute model has been used in order to determine the gap conductance effects on the fuel centerline temperature. In this model, the total heat transfer through the gap is calculated as the sum of the three aforementioned terms as represented in Eq. (21):

$$h_{total} = h_g + h_c + h_r \tag{21}$$

The heat transfer through the gas in the fuel-sheath gap is by conduction because the gap width is very small. This small gap width does not allow for the development of natural convection though the gap. The heat transfer rate through the gas is calculated using Eq. (22).

$$h_g = \frac{k_g}{1.5\,(R_1 + R_2) + t_g + g} \tag{22}$$

Where, h_g is the conductance through the gas in the gap, k_g is the thermal conductivity of the gas, R_1 and R_2 are the surface roughnesses of the fuel and the sheath, and t_g is the circumferentially average fuel-sheath gap width.

The fuel-sheath gap is very small, in the range between 0 and 125 μm (Lassmann and Hohlefeld, 1987). CANDU reactors use collapsible sheath, which leads to small fuel-sheath gaps approximately 20 μm (Lewis et al., 2008). Moreover, Hu and Wilson (2010) have reported a fuel-sheath gap width of 36 μm for a proposed PV SCWR. In the present study, the fuel centerline temperature has been calculated for both 20-μm and 36-μm gaps. In Eq. (22), g is the temperature jump distance, which is calculated using Eq. (23) (Lee et al., 1995).

$$\frac{1}{g}=\sum_i\left[\frac{y_i}{g_{o,i}}\right]\left(\frac{T_g}{273.15}\right)^{s+0.5}\left(\frac{0.101}{P_g}\right) \quad (23)$$

Where, g is the temperature jump distance, y_i is the mole fraction of the i_{th} component of gas, $g_{o,i}$ is the temperature jump distance of the i_{th} component of gas at standard temperature and pressure, T_g is the gas temperature in the fuel-sheath gap, P_g is the gas pressure in the fuel-sheath gap, and s is an exponent dependent on gas type.

In reality, the fuel pellets become in contact with sheath creating contact points. These contact points are formed due to thermal expansion and volumetric swelling of fuel pellets. As a result, heat is transferred through these contact points. The conductive heat transfer rate at the contact points are calculated using Eq. (24) (Ainscough, 1982). In Eq. (24), A is a constant, P_a is the apparent interfacial pressure, H is the Mayer hardness of the softer material. A and n are equal to 10 and 0.5.

$$h_c = A\frac{2k_f\, k_{sheath}}{\left(k_f + k_{sheath}\right)\left[\left(R_f^2 + R_{sheath}^2\right)/2\right]^{1/2}}\left(\frac{P_a}{H}\right)^n \quad (24)$$

The last term in Eq. (21) is the radiative heat transfer coefficient through the gap, which is calculated using Eq. (25) (Ainscough, 1982). It should be noted that the contribution of this heat transfer mode is negligible under normal operating conditions. However, the radiative heat transfer is significant in accident scenarios. Nevertheless, the radiative heat transfer through the fuel-sheath gap has been taken into account in this study. In Eq. (25), ε_f and ε_{sheath} are surface emissivities of the fuel and the sheath respectively; and temperatures are in degrees Kelvin.

$$h_r = \frac{\sigma\varepsilon_f\,\varepsilon_{sheath}}{\varepsilon_f + \varepsilon_{sheath} - \varepsilon_f\,\varepsilon_{sheath}}\cdot\frac{\left(T_{f,o}^4 - T_{sheath,i}^4\right)}{\left(T_{f,o} - T_{sheath,i}\right)} \quad (25)$$

6.4 Fuel centerline temperature

Equation (26) can be used to calculate the fuel centerline temperature. The thermal conductivity in Eq. (26) is the average thermal conductivity, which varies as a function of temperature. In order to increase the accuracy of the analysis, the radius of the fuel pellet has been divided into 20 rings. Initially, the inner-surface temperature is not known, therefore, an iteration loop should be created to calculate the outer-surface temperature of the fuel and the thermal conductivity of the fuel based on corresponding average temperatures.

$$T_{r,i+1} = \frac{Q_{gen}\left(r_i^2 - r_{i+1}^2\right)}{4 \cdot k_{avg}} + T_{r,i} \tag{26}$$

7. Results: Fuel centerline and sheath temperatures

There are two temperature limits that a fuel and a fuel bundle must meet. First, the sheath temperature must not exceed the design limit of 850°C (Chow and Khartabil, 2008). Second, when UO_2 fuel is used, the fuel centerline temperature must be below the industry accepted limit of 1850°C (Reisch, 2009) at all normal operating conditions.

Previously, it was mentioned that the industry accepted temperature limit for UO_2 fuel is 1850°C; however, this temperature limit might be different for fuels other than UO_2. There are several factors that may affect a fuel centerline temperature limit for a fuel. These factors include melting point, high-temperature stability, and phase change of the fuel. For instance, the accepted fuel centerline temperature limit of UO_2 fuel is approximately 1000°C below its melting point. As a result, the same fuel centerline temperature limit has been established for the other low thermal-conductivity fuels and enhanced thermal-conductivity fuels. In regards to ThO_2, the melting point is higher than that of UO_2, but a high uncertainty is associated with its melting point. Therefore, as a conservative approach, the same temperature limit has been established for ThO_2. Similarly, the corresponding limit for UC fuel would be 1500°C, because the melting point of UC is approximately 2505°C. UN fuel decomposes to uranium and gaseous nitrogen at temperatures above 1600°C. Therefore, the fuel centerline temperature limit for UN should be lower than that of UO_2 under normal operating conditions. Ma (1983) recommends a temperature limit of 1500°C for UN.

A steady-state one-dimensional heat transfer analysis was conducted in order to calculate the fuel centerline temperature at SCW fuel channels. Based on the proposed core configuration SCW fuel channels are located at the center of the core. Consequently, the thermal power in some of these fuel channels might be by a factor higher than the average channel power of 8.5 MW_{th}. Therefore, in the present study, a thermal power per channel of 9.8 MW_{th} has been considered for the SCW fuel channels with the maximum thermal power. This thermal power is approximately 15% (i. e. 10% above the average power and 5% uncertainty) above the average thermal power per channel. The conditions based on which the calculations have been conducted are as follows: an average mass flow rate of 4.4 kg/s, a constant pressure of 25 MPa, a coolant inlet temperature of 350°C, a thermal power per channel of 9.8 MW_{th}.

The presented analysis does not take into account the pressure drop of the coolant. The main reason for not taking the pressure drop into consideration is that the pressure drop is inversely proportional to the square of mass flux. In a CANDU fuel channel, the pressure drop is approximately 1.75 MPa (AECL, 2005). In addition, the mass flux in an SCWR fuel channel is approximately 5 times lower than that of a CANDU reactor. Therefore, the pressure drop of a SCWR fuel channel should be significantly lower than 1.75 MPa. As a result, the pressure drop has not been taken into consideration.

In addition, this study does not determine the sheath and the fuel centerline temperatures for the SRH fuel channels mainly due to the fact that the average thermal power in SRH channels is 5.5 MW_{th} (see Table 1). Since the thermal power in SRH channels is approximately 35% less that of the SCW channels, the sheath and the fuel centerline temperatures will be definitely lower than those of the SCW channels. As a result, if a fuel and sheath meet their corresponding temperature limits under the operating conditions of the SCW channels with the maximum thermal power, they will be suitable for the SRH channels as well.

For the SCW fuel channels, the fuel centreline temperature has been calculated at cosine, upstream-skewed cosine, downstream-skewed cosine, and uniform axial heat flux profiles. These heat flux profiles have been calculated based on the Variant-20 fuel bundle. Each of the 42 fuel elements of the Variant-20 fuel bundle has an outer diameter of 11.5 mm while the minimum required thickness of the sheath has been determined to be 0.48 mm. Therefore, the inner diameter of the sheath is 10.54 mm. Inconel-600 was chosen as the material of the sheath.

The examined fuels were UO_2, MOX, ThO_2, UC, UN, UO_2-SiC, UO_2-C, and UO_2-BeO. For each fuel, the fuel centerline temperature was analysed at the aforementioned AHFPs. Since the maximum fuel centerline temperature was reached at downstream-skewed cosine AHFP for all the examined fuels, only the results associated with this AHFP have been presented in this section. Figures 17 through 19 show the coolant, sheath, and fuel centerline temperature profiles as well as the heat transfer coefficient profile along the heated length of the fuel channel for UO_2, UC, and UO_2-BeO fuels. Each of these three fuels represents one fuel category (i.e., low, enhanced, high thermal-conductivity fuels). It should be noted that the results presented in Figs. 17 through 19 are based on a 20-μm fuel-sheath gap.

In addition, Figure 20 shows the maximum fuel centerline temperatures of all the examined fuels. As shown in Figure 20, the maximum fuel centerline temperatures of all examined low thermal-conductivity fuels exceed the temperature limit of 1850°C. On the other hand, enhanced thermal-conductivity fuels and high thermal-conductivity fuels show fuel centerline temperatures below the established temperature limits of 1850°C and 1500°C, respectively.

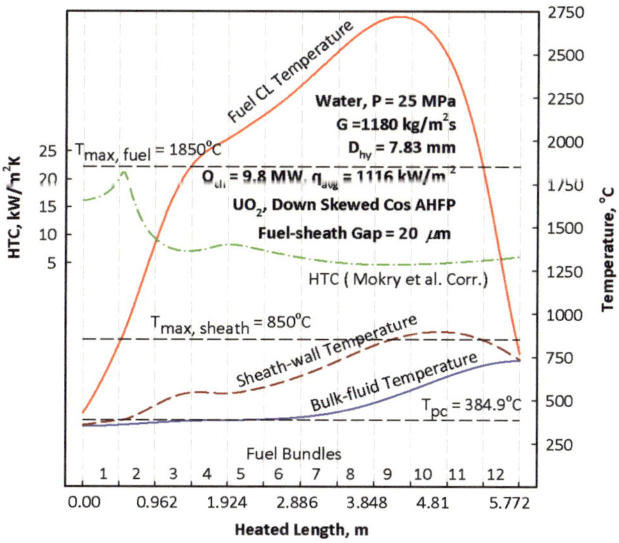

Fig. 17. Temperature and HTC profiles for UO_2 at downstream-skewed cosine AHFP.

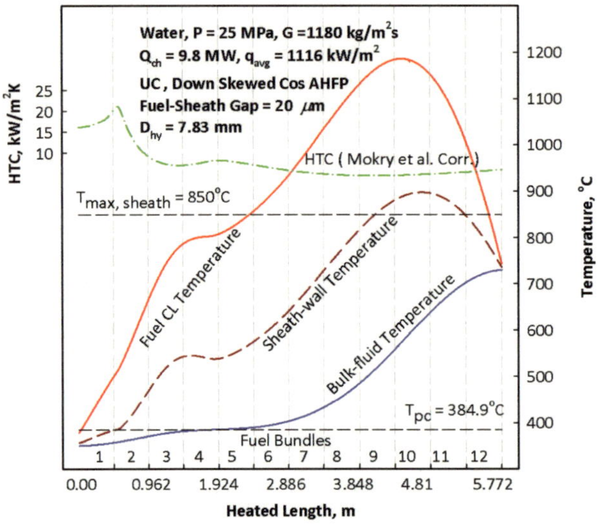

Fig. 18. Temperature and HTC profiles for UC at downstream-skewed cosine AHFP.

In regards to sheath temperature, the sheath temperature reached its maximum at downstream-skewed cosine AHFP. Figure 21 provides a comparison between the sheath temperature profiles for the four studied AHFPs. Figure 21 also shows the HTC profiles corresponding to each examined AHFPs. As shown in Fig. 21, unlike uniform AHFP, HTC reaches its maximum value in the beginning of the fuel channel for non-uniform AHFPs (i.e., downstream-skewed cosine, cosine, and upstream-skewed cosine AHFPs). This increase in HTC is due to the fact the sheath temperature reaches the pseudocritical temperature. In contrast, with uniform AHFP, the sheath temperature is above the pseudocritical temperature from the inlet of the fuel channel. Consequently, the peak in HTC at uniform AHFP occurs when the coolant reaches the pseudocritical temperature.

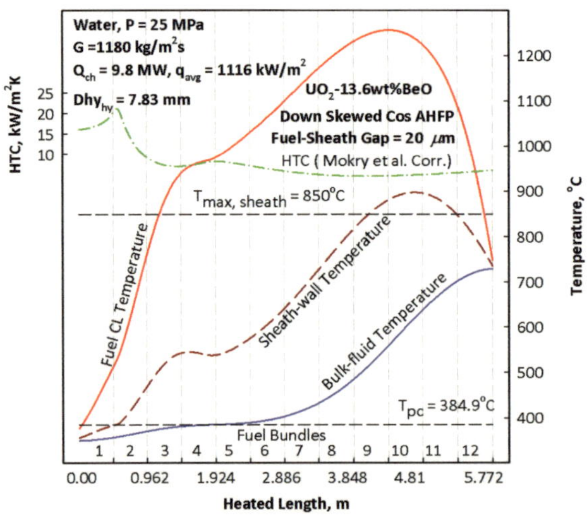

Fig. 19. Temperature and HTC profiles for UO_2–BeO at downstream-skewed cosine AHFP.

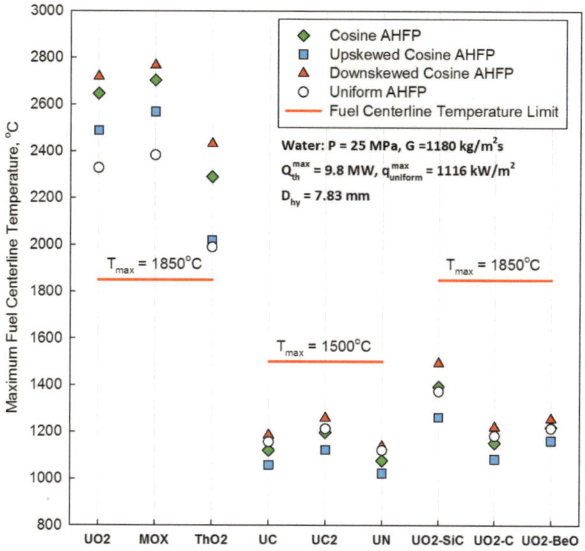

Fig. 20. Maximum fuel centerline temperatures of examined fuels based on a 20–μm fuel-sheath gap width.

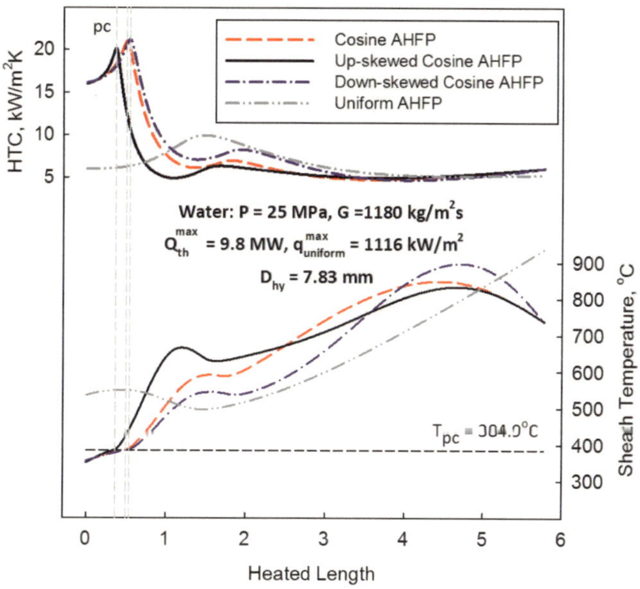

Fig. 21. HTC and sheath-wall temperature profiles as function of AHPF.

A comparison between the examined non-uniform AHFPs shows that in terms of the sheath and fuel centerline temperatures, upstream-skewed cosine AHFP is the most ideal heat flux profile. On the other hand, the downstream-skewed cosine AHFP results in the highest temperatures. Thus, for design purposes, it is a conservative approach to determine the sheath and fuel centerline temperatures based on a downstream-skewed AHFP.

8. Conclusion

Since the development of SCWRs is still in the conceptual design stage, it is worth to further investigate heat transfer and neutronic aspects of high and enhanced thermal-conductivity fuels. In regards to high thermal-conductivity and enhanced thermal-conductivity fuels, this study recommends the use of UC and UO_2-BeO, respectively. This use is conditional on the assurance of chemical compatibility, mechanical behavior, and irradiation behavior of these fuels under the SCWR conditions. In addition, the development of new fuel bundle designs, which will comply with the design temperature limits on the fuel and the sheath, is necessary. New fuel-bundle designs, which would result in lower fuel centerline temperatures, also allow for the use of low thermal-conductivity fuels.

Heat transfer at supercritical conditions has been studied by many researchers; however, still there is a need to improve the correlations used to predict the heat transfer coefficient. To the knowledge of the authors, none of the available heat-transfer correlations predicts the deteriorated heat transfer regime. The lack of capability to predict such phenomenon may result in melting of the sheath. Thus, it is significantly important to develop either look-up tables or heat transfer correlations that would predict the deteriorated heat transfer regime.

9. Acknowledgment

Financial supports from the NSERC/NRCan/AECL Generation IV Energy Technologies Program and NSERC Discovery Grant are gratefully acknowledged.

10. References

AECL (2005). CANDU 6 Technical Summary. CANDU 6 Program Team Reactor Development Business Unit.

Ainscough, J. B. (1982). Gap conductance in Zircaloy-Clad LWR Fuel Rods.United Kingdom Atomic Energy Authority.

Bae, Y.-Y., Kim, H.-Y., and Kang, D-J (2010). Forced and Mixed Convective Heat Transfer to CO_2 at a Supercritical Pressure Vertically Flowing in a Uniformly-Heated Circular Tubes. Experimental Thermal and Fluid Science, Vol. 34, 1295-1308.

Bae, Y.-Y., and Kim, H.-Y. (2009). Convective Heat Transfer to CO_2 at a Supercritical Pressure Flowing Vertically Upward in Tubes and an Annular Channel. Experimental Thermal and Fluid Science, Vol. 33, No. 2, 329-339.

Balankin, S. A., Loshmanov, L. P., Skorov, D. M., and Skolov, V. S. (1978).Thermodynamic Stability of Uranium Nitride. J. Atomic Energy 44, No. 4, 327-329.

Bishop, A. A., Sandberg, R. O., and Tong, L.S. (1965), Forced Convection Heat Transfer to Water at Near-critical Temperatures and Supercritical Pressures, A.I.Ch.E.-I.Chem.E Symposium Series No. 2, 77−85.

Bowman, A. L., Arnold, G. P., Witteman, W. G., and Wallace, T. C. (1966). Thermal Expansion of Uranium Dicarbide and Uranium Sesquicarbide. J. Nuclear Materials 19, 111-112.

Carbajo, J. J., Yoder, G. L., Popov, S. G., and Ivanov, V. K. (2001). A Review of the Thermophysical Properties of MOX and UO2 Fuels. J. Nuclear Materials 299, 181-198.

Chow, C. K., and Khartabil, H.F. (2008). Conceptual Fuel Channel Designs for CANDU-SCWR. J. Nuclear Engineering and Technology 40, 1−8.

Cochran, R. G., and Tsoulfanidis, N. (1999). The Nuclear Fuel Cycle: Analysis and Management. Second ed. American Nuclear Society, Illinois, USA.

Coninck, R. D., Lierde, W. V., and Gijs, A. (1975). Uranium Carbide: Thermal Diffusivity, Thermal Conductivity Spectral Emissivity at High Temperature. J. Nuclear Materials 57, 69-76.

Coninck, R. D., Batist, R. D., and Gijs, A. (1976). Thermal Diffusivity, Thermal Conductivity and Spectral Emissivity of Uranium Dicarbide at High Temperatures. J. Nuclear Materials 8, 167-176.

Cox, D., and Cronenberg, A. (1977). A Theoretical Prediction of the Thermal Conductivity of Uranium Carbide Vapor. J. Nuclear Materials 67, 326-331.

Diamond, W. T. (2010). Development of Out-of-Core Concepts for a Supercritical-Water, Pressure-Tube Reactor. Proc. 2nd Canada-China Joint Workshop on Supercritical Water-Cooled Reactors (CCSC-2010), Toronto, Ontario, Canada: Canadian Nuclear Society, April 25-28, 15 pages.

Farah, A., King, K., Gupta, S., Mokry, S., and Pioro, I. (2010). Comparison of Selected Heat-Transfer Correlations for Supercritical Water Flowing Upward in Vertical Bare Tubes. Proc. 2nd Canada-China Joint Workshop on Supercritical Water-Cooled Reactors (CCSC-2010), Toronto, Ontario, Canada: Canadian Nuclear Society, April 25-28, 12 pages.

Frost, B. R. (1963). The Carbides of Uranium. J. Nuclear Materials 10, 265-300.

Gabaraev, B.A., Vikulov, V.K., and Yermoshin, F.Ye. et al. (2004). Pressure tube once-through reactor with supercritical coolant pressure, (In Russian). Proceedings of the International Scientific-Technical Conference "Channel Reactors: Problems and Solutions", Moscow, Russia, October 19–22, Paper #42.

Gingerich, K. A. (1969). Vaporization of Uranium.Mononitride and Heat of Sublimation of Uranium. J. Chemical Physics 51, No. 10, 4433-4439.

Gnielinski, V. (1976). New Equation for Heat and Mass Transfer in Turbulent Pipe and Channel Flow, Intern. Chem. Eng., Vol. 16, No. 2, 359-366.

Grabezhnaya, V. A., and Kirillov, P. L. (2006). Heat Transfer at Supercritical Pressuresand Heat Transfer Deterioration Boundaries. Thermal Engineering , Vol. 53, No. 4, 296–301.

Grande, L., Mikhael, S., Villamere, B., Rodriguez-Prado, A., Allison, L., Peiman W. and Pioro, I. (2010). Thermal Aspects of Using Uranium Nitride in Supercritical Water-Cooled Nuclear Reactors, Proceedings of the 18th International Conference On Nuclear Engineering (ICONE-18), Xi'an, China.

Griem, H. (1996). A New Procedure for the Near-and Supercritical Prediction Pressure of Forced Convection Heat Transfer, Heat Mass Trans., Vol. 3, 301–305.

Groom (2003). Annulus Gas Chemistry Control. International Atomic Energy Agency. Karachi Nuclear Power Plant (KANUPP), Karachi. 19 May 2008 <http://canteach.candu.org/library/20032103.pdf>.

Harrison, J. W. (1969). The Irradiation-Induced Swelling of Uranium Carbide. J. Nuclear Materials 30, 319-323.

Hayes, S., Thomas, J., and Peddicord, K. (1990a). Material Properties of Uranium Mononitride-III Transport Properties. J. Nuclear Materials 171, 289-299.

Hayes, S., Thomas, J., and Peddicord, K. (1990b). Material Property Correlations for Uranium Mononitride-II Mechanical Properties. J. Nuclear Materials 171, 271-288.

Hayes, S., Thomas, J., Peddicord, K. (1990c). Material Property Correlations for Uranium Mononitride-IV Thermodynamic Properties. J. Nuclear Materials 171, 300-318.

Hollenbach, D. F., and Ott, L. J. (2010). Improving the Thermal Conductivity of UO2 Fuel with the Addition of Graphite Fibers, Transactions of the American Nuclear Society and Embedded Topical Meeting "Nuclear Fuels and Structural Materials for the Next Generation Nuclear Reactors", San Diego, California, June 13-17, 485-487.

Hu, L., Wang, C., Huang, Y. (2010). Porous Yttria-Stabilized Zirconia Ceramics with Ultra-Low Thermal Conductivity. J. of Materials Science 45, 3242-3246.

Hu, P. and Wilson, P. P. H. (2010). Supercritical Water Reactor Steady-State, Burn-up, and Transient Analysis with Extended PARCS/RELAP5. Journal of Nuclear Technology, Vol. 172, 143-156.

IAEA (2005). Thorium Fuel Cycle – Potential Benefits and Challenges. Retrieved March 9, 2011, from IAEA: http://www-pub.iaea.org/mtcd/publications/pdf/te_1450_web.pdf .

IAEA (2006). Thermophysical properties database of materials for light water reactors and heavy water reactors. Vienna, Austria.

IAEA (2008). Thermophysical Properties of Materials for Nuclear Engineering: A Tutorial and Collection of Data. Vienna, Austria.

Incropera, F. P., Dewitt, D. P., Bergman, T. L., and Lavine, A. S. (2006). Fundamentals of Heat and Mass Transfer, sixth ed. Wiley, USA.

INSC. Thermal Conductivity of Solid UO_2. Retrieved June 28, 2010, from International Nuclear Safety Center (INSC, 2010): http://www.insc.anl.gov/matprop/uo2/cond/solid/index.php.

Jackson, J. D. (2002). Consideration of the Heat Transfer Properties of Supercritical Pressure Water in Connection With the Cooling of Advanced Nuclear Reactors, Proc. of the 13th Pacific Basin Nuclear Conference Shenzhen City, China, 21–25 October.

Jain, D., Pillai, C., Rao, B. K., and Sahoo, K. (2006). Thermal Diffusivity and Thermal Conductivity of Thoria-Lanthana Solid Solutions up to 10 mol.% LaO (1.5), J. of Nuclear Materials 353, 35-41.

Khan, J. A., Knight, T. W., Pakala, S. B., Jiang, W., and Fang, R. (2010). Enhanced Thermal Conductivity for LWR Fuel. J. Nuclear Technology 169, 61-72.

Kikuchi, T., Takahashi, T., and Nasu, S. (1972). Porosity Dependence of Thermal Conductivity of Uranium Mononitride. J. Nuclear Materials 45, 284-292.

Kirillov, P.L., Terent'eva, M.I., and Deniskina, N.B. (2007). Thermophysical Properties of Nuclear Engineering Materials, third ed. revised and augmented, IzdAT Publ. House, Moscow, Russia, 194 pages.

Krasnoshchekov, E. A., and Protopopov, V. S. (1966). Experimental Study of Heat Exchange in Carbon Dioxide in the Supercritical Range at High Temperature Drops, Teplofizika Vysokikh Tempraturi (High Temp.), Vol. 4, No. 3.

Kuang, B., Zhang, Y., and Cheng, X. (2008). A New, Wide-Ranged Heat Transfer Correlation of Water at Supercritical Pressures in Vertical Upward Ducts, NUTHOS-7, Seoul, Korea, October 5–9.

Kutz, M. (2005). Mechanical Engineers' Handbook, Materials and Mechanical Design, 3rd Edition, John Wiley and Sons.

Lassmann, K. and Hohlefeld, F. (1987). The Revised Gap Model to Describe the Gap Conductance between Fuel and Cladding. Journal of Nuclear Engineering and Design 103, 215-221.

Lee, K. M., Ohn, M. Y., Lim, H. S., Choi, J. H., and Hwang, S. T. (1995). Study on Models for Gap Conductance between Fuel and Sheath for CANDU Reactors. Journal of Ann. Nucl. Energy, Vol. 22, No. 9, 601-610.

Leitnaker, J. M. and Godfrey, T. G. (1967). Thermodynamic Properties of Uranium Carbide. J. Nuclear Materials 21, 175-189.

Leung, L. K. H. (2008). Effect of CANDNU Bundle-Geometry Variation on Dryout Power. Proc. ICONE-16, Orlando, USA, Paper #48827, 8 pages.

Lewis, B.J., Iglesias, F.C., Dickson, R.S., and Williams, A. (2008). Overview of High Temperature Fuel Behaviour. 10th CNS International Conference on CANDU Fuel, Ottawa, Canada, October 5-8.

Liu, X.J., Yang, T., and Cheng X. (2010). Core and Sub-Channel Analysis of SCWR with Mixed Spectrum Core. Ann. Nucl. Energy, doi: 10.1016/ j.anucene.2010.07.014.

Lundberg, L. B., and Hobbins, R. R. (1992). Nuclear Fuels for Very High Temperature Applications. Intersociety Energy Conversion Engineering Conf., San Diego, 9 pages.

Ma, B. M. (1983). Nuclear Reactor Materials and Applications. Van Nostrand Reinhold Company Inc., New York.

Maxwell, J. C. (1954). Treatise on electricity and magnetism. Dover Publications.

Matthews, R. B., Chidester, K. M., Hoth, C. W., Mason, R. E., and Petty, R. L. (1988). Fabrication and Testing of Uranium Nitride Fuel for Space Power Reactors. J. Nuclear Materials 151, 334–344.

Mokry, S., Pioro, I.L., Farah, A., King, K., Gupta, S., Peiman, W. and Kirillov, P. (2011). Development of Supercritical Water Heat-Transfer Correlation for Vertical Bare Tubes, Nuclear Engineering and Design, Vol. 241, 1126-1136.

Mokry, S., Gospodinov, Ye., Pioro, I. and Kirillov, P.L. (2009). Supercritical Water Heat-Transfer Correlation for Vertical Bare Tubes. Proc. ICONE-17, July 12-16, Brussels, Belgium, Paper #76010, 8 pages.

Naidin, M., Monichan, R., Zirn, U., Gabriel, K. and Pioro, I. (2009). Thermodynamic Considerations for a Single-Reheat Cycle SCWR, Proc. ICONE-17, July 12-16, Brussels, Belgium, Paper #75984, 8 pages.

Naidin, M., Pioro, I., Duffey, R., Zirn, U., and Mokry, S. (2009). SCW NPPs: Layouts and Thermodynamic Cycles Options. Proc. Int. Conf. "Nuclear Energy for New Europe", Bled, Slovenia, Sep. 14-17, Paper #704, 12 pages.

Naterer, G., Suppiah, S., Lewis, M. (2009). Recent Canadian Advances in Nuclear-Based Hydrogen Production and the Thermochemical Cu-Cl Cycle, Int. J. of Hydrogen Energy (IJHE), Vol. 34, 2901-2917.

Naterer, G.F., Suppiah, S., Stolberg, L., et al. (2010). Canada's Program on Nuclear Hydrogen Production and the Thermochemical Cu-Cl Cycle, Int. J. of Hydrogen Energy (IJHE), Vol. 35, 10905-10926.

National Institute of Standards and Technology (2007). NIST Reference Fluid Thermodynamic and Transport Properties-REFPROP. NIST Standard Reference Database 23, Ver. 8.0. Boulder, CO, U.S.: Department of Commerce.

OECD Nuclear Energy Agency. (2010). Lead-Cooled Fast Reactor. Retrieved April 2, 2011, from Generation IV Technologies: http://www.gen-4.org/Technology/systems/lfr.htm.

Oggianu, S. M., No, H. C., and Kazimi, M. (2003). Analysis of Burnup and Economic Potential of Alternative Fuel Materials in Thermal Reactors. j. Nuclear Technology 143, 256-269.

Oka, Y., Koshizuka, S., Ishiwatari, Y., Yamaji, A. (2010). Super Light Water Reactors and Super Fast Reactors: Supercritical-Pressure Light Water Cooled Reactors. Springer, New York.

Pioro, I. L., and Duffey, R. B. (2007). Heat Transfer and Hydraulic Resistance at Supercritical Pressure in Power-Engineering Applications. ASME, New York.

Pioro, I., Mokry, S., Peiman, W., Grande, L., and Saltanov, Eu. (2010). Supercritical Water-Cooled Nuclear Reactors: NPP Layouts and Thermal Design Options of Pressure Channels. Proceedings of the 17th Pacific Basin Nuclear Conference (PBNC-2010), Cancun, Mexico.

Reisch, F. (2009). High Pressure Boiling Water Reactor, HP-BWR. Royal Institute of Technology, Nuclear Power Safety, Stockholm, Sweden.

Ross, S. B., El-Genk, M. S., and Matthews, R. B. (1988). Thermal Conductivity Correlation for Uranium Nitride Fuel. J. Nuclear Materials 151, 313-317.

Ross, S. B., El-Genk, M. S., and Matthews, R. B. (1990). Uranium nitride fuel swelling correlation. J. Nuclear Materials 170, 169-177.

Routbort, J. L. (1972). High-Temperature Deformation of Polycrystalline Uranium Carbide. J. Nuclear Materials 44, 24-30.

Routbort, J. L., and Singh, R. N. (1975). Carbide and Nitride Nuclear Fuels. J. Nuclear Materials 58, 78-114.

Saltanov, E., Monichan, R., Tchernyavskaya, E., and Pioro, I. (2009). Steam-Reheat Option for SCWRs. Proc. ICONE-17, July 12-16, Brussels, Belgium, Paper #76061, 10 pages.

Schlichting, K. W., Padture, N. P., Klemens, P. G. (2001). Thermal conductivity of dense and porous yttria-stabilized zirconia.J. of Materials Science 31, 3003-3010.

Seltzer, M. S., Wright, T. R., and Moak, D. P. (1975). Creep Behavior of Uranium Carbide-Based Alloys. J. American Ceramic Society 58, 138-142.

Shoup, R. D., and Grace, W. R., (1977). Process Variables in the Preparation of UN Microspheres. J. American Ceramic Society 60, No. 7-8, 332-335.

Special Metals.(n.d.). Inconel alloy 600. Retrieved December 2, 2008, from http://www.specialmetals.com/products/inconelalloy600.php.

Stellrecht, D. E., Farkas, M. S., and Moak, D. P. (1968). Compressive Creep of Uranium Carbide. J. American Ceramic Society 51, No. 8, 455-458.

Tokar, M., Nutt, A. W., and Leary, J. A., (1970). Mechanical Properties of Carbide and Nitride Reactor Fuels. J. Reactor Technology. http://www.osti.gov/bridge/purl.cover.jsp?purl=/4100394-tE8dXk/.

U.S. DOE Nuclear Energy Research Advisory Committee (2002). A Technology Roadmap for Generation IV Nuclear Energy Systems. Retrieved July 12, 2010, from The Generation IV International Forum: http://www.ne.doe.gov/genIV/documents/gen_iv_roadmap.pdf.

Villamere, B., Allison, L., Grande, L., Mikhael, S., Rodriguez-Prado, A., and Pioro, I. (2009). Thermal Aspects for Uranium Carbide and Uranium Dicarbide Fuels in Supercritical Water-cooled Nuclear Reactors. Proc. ICONE-17, Brussels, Belgium, July 12-16, Paper #75990, 12 pages.

Wang, S., Yuan, L. Q., and Leung, L. K. (2010). Assessment of Supercritical Heat-Transfer Correlations against AECL Database for Tubes. Proc. 2nd Canada-China Joint Workshop on Supercritical Water-Cooled Reactors (CCSC-2010), Toronto, Ontario, Canada: Canadian Nuclear Society, April 25-28, 12 pages.

Wang, Z., Naterer, G. F., and Gabriel, K. S. (2010). Thermal Integration of SCWR Nuclear and Thermochemical Hydrogen Plants. Proc. 2nd Canada-China Joint Workshop on Supercritical Water-Cooled Reactors (CCSC-2010), Toronto, Ontario, Canada: Canadian Nuclear Society, April 25-28, 12 pages.

Watts, M. J., and Chou, C-T. (1982), Mixed Convection Heat Transfer to Supercritical Pressure Water, Proc. 7th, IHTC, Munich, Germany, 495–500.

Wheeler, M. J. (1965). Thermal Diffusivity at Incandescent Temperatures by a Modulated Electron Beam Technique. Brit. J. Appl. Phys., Vol. 16, 365-376.

Zahlan, H., Groeneveld, D., and Tavoularis, S., Mokry, S., Pioro, I. (2011). Assessment of Supercritical Heat Transfer Prediction Methods. Proc. of the 5th International Symposium on Supercritical Water-Cooled Reactors (ISSCWR-5), Vancouver, British Colombia, Canada, March 13-16.

8

Development of an Analytical Method on Water-Vapor Boiling Two-Phase Flow Characteristics in BWR Fuel Assemblies Under Earthquake Condition

Takeharu Misawa, Hiroyuki Yoshida
and Kazuyuki Takase
Japan Atomic Energy Agency
Japan

1. Introduction

Safe operation of nuclear reactors under earthquake conditions cannot be guaranteed because the behavior of thermal fluids under such conditions is not yet known. For instance, the behavior of gas-liquid two-phase flow during earthquakes is unknown. In particular, fluctuation in the void fraction is an important consideration for the safe operation of a nuclear reactor, especially for a boiling water reactor (BWR). The void fraction in the coolant is one of the physical parameters important in determining the thermal power of the reactor core, and fluctuations in the void fraction are expected to affect the power of the plant.

To evaluate fluctuation in the void fraction, numerical simulation is the most effective and realistic approach. In this study, we have developed a numerical simulation technique to predict boiling two-phase flow behavior, including fluctuation in the void fraction, in a fuel assembly under earthquake conditions.

In developing this simulation technique, we selected a three-dimensional two-fluid model as an analytical method to simulate boiling two-phase flow in a fuel assembly because this model can calculate the three-dimensional time variation in boiling two-phase flow in a large-scale channel such as a fuel assembly while incurring only a realistic computational cost. In addition, this model has been used to successfully predict the void fraction for a steady-state boiling two-phase flow simulation (Misawa, et al., 2008). We expect that the development of the boiling two-phase flow analysis method for a fuel assembly under earthquake conditions can be achieved by improving the three-dimensional two-fluid model analysis code ACE-3D (Ohnuki, et al., 2001; Misawa, et al., 2008), which has been developed by the Japan Atomic Energy Agency.

This paper describes an analytical method for boiling two-phase flow in a fuel assembly under earthquake conditions by improving ACE-3D and shows how the three-dimensional behavior of boiling two-phase flow under these conditions is evaluated by the improved ACE-3D.

2. Development of an analytical method of boiling two-phase flow in a fuel assembly under earthquake conditions

2.1 Overview of ACE-3D

ACE-3D has been developed by the Japan Atomic Energy Agency (JAEA) to simulate water–vapor or water–air two-phase flows at subcritical pressures. The basic equations of ACE-3D are shown below.

Mass conservation for vapor and liquid phases:

$$\frac{\partial}{\partial t}\left(\alpha_g \rho_g\right)+\frac{\partial}{\partial x_j}\left(\alpha_g \rho_g U_{g,j}\right)=\Gamma \tag{1}$$

$$\frac{\partial}{\partial t}\left(\alpha_l \rho_l\right)+\frac{\partial}{\partial x_j}\left(\alpha_l \rho_l U_{l,j}\right)=-\Gamma \tag{2}$$

Momentum conservation for vapor and liquid phases:

$$\frac{\partial U_{g,i}}{\partial t}+U_{g,j}\frac{\partial U_{g,i}}{\partial x_j}=-\frac{1}{\rho_g}\frac{\partial P}{\partial x_i}-\frac{M_{g,i}^{\text{int}}}{\alpha_g \rho_g}-\frac{\Gamma^{+}}{\alpha_g \rho_g}\left(U_{g,i}-U_{l,i}\right)+\frac{1}{\alpha_g \rho_g}\frac{\partial \tau_{g,ij}}{\partial x_j}+g_i \tag{3}$$

$$\frac{\partial U_{l,i}}{\partial t}+U_{l,j}\frac{\partial U_{l,i}}{\partial x_j}=-\frac{1}{\rho_l}\frac{\partial P}{\partial x_i}-\frac{M_{l,i}^{\text{int}}}{\alpha_l \rho_l}-\frac{\Gamma^{-}}{\alpha_l \rho_l}\left(U_{g,i}-U_{l,i}\right)+\frac{1}{\alpha_l \rho_l}\frac{\partial \tau_{l,ij}}{\partial x_j}+g_i \tag{4}$$

Internal energy conservation for vapor and liquid phases:

$$\frac{\partial}{\partial t}\left(\alpha_g \rho_g e_g\right)+\frac{\partial\left(\alpha_g \rho_g e_g U_{g,j}\right)}{\partial x_j}=-P\left[\frac{\partial \alpha_g}{\partial t}+\frac{\partial\left(\alpha_g U_{g,j}\right)}{\partial x_j}\right]+q_g^{w}+q_g^{\text{int}}+\Gamma\cdot h_g^{sat} \tag{5}$$

$$\frac{\partial}{\partial t}\left(\alpha_l \rho_l e_l\right)+\frac{\partial\left(\alpha_l \rho_l e_l U_{l,j}\right)}{\partial x_j}=-P\left[\frac{\partial \alpha_l}{\partial t}+\frac{\partial\left(\alpha_l U_{l,j}\right)}{\partial x_j}\right]+q_l^{w}+q_l^{\text{int}}-\Gamma\cdot h_l^{sat} \tag{6}$$

Here, t represents time; x, a spatial coordinate; U, velocity; P, pressure; g, acceleration due to gravity; e, internal energy; and ρ, density. Subscripts g and l indicate vapor and liquid phases, respectively. Subscripts i and j indicate spatial coordinate components. If the subscripts of the spatial coordinate components are duplicative, the summation convention is applied to the term. The terms q^w and q^{int} indicate wall heat flux and interfacial heat flux, respectively, and h^{sat} represents saturation enthalpy. Summation of volume ratios α_g and α_l is equal to one. Γ^+, Γ^-, and Γ represent vapor generation rates. Γ^+ in Eq. (3) is equal to Γ if Γ is positive and is zero if Γ is negative. On the other hand, Γ^- is equal to Γ if Γ is negative and is zero if Γ is positive. $\tau_{g,ij}$ and $\tau_{l,ij}$ in Eqs. (3) and (4), respectively, are shear stress tensors. The liquid shear stress tensor, $\tau_{l,ij}$, is a summation of shear induced turbulence obtained and bubble induced turbulence (Bertodano, 1994).

Summation of interface stress, M^{int}, of the vapor and liquid phases is equal to zero. The interface stress is determined by correlations between turbulent diffusion force (Lehey,

1991), lift force (Tomiyama, 1995), interface friction force, and bubble diameter (Lilies, 1988). The lift force is calculated as follows.

$$M_{lift}^{int} = -c_{lift}\alpha_g \rho_g \left(\vec{U}_g - \vec{U}_l\right) \times rot\vec{U}_l$$

$$c_{lift} = c_{lift0} + c_{wake}$$

$$c_{lift0} = 0.288 \frac{1-\exp(-0.242\,\mathrm{Re}_{bubble})}{1+\exp(-0.242\,\mathrm{Re}_{bubble})}$$

$$c_{wake} = \begin{cases} 0 & (E_t < 4) \\ -0.096E_t + 0.384 & (4 \le E_t \le 10) \\ -0.576 & (E_t > 10) \end{cases} \tag{7}$$

where

$$E_t = \frac{g\left(\rho_l - \rho_g\right)D_b^2}{\sigma}, \quad \mathrm{Re}_{bubble} = \frac{\rho_l \left|U_g - U_l\right| D_b}{\nu}$$

In Eq. (7), σ is the surface tension coefficient; E_t, Eotvos number; Re_{bubble}, the bubble Reynolds number; and D_b, bubble diameter. c_{lift0} in Eq. (7) describes the effect of shear flow and is a positive value. c_{wake} describes the effect of bubble deformation and significantly depends on Eotvos number of bubble diameter. The coefficient of lift force c_{lift} is estimated by the summation of c_{lift0} and c_{wake}. If the Eotvos number of bubble diameter is small, c_{lift} is positive. However, if the Eotvos number of bubble diameter increases, c_{lift} is negative. Therefore, the lift force corresponding to a small bubble diameter acts in a direction opposite to that corresponding to a large bubble diameter.

Bubble diameter is evaluated by the following equations.

$$D_b = \left(1 - X_{slug}\right) D_{bubble} + X_{slug} D_{slug}$$

$$X_{slug} = 3X_s^2 - 2X_s^3, \quad X_s = 4\left(\alpha_g - 0.25\right)$$

$$D_{bubble} = \frac{\sigma We}{\rho_l \left|U_g - U_l\right|^2}, \quad D_{slug} = \min\left(D_{b\max}, 30\sqrt{\frac{\sigma}{g\left(\rho_g - \rho_l\right)}}\right) \tag{8}$$

where σ is the surface tension coefficient; We, a critical Weber number of 5.0; and D_{bmax}, the maximum bubble diameter, which is an input parameter. Bubble diameter, D_b, is estimated by linear interpolation of small bubble diameter, D_{bubble}, and slug diameter, D_{slug}, with a coefficient, X_{slug}, which is dependent on the void fraction. Therefore, D_b increases with an increase in the void fraction.

In ACE-3D, the two-phase flow turbulent model based on the standard k-ε model (Bertodano, 1994) is introduced below.

Turbulent energy conservation for vapor and liquid phases:

$$\alpha_g \frac{\partial k_g}{\partial t} + \alpha_g U_{g,j} \frac{\partial k_g}{\partial x_j} = \frac{\partial}{\partial x_j}\left(\alpha_g \frac{\nu_g^t}{\sigma_k} \frac{\partial k_g}{\partial x_j} \right) + \alpha_g \left(\Phi_g - \varepsilon_g \right) \quad (9)$$

$$\alpha_l \frac{\partial k_l}{\partial t} + \alpha_l U_{l,j} \frac{\partial k_l}{\partial x_j} = \frac{\partial}{\partial x_j}\left(\alpha_l \frac{\nu_l^t}{\sigma_k} \frac{\partial k_l}{\partial x_j} \right) + \alpha_l \left(\Phi_l - \varepsilon_l \right) \quad (10)$$

Turbulent energy dissipation rate conservation for vapor and liquid phases:

$$\alpha_g \frac{\partial \varepsilon_g}{\partial t} + \alpha_g U_{g,j} \frac{\partial \varepsilon_g}{\partial x_j} = \frac{\partial}{\partial x_j}\left(\alpha_g \frac{\nu_g^t}{\sigma_\varepsilon} \frac{\partial \varepsilon_g}{\partial x_j} \right) + \alpha_g \left(C_{\varepsilon 1} \frac{\Phi_g \varepsilon_g}{k_g} - C_{\varepsilon 2} \frac{\varepsilon_g^2}{k_g} \right) \quad (11)$$

$$\alpha_l \frac{\partial \varepsilon_l}{\partial t} + \alpha_l U_{l,j} \frac{\partial \varepsilon_l}{\partial x_j} = \frac{\partial}{\partial x_j}\left(\alpha_l \frac{\nu_l^t}{\sigma_\varepsilon} \frac{\partial \varepsilon_l}{\partial x_j} \right) + \alpha_l \left(C_{\varepsilon 1} \frac{\Phi_l \varepsilon_l}{k_l} - C_{\varepsilon 2} \frac{\varepsilon_l^2}{k_l} \right) \quad (12)$$

The basic equations represented in Eqs. (1) to (12) are expanded to a boundary fitted coordinate system (Yang, 1994). ACE-3D adopts the finite difference method using constructed grids, although it is difficult to construct fuel assembly geometry by using only constructed grids. Therefore, a computational domain, which consists of constructed grids, is regarded as one block, and complex geometry such as that of a fuel assembly is divided into more than one block. An analysis of complex geometry can be performed by a calculation that takes the interaction between blocks into consideration. Parallelization based on the Message Passing Interface (MPI) was also introduced to ACE-3D to enable the analysis of large-scale domains.

From a previous experiment in which the three-dimensional distribution of the void fraction (vapor volume ratio) in a tight-lattice fuel assembly was measured (Kureta, 1994), it is known that the vapor void (void fraction) is concentrated in the narrowest region between adjacent fuel rods near the starting point of boiling and as the elevation of the flow channel increases, the vapor void spreads over a wide region surrounding the fuel rods. This tendency of the vapor void to redistribute has been described by a past analysis using ACE-3D (Misawa, et al., 2008). Therefore, it is confirmed that the boiling two-phase flow analysis can be carried out using ACE-3D under steady-state conditions and with no oscillation.

2.2 Improvement of three-dimensional two-fluid model for earthquake conditions

In order to simulate the boiling two-phase flow in a fuel assembly under earthquake conditions, it is necessary to consider the influence of structural oscillation of reactor equipment on boiling two-phase flow. If the coordinate system for an analysis is fixed to an oscillating fuel assembly under earthquake conditions, it can be seen that a fictitious force acts on the boiling two-phase flow in the fuel assembly. Therefore, a new external force term, f_i, which simulates the acceleration of oscillation, was added to the momentum conservation equations (Eqs. (3) and (4)).

$$\frac{\partial U_{g,i}}{\partial t}+U_{g,j}\frac{\partial U_{g,i}}{\partial x_j}=-\frac{1}{\rho_g}\frac{\partial P}{\partial x_i}-\frac{M_{g,i}^{\text{int}}}{\alpha_g\rho_g}-\frac{\Gamma^+}{\alpha_g\rho_g}\left(U_{g,i}-U_{l,i}\right)+\frac{1}{\alpha_g\rho_g}\frac{\partial\tau_{g,ij}}{\partial x_j}+g_i+f_i \tag{13}$$

$$\frac{\partial U_{l,i}}{\partial t}+U_{l,j}\frac{\partial U_{l,i}}{\partial x_j}=-\frac{1}{\rho_l}\frac{\partial P}{\partial x_i}-\frac{M_{l,i}^{\text{int}}}{\alpha_l\rho_l}-\frac{\Gamma^-}{\alpha_l\rho_l}\left(U_{g,i}-U_{l,i}\right)+\frac{1}{\alpha_l\rho_l}\frac{\partial\tau_{l,ij}}{\partial x_j}+g_i+f_i \tag{14}$$

We assume that the analysis of boiling two-phase flow in a fuel assembly under earthquake conditions can be performed by using time-series data as an input if the time-series data of oscillation acceleration can be obtained from structural analysis results for a reactor (Yoshimura, et al., 2002) or if the measurement data of actual earthquakes can be obtained by seismographic observation.

In order to apply this improved method to the analysis of boiling two-phase flow in a fuel assembly under earthquake conditions, it is necessary to confirm that the simulation of boiling two-phase flow under oscillation conditions can be performed using the interface stress models shown in the preceding section; these stress models are empirical correlations and are based on experimental results under steady-state conditions. In the case of boiling two-phase flow analysis under oscillation conditions, these interface stress models may cause instability in simulation results.

In addition, it is necessary that large-scale analysis be performed within limited computable physical time and that it be consistent with the time-series data of oscillation acceleration obtained from the results of structural analysis in a reactor or with the measurement data from actual earthquakes. In structural analysis in a reactor (Yoshimura, et al., 2002), the minimum time interval of the analysis is limited to 0.01 s (100 Hz). Seismographic observation is also frequently performed with a sampling period of 100 Hz. If a high-frequency oscillation acceleration of over 100 Hz influences boiling two-phase flow in a fuel assembly, boiling two-phase flow analysis, which is consistent with the structural analysis in a reactor, cannot be performed. Therefore, it is necessary to evaluate the highest frequency necessary for this improved method to be consistent with the time-series data of oscillation acceleration.

A computable physical time of about 1 s is preferred for the boiling two-phase flow analysis in a fuel assembly because this analysis requires a large number of computational grids in order to simulate a large-scale domain such as a fuel assembly. If the results of the boiling two-phase flow analysis show quasi-steady time variation for long-period oscillation acceleration, it is not efficient to perform the analysis with a computable physical time span longer than the long period. Effective analysis can be performed if the analysis with a time span subequal to the shortest period of oscillation acceleration, for which the boiling two-phase flow shows quasi-steady time variation, by extracting earthquake motion at any time during the earthquake. Therefore, it is necessary to evaluate the shortest period of oscillation acceleration for which the boiling two-phase flow shows quasi-steady time variation.

The boiling two-phase flow was simulated in a heated parallel-plate channel, which is a simplification of a single subchannel in a fuel assembly. The channel was excited by vertical and horizontal oscillation to simulate an earthquake in order to confirm that the boiling two-phase flow simulation can be performed under oscillation conditions.

In addition, the influence of the oscillation period on the boiling two-phase flow behavior in a fuel assembly was investigated in order to evaluate the highest frequency necessary for the improved method to be consistent with the time-series data of oscillation acceleration and the shortest period of oscillation acceleration for which the boiling two-phase flow shows quasi-steady time variation.

2.3 Confirmation of stability of the boiling two-phase flow simulation under oscillation conditions

The parallel-plate channel, which simulates a single subchannel in a fuel assembly, was adopted as the computational domain, as shown in Fig. 1. Both plates were heated with a uniform heat flux of 270 kW/m^2. The single-phase water flows into the parallel-plate channel vertically from the inlet. The hydraulic diameter and heated length of the computational domain are equal to those of the single subchannel in the fuel assembly of a current BWR. The outlet pressure of 7.1 MPa, the inlet velocity of 2.2 m/s, and the inlet temperature of 549.15 K also reflect the operating conditions in the current BWR. The adiabatic wall region is set up on the top of the heated region in order to eliminate the influence of the outlet boundary condition. In this analysis, the maximum bubble diameter in Eq. (8) is set to the channel width of 8.2 mm.

First, an analysis was performed without applying oscillation acceleration. After a steady boiling flow was attained, oscillation acceleration was applied. The time when the oscillation acceleration was applied is regarded as t = 0 s.

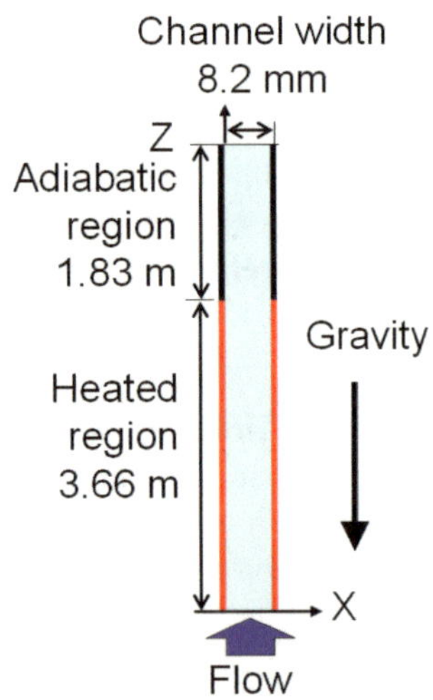

Fig. 1. Computational domain

Two cases of oscillation acceleration, in the vertical direction (Z axis) and in the horizontal direction (X axis), were applied. In both cases, the oscillation acceleration was a sine wave with a magnitude of 400 Gal and a period of 0.3 s, as shown in Fig. 2. The magnitude and period of the oscillation accelerations were taken from actual earthquake acceleration data.

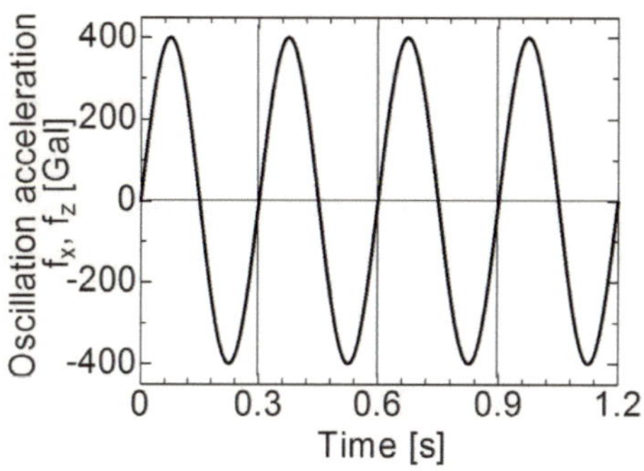

Fig. 2. Time variation in oscillation acceleration

Figure 3 shows distribution of the void fraction at t = 0 s. Much of the void fraction was distributed near the wall at Z = 2.0 m, because the effect of the lift force was dominant at this time, and the lift force acted toward the wall. On the other hand, much of the void fraction was distributed in the center of the channel at Z = 3.66 m, because the effect of bubble deformation on evaluation of lift force was dominant, and the lift force acted toward the center of the channel.

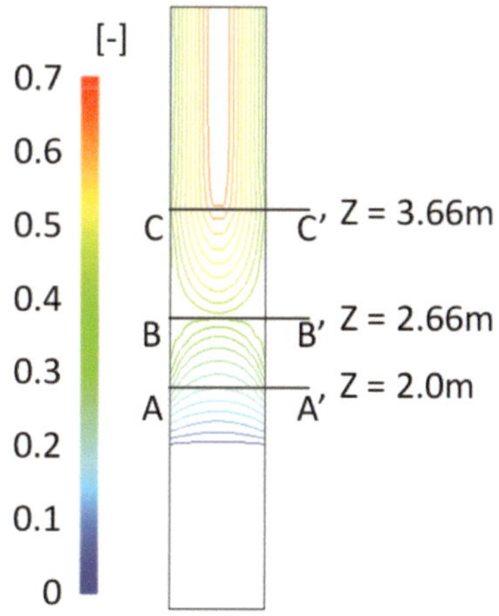

Fig. 3. Void fraction distribution at t = 0 s

The case of horizontal oscillation acceleration is shown in Fig. 4, which shows the time variation in the void fraction at Z = 2.0 m and Z = 3.66 m. The void fraction fluctuated in the horizontal direction with the same period as the oscillation acceleration; however, it moved in the direction opposite to the oscillation acceleration at both Z = 2.0 m and Z = 3.66 m.

Figure 5 shows the time variation in the horizontal velocity of liquid and vapor at Z = 2.0 m, where a positive value of velocity corresponds to the positive direction along the X axis, and a negative value of velocity corresponds to a negative direction along the X axis. The liquid velocity fluctuated in the same direction as the oscillation acceleration, while the vapor velocity fluctuated in the opposite direction. These tendencies in liquid and vapor velocities at Z = 2.0 m can also be seen at Z = 3.66 m. If oscillation acceleration is applied in the horizontal direction, a horizontal pressure gradient arises in a direction opposite to that of the oscillation acceleration in boiling flow. In this case, the liquid phase is driven by the oscillation acceleration because the influence of the oscillation acceleration is relatively large owing to a high liquid density; on the other hand, the vapor phase is driven by the horizontal pressure gradient because the influence of the oscillation acceleration is less than that of the horizontal pressure gradient owing to the low vapor density. This explains why the vapor velocity and the void fraction moved in a direction opposite to that of the oscillation acceleration.

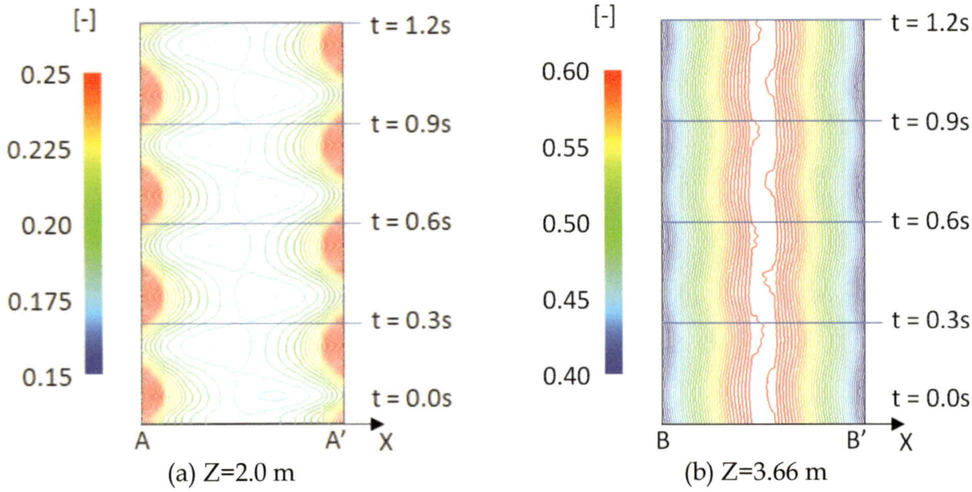

Fig. 4. Time variation in void fraction in the horizontal oscillation acceleration case

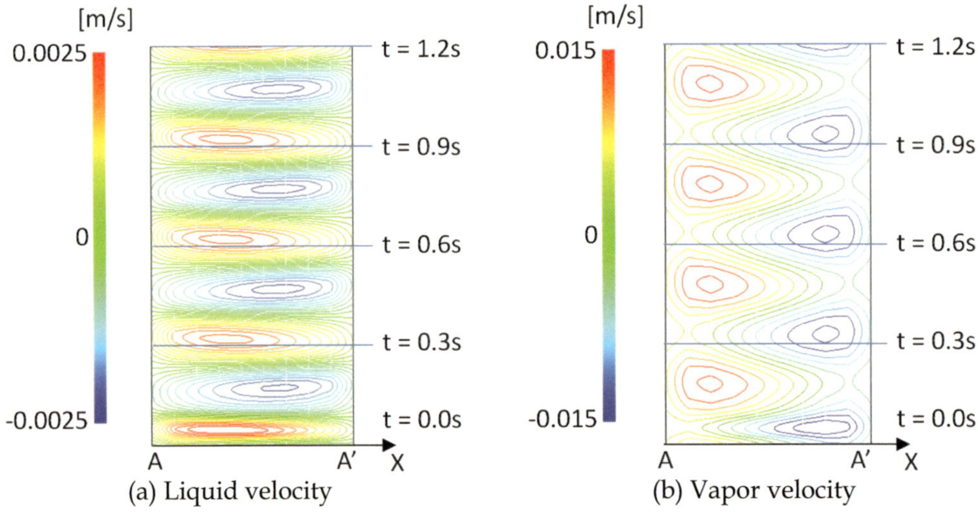

Fig. 5. Time variation in liquid and vapor velocities in the horizontal oscillation acceleration case

The case of vertical oscillation acceleration is shown in Fig. 6, which also shows the time variation in the void fraction at Z = 2.0 m and Z = 3.66 m. The distribution of the void fraction at Z = 2.0m and Z = 3.66 m fluctuated with the same period as that of the oscillation acceleration. The vertical oscillation acceleration caused fluctuations in the pressure in the channel, causing expansion and contraction of the vapor phase. This explains why the void fraction fluctuated with the same period as that of the oscillation acceleration.

A comparison between Fig. 4 and Fig. 6 indicates that the magnitude of the void fraction fluctuation for the horizontal oscillation acceleration case was greater than that for the vertical oscillation acceleration case at any vertical position.

It can therefore be confirmed that the fluctuation of the void fraction with the same period as the oscillation acceleration can be calculated in the case of both horizontal and vertical oscillation acceleration.

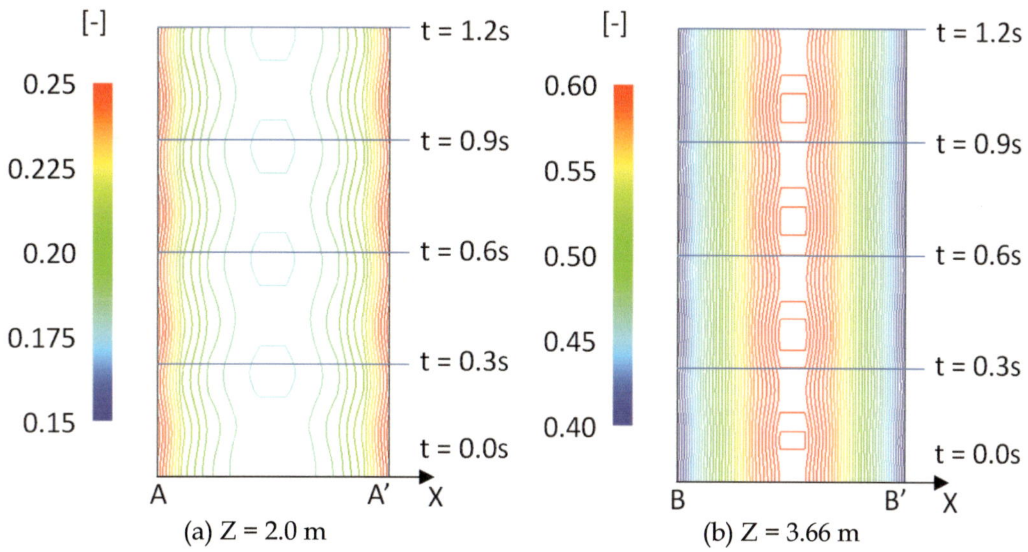

Fig. 6. Time variation in void fraction in the vertical oscillation acceleration case

2.4 Investigation of the effect of oscillation period on boiling two-phase flow behavior

The computational domain and thermal hydraulic conditions are the same as those for boiling two-phase flow in the parallel-plate channel, as described in the preceding section. The oscillation acceleration was applied at $t = 0$ s, after steady boiling flow was obtained.

Nine cases of oscillation acceleration, as shown in Table 1, were applied in order to investigate the influence of the oscillation period of the oscillation acceleration upon the boiling two-phase flow behavior. As shown in the preceding section, the influence of the horizontal oscillation acceleration upon boiling flow was greater than the influence of the vertical oscillation acceleration. Therefore, only the horizontal oscillation acceleration was investigated in these analyses. The minimum oscillation period of 0.005 s, as listed in Table 1, is equal to half of the minimum time interval of structural analysis in a reactor. The maximum oscillation period of 1.2 s is almost equal to the computable physical time of about 1 s. In all cases, magnitude of the oscillation acceleration was set to 400 Gal. Case G in Table 1 is the same as the horizontal oscillation acceleration case shown in section 2.3.

Case	Oscillation period
A	0.005 s
B	0.01 s
C	0.02 s
D	0.04 s
E	0.08 s
F	0.15 s
G	0.3 s
H	0.6 s
I	1.2 s

Table 1. Computational cases

Figure 7 shows the time-averaged distribution of the void fraction. In spite of the different oscillation periods, the time-averaged distribution of the void fraction hardly changed. Therefore, the influence of oscillation period upon the time-averaged void fraction could not be detected.

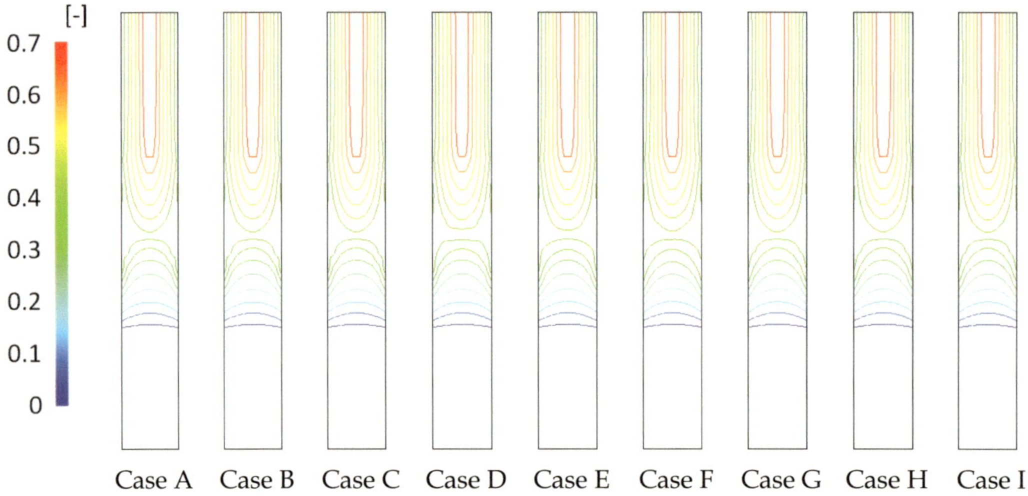

Fig. 7. Time-averaged void fraction distribution

Figure 8 shows the standard deviation distribution of void fraction fluctuation. In cases where the oscillation period is less than 0.01 s, the influence of the oscillation acceleration is small because the magnitude of the void fraction fluctuation is very small compared to that in the cases where the oscillation period is greater than 0.02 s. When the oscillation period is greater than 0.02 s, although the magnitude of the void fraction fluctuation increases with elevation, it decreases near the top of the heated region.

In cases where the oscillation period is between 0.02 s and 0.30 s, the standard deviation distributions varied significantly with the variation in the oscillation period. In Case F, the magnitude of the void fraction fluctuation was highest locally. Therefore, the distribution of void fraction fluctuation was significantly dependent on the oscillation period in this range.

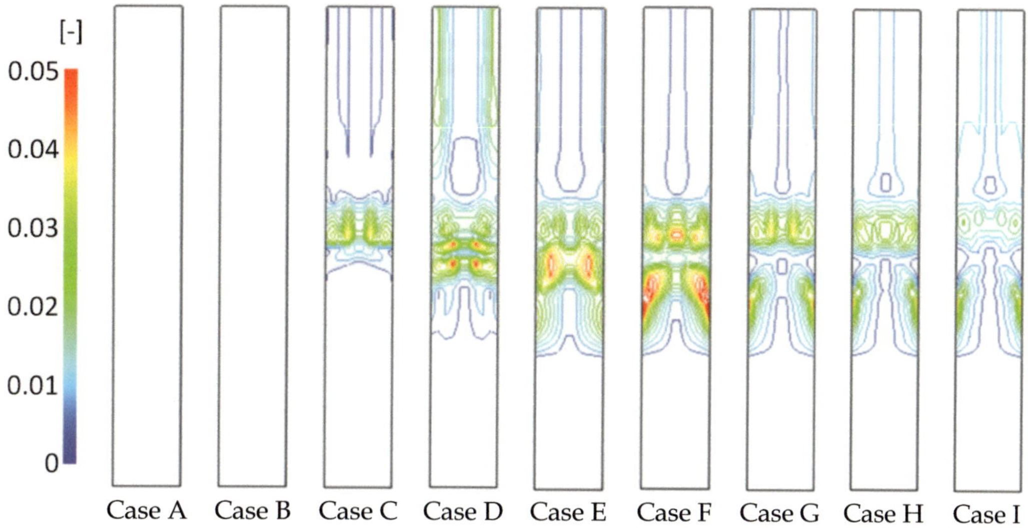

Fig. 8. Standard deviation distribution of void fraction distribution

On the other hand, in cases where the oscillation period was greater than 0.30 s, the standard deviation distributions hardly varied with the variation in the oscillation period. Therefore, the influence of the oscillation acceleration is small in this range.

From the information above, it can be confirmed that the boiling two-phase flow analysis, which is consistent with the time-series data of oscillation acceleration and has a time period greater than 0.01 s, can be performed. This is because oscillation acceleration with an oscillation period of less than 0.01 s has very little influence on the boiling two-phase flow. In addition, the time variations in the void fraction in cases where the oscillation period is greater than 0.30 s are close to quasi-steady variation. This means that the computable physical time of about 1 s is enough to evaluate the response of the boiling two-phase flow to the oscillation acceleration. Therefore, it can be confirmed that effective analysis can be performed by extracting an earthquake motion of about 1 s at any time during an earthquake.

3. Application to the boiling two-phase flow analysis in a simulated fuel assembly excited by oscillation acceleration

Boiling two-phase flow in a simulated fuel assembly excited by oscillation acceleration was performed by the improved ACE-3D in order to investigate how the three-dimensional behavior of boiling two-phase flow in a fuel assembly under oscillation conditions is evaluated by the improved ACE-3D.

3.1 Computational condition

In this analysis, a 7 × 7 fuel assembly in a current BWR core is simulated, as shown in Fig. 9. Fuel rod diameter is 10.8 mm; the narrowest gap between fuel rods is 4.4 mm, and the axial heat length is 3.66 m.

Four subchannels surrounded by nine fuel rods without channel boxes are adopted as the computational domain shown in Fig. 9; this is the smallest domain that can describe the three-dimensional behavior of boiling two-phase flow. This computational domain was determined to reflect the basic thermal-hydraulic characteristics in fuel assemblies under earthquake conditions.

In this domain, single-phase water flows in from the bottom of the channel with a mass velocity of 1673 kg/m²s and inlet temperature of 549.15 K. At the exit of the computational domain, pressure was fixed at 7.1 MPa. The mass velocity, inlet temperature, and exit pressure reflect the operating conditions in a current BWR core. The core thermal power is 351.9 W. The axial power distribution of the fuel-rod surfaces is shown in Fig. 9 and it simulates the power distribution in a current BWR core.

Figure 10 shows the boundary conditions and the computational block divisions. Here, the non-slip condition is set for each fuel-rod surface, and the slip condition is set for each symmetric boundary. In this analysis, the computational domain was divided into 9 blocks. The computational grids in each block have 10 and 256 grids in the radial and axial directions, respectively. The number of grids in the peripheral direction is as follows: 30 in block 1, block 3, block 7, and block 9; 60 in block 2, block 4, block 6, and block 8; and 120 in block 5.

In this study, the boiling two-phase flow analysis was performed under steady-state conditions to obtain a steady boiling two-phase flow. Subsequently, oscillation acceleration was applied. The time when the oscillation acceleration was applied is regarded as $t = 0$ s.

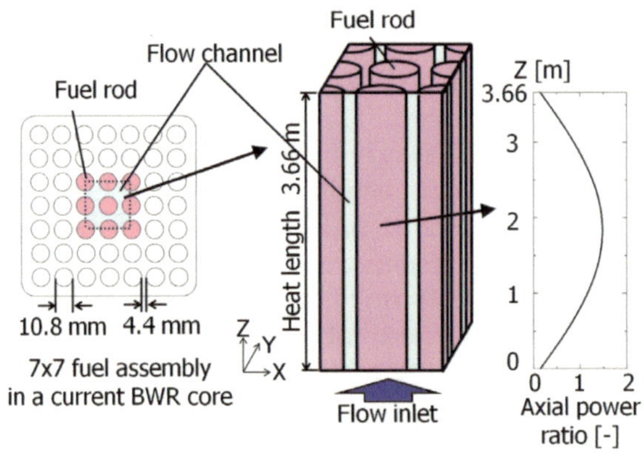

Fig. 9. Computational domain and axial power ratio

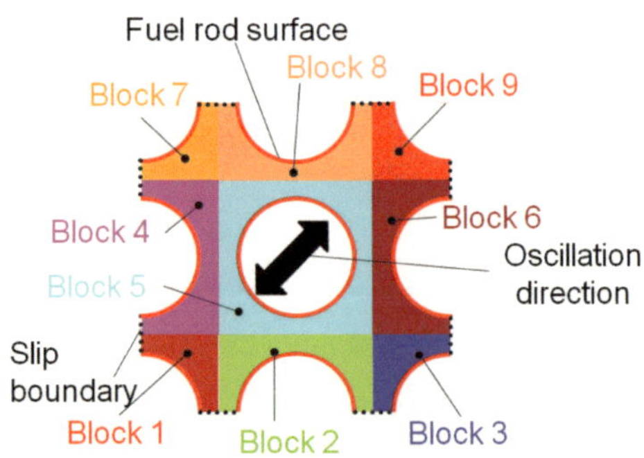

Fig. 10. Boundary conditions and computational block division

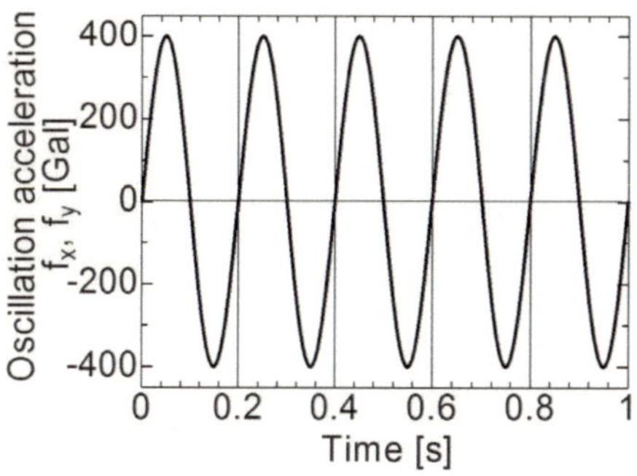

Fig. 11. Time variation in oscillation acceleration

In this analysis, in-phase sine wave acceleration was applied in the X and Y directions as shown by the black arrow in Fig. 10. The magnitude and oscillation period of the oscillation acceleration in the X and Y directions were 400 Gal and 0.2 s, respectively, as shown in Fig. 11; these values are based on actual earthquake data measured in the Kashiwazaki-Kariwa nuclear power plant. The computable physical time in this analysis after applying the oscillation acceleration was 1 s based on the results described in section 2.

3.2 Results and discussions

After applying the oscillation acceleration, the void fraction distribution fluctuated with the same period as the oscillation acceleration. Figure 12 shows the isosurfaces of the void fraction at t = 0.8 s. In the whole area where boiling occurs, the void fraction in the center of the subchannel was relatively low, and the void fraction concentrated in the positive directions along the X and Y axes was high.

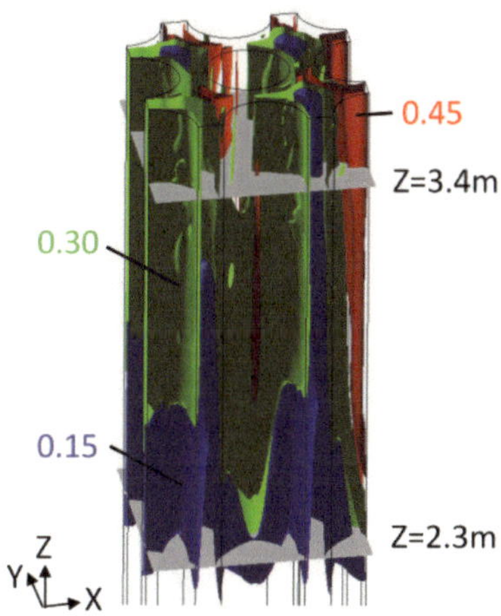

Fig. 12. Isosurface distribution of void fraction

Figure 13 shows the time variation in the void fraction at Z = 2.3 m in the upstream region of Fig. 12. The oscillation acceleration did not act at t = 0.8 s and t = 0.9 s and acted in the direction of the black arrow shown in Fig. 13(b). At t = 0.8 s, a high void fraction could be seen near the fuel-rod surface in the narrowest region between the fuel rods, as indicated by red circles in Fig. 13(a). At t = 0.85 s, a high void fraction moved in a direction opposite to the oscillation acceleration as shown in Fig. 13(b). At t = 0.9 s, a high void fraction could be seen near the fuel-rod surface in the regions marked by red circles in Fig. 13(c). This indicates that the magnitude of void fraction fluctuation at Z = 2.3 m is particularly large near the fuel-rod surface. This tendency of void fraction fluctuation is the same between t = 0.9 s and 1.0 s, when the oscillation acceleration acted in a direction opposite to that of the black arrow.

Figure 14 shows the time variation in the void fraction at Z = 3.4 m in the downstream region of Fig. 12. The black arrow shows the direction in which the oscillation acceleration acts. At t = 0.78 s, the vapor phase moved in a direction opposite to the black arrow and concentrated in the regions marked by red circles, shown in Fig. 14(a). At t = 0.8 s, the void fraction in the region marked by the red circles in Fig. 14(b) increased. In addition, a high

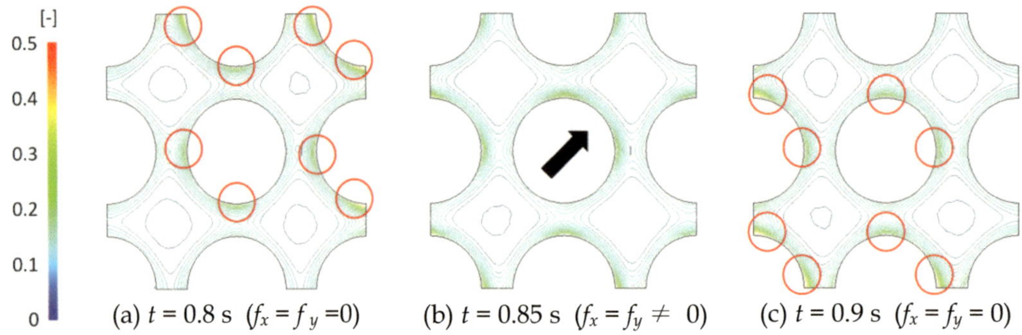

Fig. 13. Time variation in void fraction at Z = 2.3m

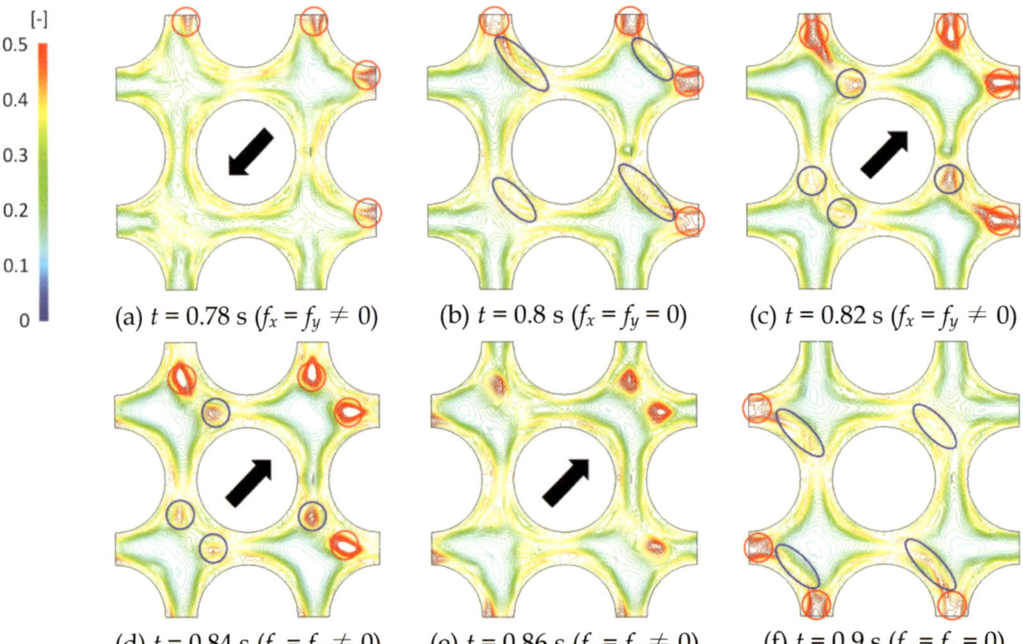

Fig. 14. Time variation in the void fraction at Z = 3.4 m

void fraction could also be seen away from the fuel rod surface, as shown by the blue circles in Fig. 14(b). The high void fraction in the regions marked by the blue circles in Fig. 14(b) split, and the high void fraction in the blue circles in Fig. 14(c) was formed at t = 0.82 s. The high void fraction regions represented by red and blue circles in Fig. 14(c) moved in a direction opposite to the black arrow as shown Fig. 14(d). While the void fraction regions indicated by the red and blue circles in Fig. 14(d) decreased as shown in Fig. 14(e), high void fraction was concentrated in the regions marked by red circles; high void fraction could also be seen in the regions away from the fuel rod surface, such as the regions indicated by the blue circles at t = 0.9 s, as shown in Fig. 14(f). Near the fuel rod surface, void fluctuation with a different period to that of the oscillation acceleration was seen while the magnitude of the void fraction was relatively small.

Figure 15 shows the time variation in vapor velocity at Z = 3.4 m. The black arrow shows the direction in which the oscillation acceleration acts at each time. At t = 0.78 s, the vapor velocity acted in the direction indicated by red arrows in Fig. 15(a); this direction is opposite, but not parallel to, the black arrow. Between t = 0.8 s and t = 0.82 s, in spite of the changing direction of the oscillation acceleration, the vapor velocity decreased but still acted in the direction of the red arrows, shown in Fig. 15(b) and Fig. 15(c), because of the effect of inertia. At t = 0.8 s, the high void fraction indicated by blue circles in Fig. 14(b) was moved by the vapor velocity. This caused the high void fraction shown in Fig. 15(b) to split, and the high void fraction represented by blue circles in Fig. 14(c) and Fig. 14(d) was formed.

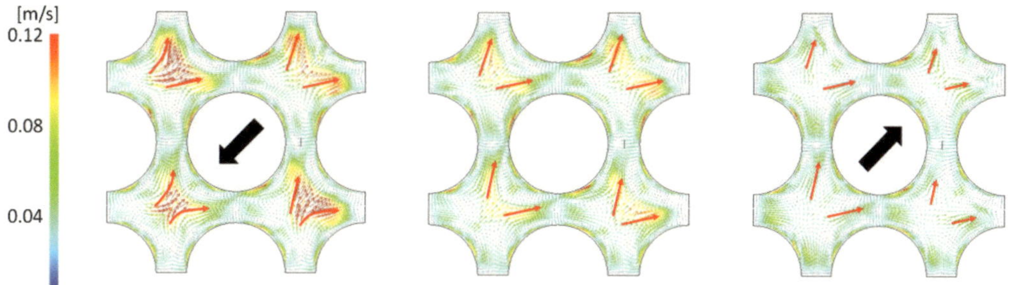

(a) t = 0.78 s ($f_x = f_y \neq 0$) (b) t = 0.8 s ($f_x = f_y = 0$) c) t = 0.82 s ($f_x = f_y \neq 0$)

Fig. 15. Time variation in vapor velocity vector at Z = 3.4 m

Figure 16 shows the time variation in the acceleration vector of the lift force at Z = 3.4 m. The black arrow shows the direction in which the oscillation acceleration acts. The lift force in the red circles in Fig. 16(a) to Fig. 16(c) acted in a direction facing away from the fuel-rod surface. Hence, the vapor velocity was directed along the red arrow, as shown in Fig. 15; this is a direction opposite, but not parallel, to the black arrow. A high void fraction could be seen away from the fuel-rod surface, as shown by blue circles in Fig. 14(b). The lift force in the regions of the blue circles in Fig. 16(b) and Fig. 16(c) also acts in a direction facing away from the fuel rod surface. Hence, a high void fraction could be seen away from the fuel rod surface, as shown by red and blue circles in Fig. 14(c) and Fig. 14(d). Near the fuel rod surface, the magnitude and direction of the lift force were not uniform along the fuel rod and fluctuated with a period different from that of the oscillation acceleration. Consequently, void fraction fluctuation at Z = 3.4 m was significantly dependent on lift force fluctuation.

Figure 17 shows the time variation in the Eotvos number at Z = 3.4 m and also shows a range of Eotvos number from 4 to 10 for which the effect of bubble deformation upon the lift force is dependent upon Eotvos number, as shown in Eq. (7). The black arrow shows the direction in which the oscillation acceleration acts. The red and blue circles in Fig. 17 correspond to regions where the magnitude of the lift force was large; the lift force acted in a direction facing away from the fuel rod surface, as shown in Fig. 16. In these regions, the effect of bubble deformation on the lift force was dominant because the Eotvos number exhibited high values. Near the fuel rod surface, the Eotvos numbers less than 4 and greater than 10 were mixed, indicating that the magnitude and direction of the lift force were not uniform near the fuel rod surface.

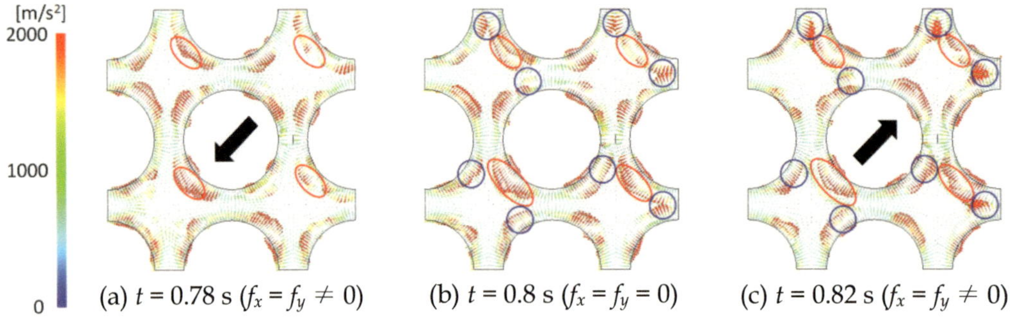

Fig. 16. Time variation in lift force vector at Z = 3.4 m

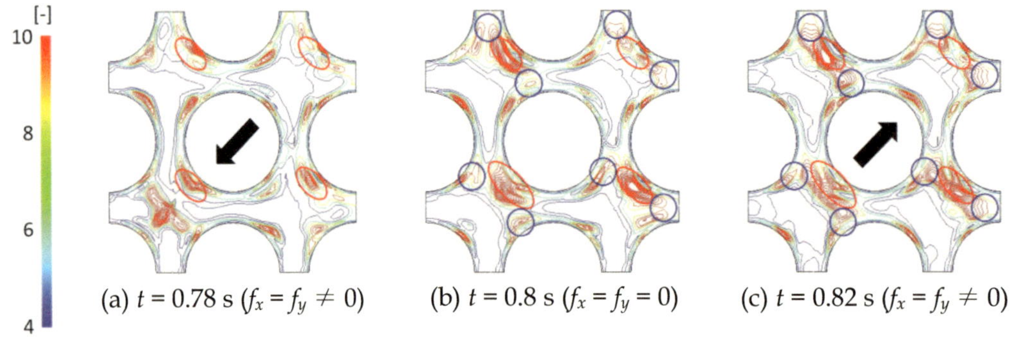

Fig. 17. Time variation in the Eotvos number at Z = 3.4 m

Figure 18 shows the variation in bubble diameter with time at Z = 3.4 m. The black arrow shows the direction in which the oscillation acceleration acts. Bubble diameters greater than 7 mm are distributed in the region where the Eotvos number is greater than 10, as shown in Fig. 17. The bubble diameter distribution shown in Fig. 18 is strongly inhomogeneous and physically invalid because large bubble diameters are mainly observed in small regions in the subchannel, while small bubble diameters of less than 3 mm are observed in the center of the subchannel. This strongly inhomogeneous bubble diameter distribution resulted in locally high Eotvos numbers and fluctuation in the direction of the lift force vectors.

The region where large bubble diameters are seen corresponds to the region of high void fraction, as shown in Fig. 14. According to Eq. (8), the bubble diameter is significantly dependent on the void fraction, and a local high void fraction results in a local large bubble diameter. Thus, a strongly inhomogeneous bubble diameter distribution results from void fraction fluctuation.

It is necessary to adequately evaluate the influence of the void fraction upon bubble diameter in order to avoid a strongly inhomogeneous bubble diameter distribution under oscillation conditions.

According to our results, void fraction fluctuation in the downstream region is significantly dependent on the lift force caused by a strongly inhomogeneous bubble diameter distribution.

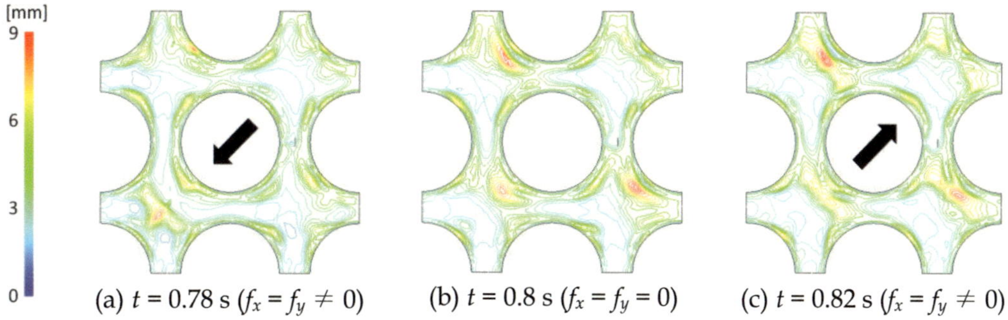

Fig. 18. Time variation in the bubble diameter at Z = 3.4 m

4. Conclusion

A new external force term, which can simulate the oscillation acceleration, was added to the momentum conservation equations in order to apply the three-dimensional two-fluid model analysis code ACE-3D under earthquake conditions.

A boiling two-phase flow excited by applying vertical and horizontal oscillation acceleration was simulated in order to confirm that the simulation can be performed under oscillation conditions. It was confirmed that the void fraction fluctuation with the same period as that of the oscillation acceleration could be calculated in the case of both horizontal and vertical oscillation acceleration.

The influence of the oscillation period of the oscillation acceleration on the boiling two-phase flow behavior in a fuel assembly was investigated in order to evaluate the highest frequency necessary for the improved method to be consistent with the time-series data of oscillation acceleration and the shortest period of oscillation acceleration for which the boiling two-phase flow shows quasi-steady time variation. It was confirmed that a boiling two-phase flow analysis consistent with the time-series data of oscillation acceleration and with a time interval greater than 0.01 s, can be performed. It was also shown that an effective analysis can be performed by extracting an earthquake motion of about 1 s at any time during the earthquake.

The three-dimensional behavior of boiling two-phase flow in a fuel assembly under oscillation conditions was evaluated using a simulated fuel assembly excited by oscillation acceleration. On the basis of this evaluation, it was confirmed that void fraction fluctuation in the downstream region is significantly dependent on the lift force caused by a strongly inhomogeneous bubble diameter distribution and that it is necessary to adequately evaluate the influence of void fraction on bubble diameter in order to avoid strongly inhomogeneous bubble diameter distribution under oscillation conditions.

5. Acknowledgment

The present study includes the result of "Research of simulation technology for estimation of quake-proof strength of nuclear power plant" conducted by the University of Tokyo as Core Research for Evolutional Science and Technology (CREST). This research was conducted using a supercomputer of the Japan Atomic Energy Agency.

6. References

Ohnuki, A. & Akimoto, H. (2001). Modeling development for bubble turbulent diffusion and bubble diameter in large vertical pipe, *Journal of Nuclear Science and Technology*, Vol. 38, No. 12, pp.1074-1080

Misawa, T.; Yoshida, H. & Akimoto, H. (2008). Development of design technology on thermal-hydraulic performance in tight-lattice rod bundle: IV Large paralleled simulation by the advanced two-fluid model Code, *Journal of Power and Energy Systems*, Vol.2, No. 1, pp.262-270

Yoshimura, S.; Shioya, R.; Noguchi, H. & Miyamura, T. (2002). Advanced general purpose computational mechanics system for large-scale analysis and design, *Journal of Computational and Applied Mathematics*, Vol. 49, pp.279-296

Satou, A.; Watanabe, T.; Maruyama, Y. & Nakamura, H. (2010). Neutron-coupled thermal hydraulic calculation of BWR under seismic acceleration, *Joint International Conference on Supercomputing in Nuclear Application and Monte Carlo 2010 (SNA+MC2010)*, Tokyo, Japan

Lehey, R. T. & Bertodano, M. L. (1991). The prediction of phase distribution using two-fluid model, *ASME/JSME Thermal Eng. Proc.*, Reno Nevada

Tomiyama, A.; Matsuoka, A.; Fukuda, T. & Sakaguchi, T. (1995). Effect of Etovos number and dimensionless liquid volumetric flux on lateral motion of a bubble in a laminar duct flow, *Advances in multiphase flow*, Elsevier, 3

Lilies, D. R.; Spore, J. W.; Knight, T. D.; Nelson, R. E.; Cappiello, M. W.; Pasamehmetaglu, K. O.; Mahaffy, J. H.; Guffee, L. A.; Stumpf, H. J.; Dotson, P. J.; Steinke, R. G.; Shire, P. R.; Greiner, S. E. & Shenwood, K. B. (1988). TRAC-PF1/MOD1 Correlations and Models, NUREC/CR-5069, LA-11208-MS

Bertodano, M. L.; Lehey Jr., R. T. & Jones, O. C. (1994). Development of a k-ε model for bubbly two-phase flow, *Journal of Fluid Engineering*, Vol.116, No. 128

Yang, H.; Habchi, S. D. & Prezekwas, A. J. (1994). General Strong Conservation Formulation of Navier-Stokes Equations in Nonorthogonal Curvilinear Coordinates, *AIAA journal*, Vol.32, No.5

Kureta, M. & Akimoto, H. (1994). Measurement of Vapor Behavior in Tight-Lattice Bundles by Neutron Radiography, *6th International Conference on Nuclear Thermal Hydraulics, Operations and Safety (NUTHOS-6)*, Nara, Japan

9

The Theoretical Simulation of a Model by SIMULINK for Surveying the Work and Dynamical Stability of Nuclear Reactors Cores

Seyed Alireza Mousavi Shirazi
Department of Physics, Islamic Azad University, South Tehran Branch, Tehran Iran

1. Introduction

According to complexity of nuclear reactor technology, applying a highly developed simulation is necessary for controlling the nuclear reactor control rods, so in this proposal the processes of a controlling model for nuclear reactors have been developed and simulated by the SIMULINK tool kit of MATLAB software and all responses, including oscillation and transient responses, have been analyzed.

In this work an arbitrary value of K_{eff} as a comparable value is purposed and attributed to input block (*H*) of diagram and then this value with the received feedback value from block diagram is compared. Since the stability of the cited simulation depends on either velocity or delay time values, therefore according to this simulation the best response and operation which a reactor can have from stability aspect, have been derived. Meantime by viewing the results, the best ranges of velocity and delay time of control rod movement (in unit per second and millisecond respectively) for stability a nuclear reactor has been deduced.

Though the highlights of this proposal are respectively the following:

- Defining a mathematical model for control rod movement
- Simulation of a mathematical model by SIMULINK of MATLAB
- Determination of the best ranges for both velocity and delay time of control rod movement (in unit per second and millisecond respectively) based on the obtained results for stability an LWR nuclear reactor

In view of the great advancing the nuclear reactors technology, the phenomenal and significant changes in evolution of made nuclear reactors is observed. Since the make of the first nuclear reactor on 1948 until modern reactors, too changes are obvious. The major of these changes to: the kind of reactor design, the percent of fuel enrichment, the kind of coolant and neutron moderator, more safety and the dimensions of core are referred.

The power control system is a key control system for a nuclear reactor, which directly affects the safe operation of a nuclear reactor. Much attention has been spent to the power control system performance of nuclear reactor in engineering (Zhao et al., 2003).

High reliability is one of the main objectives of the design and operation of control systems in nuclear power plants (Basu and Zemdegs, 1978; Stark, 1976).

Prototyping a control-rod driving mechanism (CRDM), which is a crucial safety system in the Taiwan Research Reactor (TRR-II) has been implemented, by iterative parallel procedures. Hence to ensure the mechanical integrity and substance of the prototype, a series of performance testing and design improvement have been interactively executed. Functional testing results show that the overall performance of the CRDM meets the specification requirements (Chyou and Cheng, 2004).

Also the SCK.CEN/ININ joint project, which deals with the design and application of modern/expert control and real-time simulation techniques for the secure operation of a TRIGA Mark III research nuclear reactor, has been undertaken (Dong et al., 2009).

This project has been proposed as the first of its kind under a general collaboration agreement between the Belgian Nuclear Research Centre (SCK·CEN) and the National Nuclear Research Institute (ININ) of Mexico (Benítez et al., 2005). In addition to the fuzzy proportional-integral-derivative (fuzzy-PID) control strategy has been applied recently as a nuclear reactor power control system. In the fuzzy-PID control strategy, the fuzzy logic controller (FLC) is exploited to extend the finite sets of PID gains to the possible combinations of PID gains in stable region and the genetic algorithm (Cheng et al., 2009; Park and Cho, 1992).

Until now, manual controlling systems have been used for controlling and tuning the control rods in the core of Gen II and some Gen III reactors (Tachibana et al., 2004).

But by application of this simulation that is the subject of this proposal the best response for operating and the best velocity and delay time of control rod movement in which can be caused to stability and critical state of a nuclear reactor, have been derived.

The safe situation is state in which the reactor stabilizes in the critical situation, meaning that the period is infinite and the Keff is 1 (Lamarsh, 1975).

1.1 Accidents

In which two states the positive reactivity overcomes the temperature coefficient of reactivity (α_T):

Increasing the power and temperature of reactor core might decrease concentration of boric acid. Accordingly this event might cause to inject positive reactivity.

In addition for the reactors which apply fuels including Pu, because of having a resonance for Pu in thermal neutrons range so through increasing core's temperature the related resonance is broadened and absorbs more neutrons and because of Pu is fissionable therefore fissionable absorption occurs and is caused excess reactivity.

Also either accident or unfavorable issues as a feedback can be considered. Accidents of a nuclear reactor are totally classified based on following:

1. Over power accident.
2. Under cooling accident.

Each mentioned issues are divided to other sub issues. Over power accident is due factors such as:

1. Control rod withdrawal including uncontrolled rod withdrawal at sub critical power and uncontrolled rod withdrawal at power that will cause power excursion.
2. Control rod ejection.
3. Spent fuel handling.
4. Stem line break.
5. External events such as earthquake, enemy attack and etc.

In each mentioned issues the positive reactivity to nuclear reactor core can be injected. But there is another important accident that is: under cooling accident. Under cooling accident is classified to three sub accident among: LOCA (loss of coolant accident), LOHA (loss of heat sink accident) and LOFA (loss of flow accident).

LOCA accident from loosing coolant is derived. This event in PWR reactor can occur through breaking in primary loop of reactor either hot leg or cold leg.

In this state the existent water in the primary loop along with steam are strongly leaked that blow down event occurs. But when the lost water of primary loop through RHRS (Residual and Heat Removal System) including HPIS (High Pressure Injection System), LPIS (Low Pressure Injection System) and Accumulator (passive system) are filled this process is entitled Refill. When the primary is filled and all the lost water in it is compensated then the reactor sets in the normal status.

In the blow down status the raised steam is due loss of pressure in primary loop and moving the situation of reactor's primary loop from single phase to two phase flow.

LOHA accident from loosing heat sink in reactor is derived. Heat sink is as steam generator in nuclear reactor. This event when occurs heat exchanging between primary and secondary loops are not done.

This event might occur through lacking water circulation in the secondary loop of reactor. Not circulating the water might occur through closing either block valve (which sets after demineralizer tank) or other existent valves in the secondary loop.

LOFA accident from disabling and loosing the pumps in either primary loop or secondary is derived. In case either primary loop's pump or secondary encounter with problems LOFA accident might occur.

In the secondary loop the main feed water pump has duty of circulation of feed water to steam generator and sets after condenser pump.

In view of dynamically stability of a nuclear reactor, there is a stable system so that an excess reactivity is injected to it and it able to be stabled again in shortest time. The stability of linear systems in the field of complex numbers by defining the polarity of closed loop transfer's function is determined.

In case all the polarities are in the left side of imagine page then system will be stabled. In the time field the stability definition means system's response to each input will be definitive. In the matrix form all the Eigen values of system have real negative part.

1.2 Dynamics of nuclear reactor

There are several methods for investigation of nuclear reactors dynamics.

One of the most important methods to study reactor dynamics and the stability of nuclear reactor is define of transfer's functions and application of it to analyze the closed loop function.

According to following figure a closed loop system including transfer function, feedback and related applied reactivity are shown:

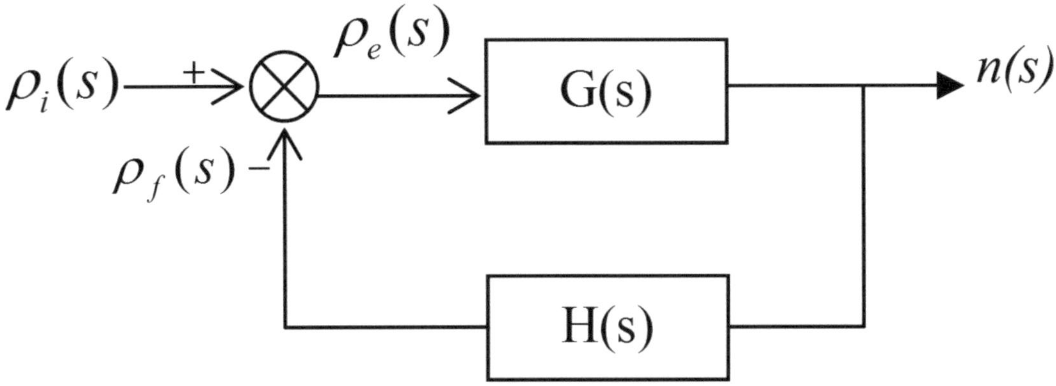

Fig. 1. The closed loop conversion function based on reactivity

Where:

$\rho_i(s)$ is: input reactivity in frequency field, $\rho_f(s)$ is: reactivity due to feedback in frequency field, $\rho_e(s)$ is: error reactivity in frequency field that is as input reactivity to transfer function, $G(s)$ is: transfer function, $H(s)$ is: feedback function and $n(s)$ is: output of closed loop conversion function that means the density of neutrons.

There is also:

$$\rho_e(s) = \rho_i(s) - \rho_f(s) \tag{1}$$

According to Fig.1 for both transfer function and feedback function existing in closed loop can write:

$$G(s) = \frac{n(s)}{\rho_e(s)}, \tag{2}$$

$$H(s) = \frac{\rho_f(s)}{n(s)} \tag{3}$$

and it can also be written:

Conversion Function: $$T(s) = \frac{n(s)}{\rho_i(s)} = \frac{G(s)}{1 + G(s).H(s)} \tag{4}$$

In order to survey the stability of a closed loop system the term of $[1+G(s).H(s)]$ must be set zero and by solving this equation, all the roots that are as zero and pole for closes loop system, will be defined. The stability condition of a closed loop system is lack of positive real part of poles. It means all the poles must be the left side of real-imagine graph.

Reactivity feedback causes the steady operation of nuclear reactor and equilibrium of its dynamical system.

A transfer function can be either linear or not. Each system variable can be affected as an input reactivity to transfer function as shown in Fig.2 :

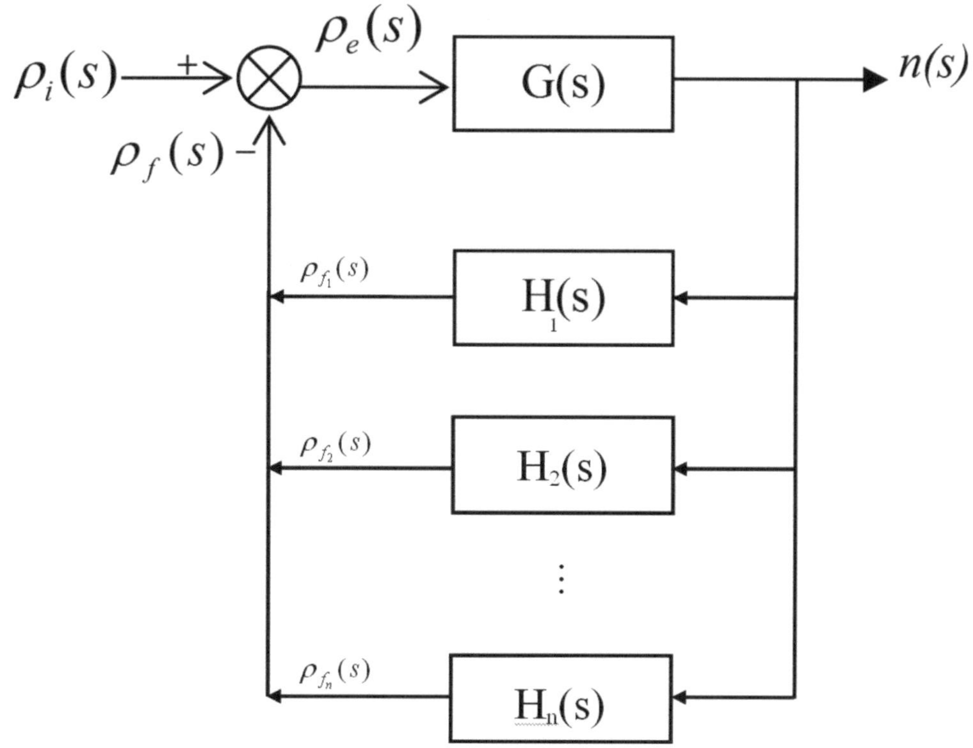

Fig. 2. The closed loop for several feedback reactivities

1.3 System's variables of nuclear reactor

There are several controlling factors in nuclear reactors such as:

1. Coolant flow rate.
2. Movement of control rods.
3. Concentration of boric acid.
4. Reaction rate.
5. Error function.
6. Temperature of core.
7. Power of reactor.
8. Core expansion.

9. Fission poisons like Xe and Sm.
10. Fission fragments and fission products.
11. Burn up.
12. Power demand.
13. The kind of fuel.
14. Energy of neutrons.
15. Doppler's effect.
16. Value of β.

If each system variable as a mathematics variable is considered then can write:

$$\dot{x}(t) = Ax(t) + Bu(t) \tag{5}$$

and:

$$y(t) = Cx(t) \tag{6}$$

Where:

x(t) is: variable of system, $\dot{x}(t)$ is: derivative of system's variable, *y(t)* is: output, *A* is: system's matrix, B is: control's matrix, C is: matrix of output and *u(t)* is: control's variable.

By taking the laplace conversions from two sides of above equations can write:

$$sX(s) = AX(s) + Bu(s) \tag{7}$$

and:

$$Y(s) = CX(s) \tag{8}$$

So two last equations that are based on Laplace conversions can be converted to following form:

$$(sI - A)X(s) = Bu(s) \tag{9}$$

and:

$$Y(s) = CX(s) \tag{10}$$

Therefore:

$$X(s) = (sI - A)^{-1}Bu(s) \tag{11}$$

and:

$$Y(s) = C(sI - A)^{-1}Bu(s) \tag{12}$$

In order to define the transfer function, it will be deduced as shown below:

$$G(s) = \frac{Y(s)}{u(s)} = C(sI - A)^{-1}B \tag{13}$$

Where:

$$(sI - A)^{-1} = \varphi(s) \tag{14}$$

and:

$$G(s) = C\varphi(s)B \tag{15}$$

1.4 Six factors coefficients

The effective multiplying coefficient is: the ratio of generated neutrons in every generation to generated neutrons in last generation. So to operate the nuclear reactor in steady state, this parameter should be: 1 means the generated neutrons in every generation are equal with neutrons which have absorbed or leaked in last generation that means: critical state. The minimum value of K_{eff} is: 0 and maximum of it is: υ namely: 2.43.

The effective multiplying coefficient is [5]:

$$K_{eff} = \eta.f.p.\varepsilon.P_{FNL}.P_{THNL} = \eta.f.p.\varepsilon.P_{TNL} \tag{16}$$

As the thermal fission coefficient (η) is [5, 6]:

$$\eta = \upsilon\frac{\Sigma_f^F}{\Sigma_{fm}^F} = \frac{\upsilon.N^{235}.\sigma_f^{235}.g_f^{235}}{N^{235}.\sigma_a^{235}.g_a^{235} + N^{238}.\sigma_a^{238}.g_a^{238} + N^O.\sigma_a^O} \tag{17}$$

Where:

$$N^{235} = \frac{m^{235}.A}{M^{235}}, \tag{18}$$

$$N^{238} = \frac{m^{238}.A}{M^{238}} \tag{19}$$

and:

$$N^o = 2N^u = 2\frac{m^u.A}{M^u} = 2\frac{m^u.A}{rM^{235} + (1-r)M^{238}} \tag{20}$$

Also the thermal absorption coefficient (f) is [5]:

$$f = \frac{\Sigma_a^F}{\Sigma_a^F + \Sigma_a^M} = \frac{N^{235}.\sigma_a^{235}.g_a^{235} + N^{238}.\sigma_a^{238}.g_a^{238} + N^O.\sigma_a^O}{N^{235}.\sigma_a^{235}.g_a^{235} + N^{238}.\sigma_a^{238}.g_a^{238} + N^O.\sigma_a^O + N^H.\sigma_a^H + N^{O_M}.\sigma_a^O} \tag{21}$$

The resonance escape probability for fast neutrons (p) also is calculated as following [5]:

$$p = e^{-\left(\frac{N^{238}}{\zeta.\Sigma_s}\right)\times I_{eff}} \tag{22}$$

Where:

$$I_{eff} = 3.9\times\left(\frac{\Sigma_S}{N^{238}}\right)^{0.415}, \tag{23}$$

$$\xi = \frac{A}{A+\frac{2}{3}} \tag{24}$$

and:

$$\Sigma_S = N^{235}.\sigma_s^{235} + N^{238}.\sigma_s^{238} + N^O.\sigma^O + N^H.\sigma^H + N^{O_M}.\sigma^O + N^{Zr}.\sigma^{Zr} \tag{25}$$

If the enrichment of fuel is 100% then the resonance escape probability for fast neutrons (p) will be maximum value.

According to enrichment of applied fuel and its mass can write [7]:

$$r = \frac{m_{ff}}{m_f}, \tag{26}$$

$$f_{fm} = \frac{m_f}{m_{fm}} = \frac{rM_{ff} + (1-r)M_{nf}}{rM_{ff} + (1-r)M_{nf} + M_{O_2}} \tag{27}$$

Where [8]:

$$m_{UO_2} = m_{fm} = \frac{N_{fm}.M_{fm}}{A} = \rho_{fm}.V_{fm}, \tag{28}$$

$$m_U = m_f = m_{fm}.f_{fm} = \rho_{fm}.V_{fm}.\frac{rM_{ff} + (1-r)M_{nf}}{rM_{ff} + (1-r)M_{nf} + M_{O_2}}, \tag{29}$$

and:

$$m_{U^{235}} = m_{ff} = m_{fm}.f_{fm}.r = \rho_{fm}.V_{fm}.\frac{rM_{ff} + (1-r)M_{nf}}{rM_{ff} + (1-r)M_{nf} + M_{O_2}}.r \tag{30}$$

If the transfer function of G(s) as V function is assumes then:

$$\dot{G}(\rho) = \frac{\partial G(\rho)}{\partial \rho_1}\frac{d\rho_1}{dt} + \frac{\partial G(\rho)}{\partial \rho_2}\frac{d\rho_2}{dt} + \cdots + \frac{\partial G(\rho)}{\partial \rho_n}\frac{d\rho_n}{dt} \tag{31}$$

and:

$$\dot{G}(\rho) = \nabla G^T(\rho).\dot{\rho} \tag{32}$$

Firstly the transfer function is supposed. Secondly due to mostly reactor's cores are twin therefore once a signal with delay time (τ_d) from first part to second part is transmitted then that signal with a same delay time from second part to first part will be transmitted.

1.5 Neutron point kinetics (NPK)

The treatment of the neutron transport as a diffusion process has only been validated. For example, in a Light Water Reactor (LWR) the mean free path of thermal neutrons is typically around 1 cm.

The fractional model has been derived for the NPK equations with n groups of delayed neutrons, is given by:

$$\tau^K \frac{d^{k+1}n}{dt^{k+1}} + \tau^K \left[\frac{1}{l} + \frac{(1-\beta)}{\Lambda}\right]\frac{d^k n}{dt^k} + \frac{dn}{dt} = \frac{\rho-\beta}{\Lambda}n + \sum_{i=1}^{m}\lambda_i C_i + \tau^k \sum_{i=1}^{m}\left(\lambda_i \frac{d^k C_i}{dt^k}\right), \quad 0 < k \le 2 \tag{33}$$

Where:

τ is the relaxation time, k is the anomalous diffusion order (for sub-diffusion process: $0 < k < 1$; while that for super-diffusion process: $1 < k < 2$), n is the neutron density, Ci is the concentration of delayed neutron precursor, l is the prompt-neutron lifetime for finite media, K is the neutron generation time, β is the fraction of delayed neutrons, and ρ is the reactivity. When $\tau^k \to 0$, the classic NPK equation is recovered.

The fractional model includes three additional terms relating to the classic equations which are contained of fractional derivatives (Gilberto and Espinosa, 2011):

$$\frac{d^{k+1}n}{dt^{k+1}}, \tag{34}$$

$$\frac{d^k n}{dt^k} \tag{35}$$

and:

$$\frac{d^k C_i}{dt^k} \tag{36}$$

The physical meaning of above terms suggests that for sub-diffusion processes, the first term has an important contribution for rapid changes in the neutron density (for example in the turbine trip in a BWR nuclear power plant (NPP)), while the second term represents an important contribution when the changes in the neutron density is almost slow, for example during startup in a NPP that involves operational maneuvers due to movement of control rod mechanism. The importance of third term is when the reactor sets in shutdown state, it could also be important to understand the processes in the accelerator driven system (ADS), which is a subcritical system characterized by a low fraction of delayed neutrons and by a small Doppler reactivity coefficient and totally there are many interesting problems to consider under the view point of fractional differential equations (FDEs) (Gilberto and Espinosa, 2011).

The neutron point kinetics (NPK) equations are one of the most important reduced models of nuclear engineering, and they have been the subject of countless studies and applications to understand the neutron dynamics and its effects. These equations are shown below:

$$\frac{dn(t)}{dt} = \dot{n}(t) = \frac{\rho_e - \beta}{\Lambda} n(t) + \sum_{i=1}^{6} \lambda_i C_i(t) \tag{37}$$

and:

$$\frac{dC_i(t)}{dt} = \dot{C}_i(t) = \frac{\beta}{\Lambda} n(t) - \lambda_i C_i(t) \tag{38}$$

Where:

$\dot{n}(t)$ is: the variations of neutrons density, ρ_e is: injected reactivity to system's transfer function, β is: delayed neutrons fraction, Λ is: neutron generation time, λ_i is: decay coefficient per density of each group of delayed neutrons, C_i is: density of each group of fission fragments which are as delayed neutrons generators.

If the partial variations per each system and control variables including: $\dot{n}(t)$, ρ_e, $n(t)$, $\dot{C}(t)$ and $C(t)$ are considered the these can write as following:

$$\delta\dot{n}(t) = \frac{\delta\rho - \beta}{\Lambda}(n_0 + \delta n(t)) + \lambda(C_0 + \delta C(t)) \tag{39}$$

and:

$$\delta\dot{C}(t) = \frac{\beta}{\Lambda}(n_0 + \delta n(t)) - \lambda(C_0 + \delta C(t)) \tag{40}$$

If these both $\delta\dot{n}(t)$ and $\delta\dot{C}(t)$ equations to matrix format are written, then the matrix format of them is shown as below:

$$\begin{bmatrix} \delta\dot{n}(t) \\ \delta\dot{C}(t) \end{bmatrix} = \begin{bmatrix} \frac{-\beta}{\Lambda} & \lambda \\ \frac{\beta}{\Lambda} & -\lambda \end{bmatrix} \times \begin{bmatrix} \delta n(t) \\ \delta C(t) \end{bmatrix} + \begin{bmatrix} \frac{n_0}{\Lambda} \\ 0 \end{bmatrix} \delta\rho \tag{41}$$

In the linear state can write last equation as the following:

$$\delta\dot{n}(t) = -\frac{\beta}{\Lambda}\delta n(t) + \lambda\delta C(t) + \frac{n_0\delta\rho}{\Lambda} \tag{42}$$

In case the point kinetic equations of reactor from both neutron density and fission fragments aspects are considered, it can be written:

Transfer Function: $$G(s)=\frac{\delta n(s)}{\delta \rho_e(s)}=\frac{[1\quad 0]\,adj\begin{bmatrix} s+\frac{\beta}{\Lambda} & -\lambda \\ \frac{-\beta}{\Lambda} & s+\lambda \end{bmatrix}\begin{bmatrix} \frac{n_0}{\Lambda} \\ 0 \end{bmatrix}}{\begin{vmatrix} s+\frac{\beta}{\Lambda} & -\lambda \\ \frac{-\beta}{\Lambda} & s+\lambda \end{vmatrix}}=$$

$$\frac{[1\quad 0]\begin{bmatrix} s+\lambda & \lambda \\ \frac{\beta}{\Lambda} & s+\frac{\beta}{\Lambda} \end{bmatrix}\begin{bmatrix} \frac{n_0}{\Lambda} \\ 0 \end{bmatrix}}{\begin{vmatrix} s+\frac{\beta}{\Lambda} & -\lambda \\ \frac{-\beta}{\Lambda} & s+\lambda \end{vmatrix}} \tag{43}$$

Where:

$$\begin{vmatrix} s+\frac{\beta}{\Lambda} & -\lambda \\ \frac{-\beta}{\Lambda} & s+\lambda \end{vmatrix}=s^2+s(\lambda+\frac{\beta}{\Lambda})=s(s+\lambda+\frac{\beta}{\Lambda}) \tag{44}$$

Therefore the transfer function will be as shown:

$$G(s)=\frac{\delta n(s)}{\delta \rho_e(s)}=\frac{[1\quad 0]\begin{bmatrix} s+\lambda & \lambda \\ \frac{\beta}{\Lambda} & s+\frac{\beta}{\Lambda} \end{bmatrix}\begin{bmatrix} \frac{n_0}{\Lambda} \\ 0 \end{bmatrix}}{s(s+\lambda+\frac{\beta}{\Lambda})} \tag{45}$$

Where:

$$\delta \rho_e(s)=\rho_0-\delta \rho_f(s) \tag{46}$$

Also the equation of zero power transfer function is as following:

$$\frac{\delta n(s)}{\delta \rho_e(s)}=\frac{[1\quad 0]\begin{bmatrix} \frac{n_0(s+\lambda)}{\Lambda} \\ \frac{\beta n_0}{\Lambda^2} \end{bmatrix}}{s(s+\lambda+\frac{\beta}{\Lambda})}=\frac{\frac{n_0}{\Lambda}(s+\lambda)}{s(s+\frac{\beta}{\Lambda})}=\frac{n_0(s+\lambda)}{s(s\Lambda+\beta)} \tag{47}$$

and for variation of fission fragment density as per error reactivity variation, it can write:

$$\frac{\delta C(s)}{\delta \rho_e(s)} = \frac{\begin{bmatrix} 0 & 1 \end{bmatrix} \begin{bmatrix} \dfrac{n_0(s+\lambda)}{\Lambda} \\ \dfrac{\beta n_0}{\Lambda^2} \end{bmatrix}}{s(s+\lambda+\dfrac{\beta}{\Lambda})} = \frac{\beta n_0 / \Lambda^2}{s(s+\dfrac{\beta}{\Lambda})} \tag{48}$$

In case the point kinetic equations of reactor from neutron density and fission fragments aspects and also temperature aspect (according to Newton's low of cooling) are considered, it can be expressed:

$$G(s) = \frac{\delta n(s)}{\delta \rho_e(s)} = \frac{\begin{bmatrix} 1 & 0 & 0 \end{bmatrix} adj \begin{bmatrix} s+\dfrac{\beta}{\Lambda} & -\lambda & \dfrac{\alpha n_0}{\Lambda} \\ \dfrac{-\beta}{\Lambda} & s+\lambda & 0 \\ -k & 0 & s+a \end{bmatrix} \begin{bmatrix} \dfrac{n_0}{\Lambda} \\ 0 \\ 0 \end{bmatrix}}{\begin{vmatrix} s+\dfrac{\beta}{\Lambda} & -\lambda & \dfrac{\alpha n_0}{\Lambda} \\ \dfrac{-\beta}{\Lambda} & s+\lambda & 0 \\ -k & 0 & s+a \end{vmatrix}} \tag{49}$$

Where:

$$\begin{vmatrix} s+\dfrac{\beta}{\Lambda} & -\lambda & \dfrac{\alpha n_0}{\Lambda} \\ \dfrac{-\beta}{\Lambda} & s+\lambda & 0 \\ -k & 0 & s+a \end{vmatrix} = s^3 + s^2(a+\lambda+\frac{\beta}{\Lambda}) + s(a\lambda+\frac{\beta a}{\Lambda}+\frac{\alpha n_0 k}{\Lambda}) + \frac{\alpha n_0 k \lambda}{\Lambda} \tag{50}$$

In this state the equation of zero power transfer function is as following:

$$G(s) = \frac{\delta n(s)}{\delta \rho_e(s)} = \frac{(n_0/\Lambda)(s+\lambda)(s+a)}{s^3 + s^2(a+\lambda+\dfrac{\beta}{\Lambda}) + s(a\lambda+\dfrac{\beta a}{\Lambda}+\dfrac{\alpha n_0 k}{\Lambda}) + \dfrac{\alpha n_0 k \lambda}{\Lambda}} \tag{51}$$

In this stage according to the main transfer function that has mentioned in the last stage, $G(\rho)$ is defined:

$$G(\rho) = \frac{\delta n(\rho)}{\delta \rho_e(\rho)} = \frac{(n_0/\Lambda)(\rho+\lambda)(\rho+a)}{\rho^3 + \rho^2(a+\lambda+\dfrac{\beta}{\Lambda}) + \rho(a\lambda+\dfrac{\beta a}{\Lambda}+\dfrac{\alpha n_0 k}{\Lambda}) + \dfrac{\alpha n_0 k \lambda}{\Lambda}} \tag{52}$$

Therefore through $G(\rho)$, the parts of $\dot{G}(\rho)$ will be defined and the stability condition through V function is applied.

So can write:

$$G(s) = c^T \phi(s) b \tag{53}$$

In large twin reactor's cores, it can be written:

$$\dot{n}(t) = \frac{\rho_e - \beta}{\Lambda} n(t) + \sum_{i=1}^{6} \lambda_i C_i(t) + q_i(t) \tag{54}$$

It is also supposed:

$$q_i(t) = \sum_{j=1 \neq i} \frac{\alpha_{ij}}{\Lambda_i} n_j(t - T_{ij}) \tag{55}$$

and:

$$\alpha_{12} = \alpha_{21} \tag{56}$$

q_1 , q_2 and ρ_i are respectively as following:

$$q_1(t) = \frac{\alpha_{12}}{\Lambda_1} n_2(t - T_{12}), \tag{57}$$

$$q_2(t) = \frac{\alpha_{21}}{\Lambda_2} n_1(t - T_{21}) \tag{58}$$

and:

$$\rho_i = \sum_{j=1 \neq i} \alpha_{ij} \frac{n_j(t - T_{ij})}{n_i(t)} \tag{59}$$

So $q_i(t)$ divided by ρ_i equals:

$$\frac{q_i(t)}{\rho_i} = \frac{\sum_{j=1 \neq i} \frac{\alpha_{ij}}{\Lambda_i} n_j(t - T_{ij})}{\sum_{j=1 \neq i} \alpha_{ij} \frac{n_j(t - T_{ij})}{n_i(t)}} = \sum_{j=1 \neq i} \frac{n_i(t)}{\Lambda_i} \tag{60}$$

According to Newton's low of cooling can write:

$$\dot{T}(t) = \frac{1}{mc_p} n(t) - aT(t) \tag{61}$$

If the variations of neutron density than initial neutron numbers, fission fragment density than initial fission fragment density and temperature than initial temperature as system variables are considered respectively as below:

$$x(t) = \frac{n(t) - n_0}{n_0}, \tag{62}$$

$$y(t) = \frac{C(t) - C_0}{C_0} \tag{63}$$

and:

$$z(t) = \frac{T(t) - T_0}{T_0} \tag{64}$$

Then the differentials of these system variables are respectively as following:

$$\dot{x}_i(t) = \frac{\delta_i(x,t)}{\Lambda}[1 + x_i(t)] - \frac{\beta}{\Lambda}x_i(t) + \frac{\beta}{\Lambda}y_i(t) - \sum_{j=1\neq i}^{N} \frac{n_{j_0}}{n_{i_0}} \cdot \frac{\alpha_{ij}}{\Lambda} x_i(t) + \sum_{j=1\neq i}^{N} \frac{n_{j_0}}{n_{i_0}} \cdot \frac{\alpha_{ij}}{\Lambda} x_j(t - T_{ij}), \tag{65}$$

$$\dot{y}_i(t) = \lambda x_i(t) - \lambda y_i(t) \tag{66}$$

and:

$$\dot{z}_i(t) = a x_i(t) - a z_i(t) \tag{67}$$

If these system variables differentials are to matrix format are written, then can write:

$$\begin{bmatrix} \dot{x}_i(t) \\ \dot{y}_i(t) \\ \dot{z}_i(t) \end{bmatrix} = \begin{bmatrix} \frac{\delta_i - \beta}{\Lambda} - \sum_{j=1\neq i}^{N} \frac{n_{j_0}}{n_{i_0}} \cdot \frac{\alpha_{ij}}{\Lambda} & \frac{\beta}{\Lambda} & \\ \lambda & \lambda & 0 \\ a & 0 & -a \end{bmatrix} \times \begin{bmatrix} x_i(t) \\ y_i(t) \\ z_i(t) \end{bmatrix} + \begin{bmatrix} \frac{\delta_i}{\Lambda} \\ 0 \\ 0 \end{bmatrix} \tag{68}$$

1.6 Stability of reactor

There are some methods for determination of nuclear reactor stability. Among methods which are applied for this aim are as following:

Liapunov method- Lagrange method- Popov method- Pontryagin method.

There is Lyapunov's method to determine the stability of nonlinear reactor dynamics by constructing certain positive definite functions of the reactor variables and parameters (Pankaj and Vivek, 2011).

There has been calculated the Lyapunov exponents from a time series of the excess neutron population of a boiling water reactor (BWR) and used it to conclude about the stability of the steady state operation of that particular BWR (Munoz et al., 1992).

There has also discussed the application of topological methods in reactor kinetics study (Smets and Giftopoulos, 1959).

The topological and Lyapunov methods were compared with Aizermann-Rosen methods for analyzing a point reactor model (Devooght and Smets, 1967).

The Padé approximations has been used to obtain solutions for point kinetic equations (Aboanberand, 2002).

Perturbation theory has also been widely used in studying reactor dynamics. There has been obtained specific types of steady solutions to study power oscillations in a reactor results from a Hopf bifurcation (Pandey, 1996; Munoz and Verdu, 1991; Tsuji et al., 1993; Konno et al., 1994). The KBM theory has been used for the nonlinear analysis of a reactor model with the effect of time-delay in the automated control system (Konno et al, 1992).

A singular perturbation has been used to study relaxation oscillations in typical nuclear reactors (Ward and Lee, 1987).

The point kinetic equations in the presence of delayed neutrons with one temperature reactivity coefficient for a step input of reactivity have analytically been solved by applying the perturbation theory (Gupta and Trasi, 1986).

The regular perturbations to obtain an analytical solution for general reactivity have been used (Nahla, 2009).

The multiple time-scales expansion to obtain analytical solutions of the neutron kinetic equations has been applied (Merk and Cacuci, 2005).

The variation methods in conjunction with the Hopf bifurcation theory for a BWR with one group delayed neutron have also been applied (Munoz and Verdu, 1991).

A combination of the center-manifold reduction (CMR) and the method of normal forms have already been applied widely for nonlinear analysis of nuclear reactor dynamics (Pandey, 1996; Tsuji et al., 1993; Konno et al., 1994).

In the Liapunov model, the dynamically equations may be either differential or non differential equations. But the method of problem evaluation is not based on solving the equations. This method is based on energy in classic mechanical. In one of mechanical system the stability condition is when the total energy of a system decreases.

Liapunov applied this property and based the stability function. This function is entitled: *V(x)* Function.

This function has features as following:

1. $\dot{V}(x)$ has definite positive value.
2. The partial differential of $V(x)$ is continued.
3. Where the ρ is momentum, $V(\rho)$ is negative quasi relatively.
4. The $V(x)$ function can be written as following:

$$\dot{V}(x)=\frac{\partial V(x)}{\partial x_1}.\frac{dx_1}{dt}+\frac{\partial V(x)}{\partial x_2}.\frac{dx_2}{dt}+\ldots+\frac{\partial V(x)}{\partial x_n}.\frac{dx_n}{dt} \tag{69}$$

It can also write:

$$\dot{V}(x)=\nabla V^T(x)\dot{x} \tag{70}$$

In fact the Liapunov function is a function that considers either state variables or variables which cause to imbalance state. Due to the potential function is able to do a process, so the Liapunov function is as V function. One can write:

$$V = f(x_1, x_2, ..., x_n) \tag{71}$$

If: $\dot{V} = 0$ then the system will be steady state.

This method revolved about the determination of a V function, which satisfies certain requirements of the stability theorem. In an initial reactor model a point reactor with constant power removal and without delayed neutrons is considered.

Due to ρ is a variable which causes unbalancing the system, therefore it can be considered as an x variable. Then it can be written:

$$\dot{V}(\rho) = \frac{\partial V(\rho)}{\partial \rho_1}\frac{d\rho_1}{dt} + \frac{\partial V(\rho)}{\partial \rho_2}\frac{d\rho_2}{dt} + \cdots + \frac{\partial V(\rho)}{\partial \rho_n}\frac{d\rho_n}{dt} \tag{72}$$

and:

$$\dot{V}(\rho) = \nabla V^T(\rho).\dot{\rho} \tag{73}$$

2. Methodology

2.1 Analyzing the theory by mathematical model

In this work the value of K_{eff} as a comparable value is supposed and attributed to input parameter block (*H*) and then this value with the received feedback value is compared.

The unit of the control rod velocity (*v*) can be mm/s, the rate is steady, and the control rod movement is only to up and down directions, so: x(0) =0 (Shirazi et al., 2010).

Since the *sgn(x)* function is nonlinear; so conversion function can not be calculated; thus in this stage arguing the frequency response is not meaningful. Therefore the steady state must be considered for this nonlinear function; though it is rather complicated (Marie and Mokhtari, 2000).

To analyze the controlling system theory these are assumed:

If: Input=*H*; Output=*x(t)*; in the top of control rod: *x*=0; in the bottom of the control rod: $x=x_{max}$;

$$F = kHx(t) + K_0 \tag{74}$$

Where:

F: Function, k: constant coefficient, H: input parameter, $x(t)$: the control rod position, K_0: initial value of K_{eff}.

$$\Delta x = v\,\mathrm{sgn}(F - K_{sp}).\Delta t \tag{75}$$

Where Δx is: the amount of control rod movement, Δt is: time.

$$\frac{dx}{dt} = v\,\text{sgn}(F - K_{sp}) \tag{76}$$

Where $\frac{dx}{dt}$ is: the velocity of control rod from the movement aspect to up and down, K_{sp} : the secondary value of K_{eff} in the recent position of control rod.

$$x(t) = \int_0^t v sgn[kx(t)H(t - t_D) + K_0 - K_{sp}]dt \tag{77}$$

Supposition: $x(0) = 0$,

So:

$$x(t) = x_0 \pm vtsgn(t - t_D) \tag{78}$$

Where x is: absolutely descending, x_0 is: the initial value of x and t_D is: the innate delay time.

The SIMULINK of MATLAB is an appropriate software to analyze the performance of this simulation (Tewari, 2002).

The simulated model is considered according to Fig.3:

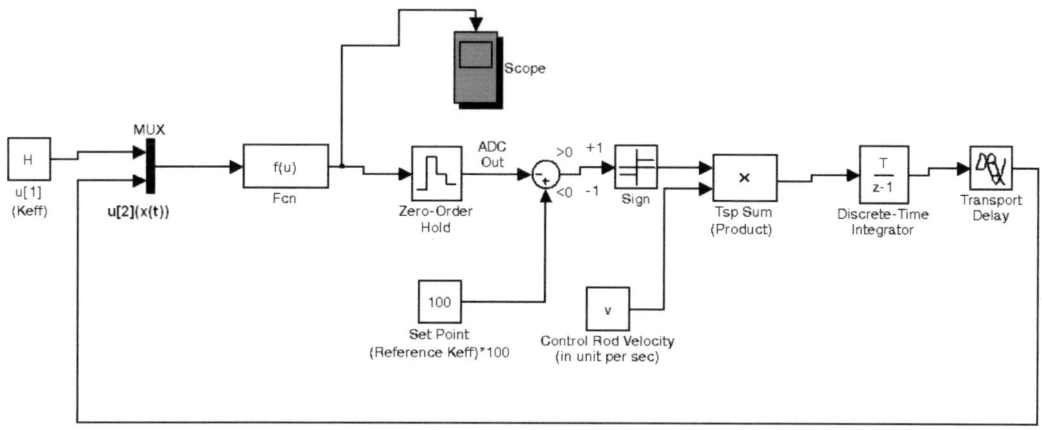

Fig. 3. The block diagram of simulation by SIMULINK

2.2 The simulation by SIMULINK and related block diagram

The work processes have been simulated by the SIMULINK of MATLAB software and all responses such as oscillation and transient responses have been analyzed by it as well. The main function *(F)* is *Fcn*. It includes two other functions that are: $u[1]$ and $u[2]$ which are defined for SIMULINK. $u[1]$ is one of two input functions of Mux that has been shown by the input block that is: *H*. This block is presenting amount of the K_{eff}. $u[2]$ equals with amount of the feedback which has been sampled as follows: *x(t)*.

Thus:

$$Fcn = F = ku[1] \times u[2] + K_0 \tag{79}$$

Because of the control rod movement is steady, in order to calculate the total amount of the discrete movements of control rod, the Discrete-time Integrator block has been used. The *Fcn* produced function has been transferred to Zero-Order Hold block which plays logic converter role. In addition the Transport Delay block is related to the inherent delay time that is: t_D. The parameters which must be adjusted are: Set Point that is: the default amount of K_{eff} as reference K_{eff} and the meaning of the Set Point=100 is: $K_{eff}=1$, the velocity of control rod (v), recent K_{eff} (block H) and the stop time that is: the innate delay time or t_D. The graphs can be observed by the oscilloscope.

3. Results

In this simulation the input parameter value (*H*) is attributed to arbitrary K_{eff}. So if the favorite K_{eff} is: 1 then the value of *H* will be defined: 100 and it is also for Set point (reference K_{eff}). So Set Point: 50 means the reference K_{eff} is: 0.50 and in this situation this K_{eff} is enumerated as the arbitrary and favorite K_{eff} that stability of reactor in this situation is based on it. This arbitrary K_{eff} with output of Zero-Order Hold block is compared for performance of simulation by SIMULINK software. Also for example velocity: 3 means the velocity of control rod in this simulation is: 3 units per second (for example: 3mm/s). In this state the velocity of control rod is increased comparing to the last stage which was: 1 unit per second. The velocity of control rod belongs to Speed block which is an input to Tsp Sum block in the block diagram.

By changing the values of the cited parameters (which were: Set Point, the velocity of control rod (v), *H* and the delay time), the different states of the graphs can be shown according to Figs.4, 5, 6 and 7:

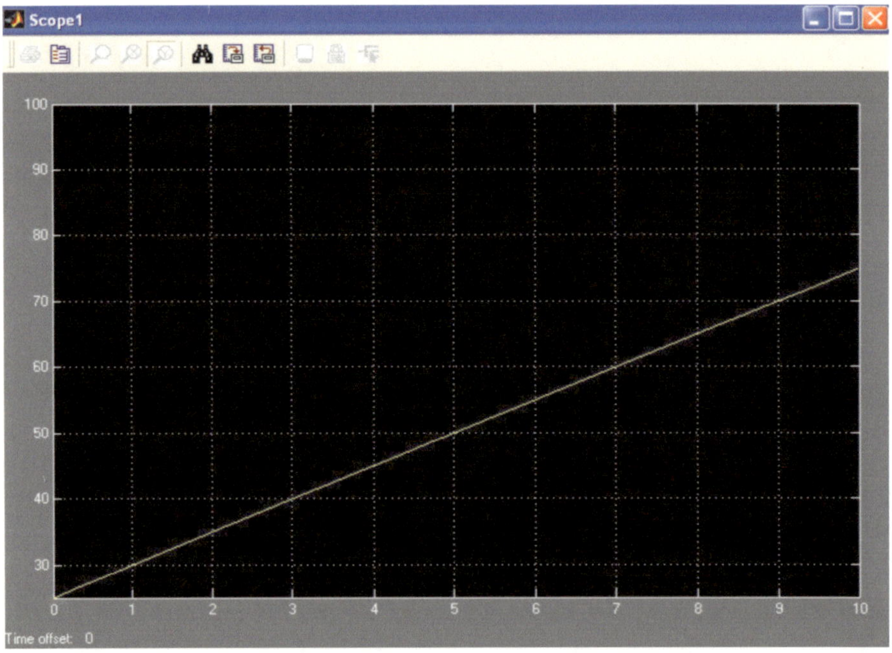

Fig. 4. Without oscillation for: Set Point: 100, v: 1, H: 50, delay time: 10 ms

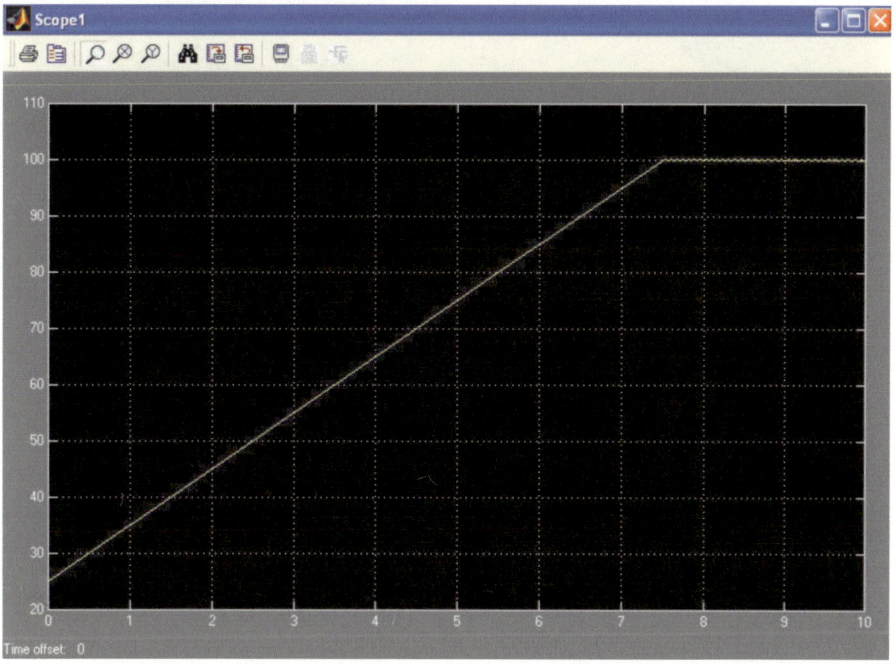

Fig. 5. The low oscillation for: Set Point: 100, v: 1, H: 100, delay time: 15 ms

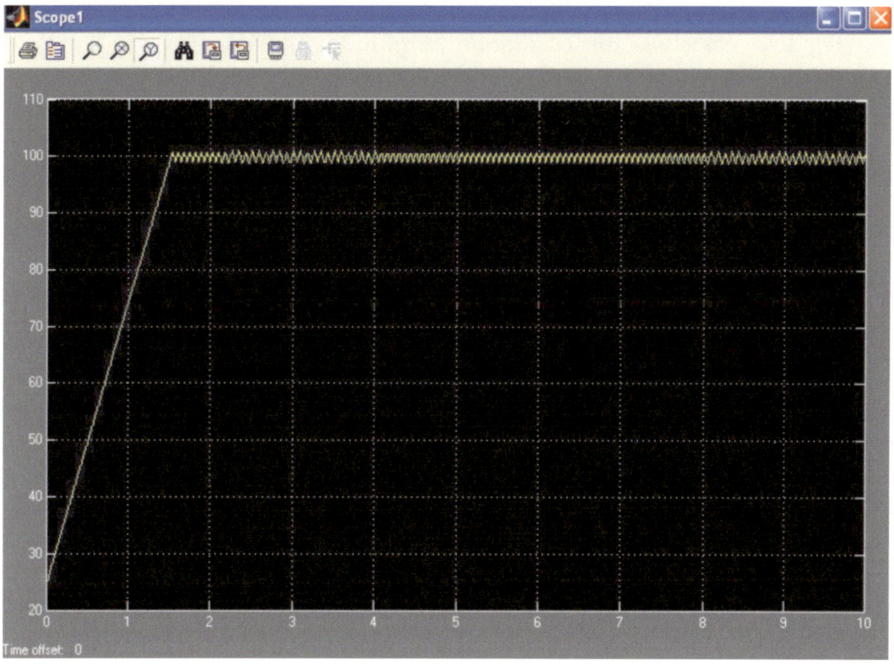

Fig. 6. The medium oscillation for: Set Point: 100, v: 5, H: 100, delay time: 15 ms

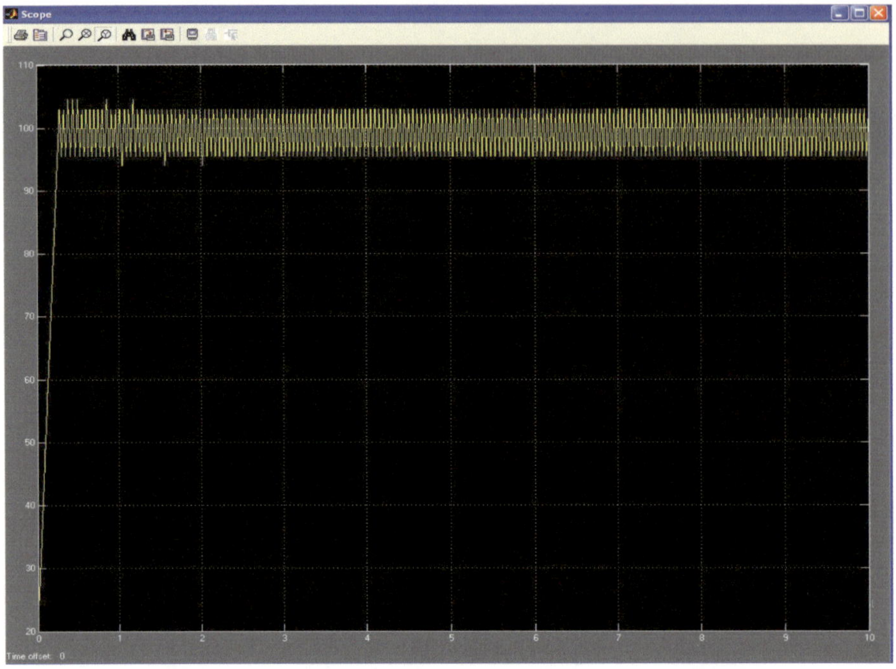

Fig. 7. The large oscillation for: Set Point: 100, v: 7, H: 100, delay time: 10 ms

The Figs.4, 5, 6 and 7 show if the velocity of control rod for upward and downward movements is increased then the K_{eff} will be pendulous surrounds of defined Set Point (reference K_{eff}) and also the oscillation amplitude will be more than lower velocity situations. For low velocities the oscillation amplitude is slight and acceptable. Another effective factor is inherent delay (such as derived delay of control rod mechanism) and its inordinate increasing can cause unstable states.

4. Conclusion

By this simulation the best response and operation which a reactor can have from stability aspect, according to its control rod velocity is derived.

According to Figs.4-8 the best status in the Fig.4 is observed in which there is no vibration in response. In this simulation the stability of reactor depends on either velocity or delay time values directly because delay time plays a key role. Therefore in this simulation the admissible ranges of velocity and delay time which can be caused to stable the reactor are respectively: low velocity of control rod around 1-3 units per second and short delay time (10ms). However in this case reach critical state (Keff=1) for nuclear reactor will be taken more than modes Figs.5-8. As in all the figures is observed, so can deduce the velocity of control rod plays the more important role than delay time of mechanism in stability of nuclear reactor. Whereas for minor and major changes of reactivity and shut down of reactor in emergency states, there are some kinds of control rods at nuclear reactors cores such as regulating rods, safety rods and shim rods, therefore this simulation can be applied for each control rod in either LWR nuclear reactor or research reactor cores which have vertical control rods. Also this simulation can be applied for each batch control rods which act in the same way as the cluster at fuel assembly at core of nuclear power reactors though the moving speed of regulating rod is much less than the safety rod.

5. Acknowledgment

This book chapter is related to a research project entitled: "The Simulation of a Model by SIMULINK of MATLAB for Determining the Best Ranges for Velocity and Delay Time of Control Rod Movement in LWR Reactors" that by financial supporting the Islamic Azad University-South Tehran Branch has been carried out.

6. References

Aboanber, A and Nahla, A., 2002. Solution of point kinetics equations in the presence of Newtonian temperature feedback by Padé approximations via the analytical inversion method. *Journal of Physics, Mathematical and General*, Vol. 35, 9609–9627.

Basu, S., Zemdegs, R., 1978. Method of reliability analysis of control systems for nuclear power plants. *Microelectronics Reliability*, Vol. 17, No.1, (January 1978), 105-116.

Benítez, J., Martínez, P and Pérez, H., 2005. International cooperation on control for safe operation of nuclear research reactors. *Progress in Nuclear Energy*, Vol. 46, No.3-4, 321-327.

Cheng, L., Feng, J and Zhao, F., 2009. Design and optimization of fuzzy-PID controller for the nuclear reactor power control. *Nuclear Engineering and Design*, Vol. 239, No.11, (November 2009), 2311-2316.

Chyou, Y., Cheng, Y., 2004. Performance validation on the prototype of control rod driving mechanism for the TRR-II project. *Nuclear Engineering and Design*, Vol. 227, No.2, (January 2004), 195-207.

Devooght, G and Smets, H., 1967. Determination of stability domains in point reactor dynamics. *Nuclear Science and Engineering*, Vol. 28, 226–236.

Dong, Z., Huang, X., Feng, J and Zhang, L., 2009. Dynamic model for control system design and simulation of a low temperature nuclear reactor. *Nuclear Engineering and Design*, Vol.239, No. 10, (October 2009), 2141-2151.

Gilberto, M., Polo, E and Espinosa, E., 2011. Fractional neutron point kinetics equations for nuclear reactor dynamics. *Annals of Nuclear Energy*, Vol. 38, No.2-3, (February-March 2011), 307-330.

Gupta, H and Trasi, M., 1986. Asymptotically stable solutions of point-reactor kinetics equations in the presence of Newtonian temperature feedback. *Annals of Nuclear Energy*, Vol. 13, No.4, 203–207.

Konno, H., Hayashi, K and Shinohara, Y., 1992. Nonlinear dynamics of reactor with time delay in automatic control system and temperature effect. *Journal of Nuclear Science and Technology*, Vol. 29, No.6, 530–546.

Lamarsh, J., 1975. *Introduction to nuclear engineering*. Addison Wesley Publishing Company, Chapter 5.

Marie, M., Mokhtari, M., 2000. *Engineering Applications of MATLAB 5.3 and SIMULINK 3*. Springer Pub, 30-75.

Merk, B and Cacuci, D., 2005. Multiple timescale expansions for neutron kinetics-illustrative application to the point kinetics model. *Nuclear Science and Engineering*, Vol. 151, 184–193.

Munoz, J and Verdu, G., 1991. Application of Hopf bifurcation theory and variational methods to the study of limit cycles in boiling water reactors. *Annals of Nuclear Energy*, Vol. 18, No.5, 269–302.

Munoz, J., Verdu, G., Pereira, C., 1992. Dynamic reconstruction and Lyapunov exponents from time series data in boiling water reactors. *Annals of Nuclear Energy*, Vol. 19, No.4, 223–235.

Nahla, A., 2009. An analytical solution for the point reactor kinetics equations with one group of delayed neutrons and the adiabatic feedback model. *Progress in Nuclear Energy*, Vol. 51, No.1, 124–128.

Pandey, M., 1996. *Nonlinear Reactivity Interactions in Fission Reactor Dynamical Systems*, Ph.D. Thesis, Indian Institute of Technology-Kanpur.

Pankaj, W and Vivek, K., 2011. Nonlinear stability analysis of a reduced order model of nuclear reactors: A parametric study relevant to the advanced heavy water reactor. *Nuclear Engineering and Design*, Vol. 241, No.1, (January), 134-143.

Park, G., Cho, N., 1992. Design of a nonlinear model-based controller with adaptive PI gains for robust control of a nuclear reactor. *Progress in Nuclear Energy*, Vol. 27, No.1, 37-49.

Shirazi, S., Aghanajafi, C., Sadoughi, S and Sharifloo, N., 2010. Design, construction and simulation of a multipurpose system for precision movement of control rods in nuclear reactors. *Annals of Nuclear Energy*, Vol. 37, No.12, (December 2010), 1659-1665.

Smets, H and Giftopoulos, E., 1959. The application of topological methods to the kinetics of homogeneous reactors. *Nuclear Science and Engineering*, Vol. 6, 341–349.

Stark, K., 1976. *Modal control of a nuclear power reactor*. Automatica 12, 613-618.

Tachibana, Y., Sawahata, H., Iyoku, T and Nakazawa, T., 2004. Reactivity control system of the high temperature engineering test reactor. *Nuclear Engineering and Design*, Vol. 233, No.1-3, (October 2004), 89-101.

Tewari, A., 2002. *Modern Control Design with MATLAB and SIMULINK*. John Wiley, Chapter 2, 3.

Ward, M and Lee, J., 1987. Singular perturbation analysis of relaxation oscillations in reactor systems. *Nuclear Science and Engineering*, Vol. 95, 47–59.

Zhao, F., Cheung, K and Yeung, R., 2003. Optimal power control system of a research nuclear reactor. *Nuclear Engineering and Design*, Vol. 219, No.3, (February 2003), 247-252.

10

Theory of Fuel Life Control Methods at Nuclear Power Plants (NPP) with Water-Water Energetic Reactor (WWER)

Sergey Pelykh and Maksim Maksimov
Odessa National Polytechnic University, Odessa
Ukraine

1. Introduction

The problem of fuel life control at nuclear power plants (NPP) with WWER-type light-water reactors (PWR) will be discussed for design (normal) loading conditions only. That is, emergency nuclear reactor (NR) operation leading to cladding material plastic deformation is not studied here, therefore the hot plasticity (stress softening) arising at the expense of yield stress decrease under emergency cladding temperature rise, will not be considered here.

Analysing the current Ukrainian energetics status it is necessary to state that on-peak regulating powers constitute 8 % of the total consolidated power system (CPS), while a stable CPS must have 15 % of on-peak regulating powers at least. More than 95 % of all thermal plants have passed their design life and the Ukrainian thermal power engineering averaged remaining life equals to about 5 years. As known, the nuclear energetics part in Ukraine is near 50 %. Hence, operation of nuclear power units of Ukraine in the variable part of electric loading schedule (variable loading mode) has become actual recently, that means there are repeated cyclic NR capacity changes during NR normal operation.

Control of fuel resource at WWER nuclear units is a complex problem consisting of a few subproblems. First of all, a physically based fuel cladding failure model, fit for all possible regimes of normal NR operation including variable loading and burnups above 50 MW·d/kg, must be worked out. This model must use a certified code developed for fuel element (FE) behaviour analysis, which was verified on available experimental data on cladding destruction.

The next condition for implementation of nuclear fuel resource control is availability of a verified code estimating distribution of power flux in the active core for any reactor normal operation mode including variable loading.

It should be noticed that calculation of nuclear fuel remaining life requires estimating change of the state of a fuel assembly (FA) rack. For instance, the state of a rack can change considerably at core disassembling (after a design accident) or at spent fuel handling. Generally speaking, the total fuel handling time period must be considered including the duration of dry/wet storage. Before designing a nuclear fuel resource control system, using probability theory and physically based FA failure criteria, the failure probability for all FA must be estimated. Having satisfied the listed conditions, a computer-based system for control of nuclear fuel remaining life can be worked out.

The FEMAXI code has been used to calculate the cladding stress/strain development for such its quality as simultaneous solution of the FE heat conduction and mechanical deformation equations using the finite element method (FEM) allowing consideration of variable loading (Suzuki, 2000). Sintered uranium dioxide was assumed to be the material of pellets while stress relieved Zircaloy-4 was assumed to be the material of cladding (Suzuki, 2010). Cladding material properties in the FEMAXI code are designated in compliance with (MATPRO-09, 1976). But the manufacturing process and the zircaloy alloy used are not specified here.

FE behaviour for UTVS (the serial FA of WWER-1000, V-320 project), TVS-A (the serial FA of WWER-1000 produced by OKBM named after I.I. Aphrikantov) and TVS-W (the serial FA produced by WESTINGHOUSE) has been analysed.

The full list of input parameters used when analyzing the PWR fuel cladding durability can be seen in (Suzuki, 2000). The NR regime and FA constructional parameters were set in compliance with Shmelev's method (Shmelev et al., 2004). The main input parameters of FE and FA used when analyzing the WWER-1000 fuel cladding durability are listed in Table 1.

Parameter	UTVS	TVS	
		TVS-A	TVS-W
Cladding outer diameter, cm	0.910	0.910	0.914
Cladding inner diameter, cm	0.773	0.773	0.800
Cladding thickness, cm	0.069	0.069	0.057
Pellet diameter, cm	0.757	0.757	0.784
Pellet centre hole diameter, cm	0.24	0.14	–
Pellet dish	–	–	each side
Equivalent coolant hydraulic diameter, cm	1.06	1.06	1.05
Total fuel weight for a FE, kg	1.385	1.487	1.554

Table 1. Different parameters of UTVS, TVS-A and TVS-W.

FE cladding rupture life control for a power-cycling nuclear unit having the WWER-1000 NR is a key task in terms of rod design and reliability. Operation of a FE is characterized by long influence of high-level temperature-power stressing leading to uncontrollable cladding material creep processes causing, after a while, its destruction, and fission products enter the circuit in the quantities exceeding both operational limits and limits of safe operation. In this connection, estimation of cladding integrity time for a NR variable loading mode, taking into account some appointed criteria, becomes one of key problems of FE designing and active core operational reliability analysis.

In accordance with the experience, there are following main characteristic cladding destruction mechanisms for the WWER-1000 varying loading mode (Suzuki, 2010): pellet-cladding mechanical interaction (PCMI), especially at low burnups and stress corrosion cracking (SCC); corrosion at high burnups (>50 MWd/kg-U); cladding failure caused by multiple cyclic and long-term static loads.

It is supposed that influence of low-burnup PCMI is eliminated by implementation of the WWER-1000 maximum linear heat rate (LHR) regulation conditions. Non-admission of cladding mechanical damage caused by SCC is ensured by control of linear heat power permissible values and jumps also. The high-burnup corrosion influence is eliminated by optimization of the alloy fabrication technique.

As all power history affects fuel cladding, it is incorrect to transfer experimental stationary and emergency operation cladding material creep data onto the FE cladding working at variable loading. Emergency NR operation leading to cladding material plastic deformation is not studied here, therefore hot plasticity (stress softening) arising at the expense of yield stress decrease under emergency cladding temperature rise, is not considered.

To solve this problem, we are to define main operating conditions affecting FE cladding durability and to study this influence mechanism. The normative safety factor K_{norm} for cladding strength criteria is defined as

$$K_{\text{norm}} = R^{\max} / R\,, \tag{1}$$

where $R^{\max}$ is the limit value of a parameter; R is the estimated value of a parameter.

The groupe of WWER-1000 cladding strength criteria includes the criteria SC1…SC5 – see Table 2 (Novikov et al., 2005). According to SC4, the WWER-1000 FE cladding total damage parameter is usually estimated by the relative service life of cladding, when steady-state operation and varying duty are considered separately:

$$\omega(\tau) = \sum_i \frac{NC_i}{NC_i^{\max}} + \int_0^{\tau} \frac{dt}{t^{\max}} < 1\,, \tag{2}$$

where $\omega(\tau)$ is the cladding material damage parameter; NC_i and $NC_i^{\max}$ are the number of i-type power-cycles and the allowable number of i-type power-cycles, respectively; t is time; $t^{\max}$ is the creep-rupture life under steady-state operation conditions.

The cladding material damage parameter can be considered as a structure parameter describing the material state (ω = 0, for the intact material and ω = 1, for the damaged material). The second possible approach is considering $\omega(\tau)$ as a characteristic of discontinuity flaw. That is when $\omega = 0$, there are no submicrocracks in the cladding material. But if $\omega = 1$, it is supposed that the submicrocracks have integrated into a macrocrack situated in some cross-section of the cladding

Criterion	Definition	K_{norm}
SC1	$\sigma_\theta^{max} \le 250$ MPa, where σ_θ^{max} is maximum circumferential stress.	1.2
SC2	$\sigma_e^{max} < \sigma_0(T,\phi)$, where σ_e^{max} is maximum equivalent stress, Pa; σ_0 is yield stress, Pa; T is temperature, K; ϕ is neutron fluence, cm^{-2} s^{-1}.	-
SC3	$P_c \le P_c^{max}$, where P_c is coolant pressure, Pa.	1.5
SC4	$\omega(\tau) = \sum_i \frac{NC_i}{NC_i^{max}} + \int_0^\tau \frac{dt}{t^{max}} < 1$.	10
SC5	$\varepsilon_{\theta,pl}^{max} \le 0.5\,\%$, where $\varepsilon_{\theta,pl}^{max}$ is cladding limit circumferential plastic strain	-

Table 2. Cladding strength criteria.

An experimental study of Zircaloy-4 cladding deformation behavior under cyclic pressurization (at 350 °C) was carried out in (Kim et al., 2007). The investigated cladding had an outer diameter and thickness of 9.5 mm and 0.57 mm, respectively. The microstructure of Zircaloy-4 was a stress-relieved state. A sawtooth pressure waveform was applied at different rates of pressurization and depressurization, where the maximum hoop stress was varied from 310 MPa to 470 MPa, while the minimum hoop stress was held constant at 78 MPa. Using the cladding stress-life diagram and analyzing the metal structure and fatigue striation appearance, it was found that when loading frequency $v < 1$ Hz, creep was the main mechanism of thin cladding deformation, while the fatigue component of strain was negligibly small.

Taking into account the experimental results (Kim et al., 2007), it can be concluded that estimation of $\omega(\tau)$ by separate consideration of NR steady-state operation and varying duty (2) has the following disadvantages: the physical mechanism (creep) of cladding damage accumulation and real stress history are not taken into account; uncertainty of the cladding durability estimate forces us into unreasonably assumption $K_{norm} = 10$; there is no public data on N_i^{max} and t^{max} for all possible loading conditions.

Now the WWER-1000 fuel cladding safety and durability requirements have not been clearly defined (Semishkin et al., 2009). As strength of fuel elements under multiple cyclic power changes is of great importance when performing validation of a NR project, a tendency to in-depth studies of this problem is observed. The well-known cladding fatigue failure criterion based on the relationship between the maximum circumferential stress amplitude σ_θ^{max} and the allowable number of power-cycles NC^{max} is most popular at present (Kim et al., 2007). Nevertheless, in case of satisfactory fit between the experimental and calculated data describing the maximum number of cycles prior to the cladding failure, still there stays the problem of disagreement between experimental conditions and real operating environment (e.g. fluence; neutron spectrum; rod internal pressure; coolant temperature conditions; cladding water-side corrosion rate; radiation growth; cladding defect distribution; algorithm of fuel pick-and-place operations; reactor control system regulating unit movement amplitude and end effects; loading cycle parameters, etc.). In connection with this problem, to ensure a satisfactory accuracy of the cladding state estimation at variable loading conditions, it is necessary to develop physically based FE cladding durability analysis methods, on the basis of verified codes available through an international data bank.

As is known, when repair time is not considered, reactor capacity factor CF is obtained as

$$\mathrm{CF} = \frac{\sum_{i=1}^{n} (\Delta\tau_i \cdot P_i)}{T \cdot P}, \tag{3}$$

where $\Delta\tau_i$ – NR operating time at the capacity of P_i; T – total NR operating time; P – maximum NR capacity (100 %).

Using (3), the number of daily cycles $N_{e,0}$ that the cladding can withstand prior to the beginning of the rapid creep stage, expressed in effective days, is defined from the following equation:

$$N_{e,0} = N_0 \cdot \mathrm{CF},$$

where N_0 – the number of calendar daily cycles prior to the beginning of the rapid creep stage.

It should be stressed that CF is a summary number taking into account only the real NR loading history. For instance, the following NR loading modes can be considered:

1. Stationary operation at 100 % NR capacity level, CF = 1.
2. The NR works at 100 % capacity level within 5 days, then the reactor is transferred to 50 % capacity level within 1 hour. Further the NR works at the capacity level of 50 % within 46 hours, then comes back to 100 % capacity level within 1 hour. Such NR operating mode will be designated as the (5 d – 100 %, 46 h – 50 %) weekly load cycle, CF = 0.860.
3. The NR works at 100 % capacity level within 16 hours, then the reactor is transferred to 75 % capacity level within 1 hour. Further the NR works at 75 % capacity level within 6 hours, then comes back to 100 % capacity level within 1 hour. Such NR operating mode will be designated as the (16 h – 100 %, 6 h – 75 %) daily load cycle, CF = 0.927.
4. The NR works at 100 % capacity level within 16 hours, then the reactor is transferred to 75 % capacity level within 1 hour. Further the NR works at 75 % capacity level within 6 hours, then comes back to 100 % capacity level within 1 hour. But the NR capacity decreases to 50 % level within last hour of every fifth day of a week. Further the reactor works during 47 hours at 50 % capacity level and, at last, within last hour of every seventh day the NR capacity rises to the level of 100 %. Such NR operating mode will be designated as the (5 d – 100 % + 75 %, 2 d – 50 %) combined load cycle, CF = 0.805.

2. The CET-method of fuel cladding durability estimation at variable loading

The new cladding durability analysis method, which is based on the creep energy theory (CET) and permits us to integrate all known cladding strength criteria within a single calculation model, is fit for any normal WWER/PWR operating conditions (Pelykh et al., 2008). The CET-model of cladding behaviour makes it possible to work out cladding rupture life control methods for a power-cycling WWER-1000 nuclear unit. As the WWER-1000 Khmelnitskiy nuclear power plant (KhNPP) is a base station for study of varying duty cycles in the National Nuclear Energy Generating Company ENERGOATOM (Ukraine), the second power unit of KhNPP will be considered.

According to CET, to estimate FE cladding running time under multiple cyclic NR power changes, it is enough to calculate the energy A_0 accumulated during the creep process, by the moment of cladding failure and spent for cladding material destruction (Sosnin and Gorev, 1986). The energy spent for FE cladding material destruction is called as specific dispersion energy (SDE) $A(\tau)$. The proposed method of FE cladding running time analysis is based on the following assumptions of CET: creep and destruction processes proceed in common and influence against each other; at any moment τ creep process intensity is estimated by specific dispersion power (SDP) $W(\tau)$, while intensity of failure is estimated by $A(\tau)$ accumulated during the creep process by the moment τ

$$A(\tau) = \int_0^{\tau} W(\tau) \cdot d\tau \,, \tag{4}$$

where SDP standing in (4) is defined by the following equation (Nemirovsky, 2001):

$$W(\tau) = \sigma_e \cdot \dot{p}_e \,, \tag{5}$$

where σ_e is equivalent stress, Pa; $\dot{p}_e$ is rate of equivalent creep strain, s^{-1}.

Equivalent stress σ_e is expressed as

$$\sigma_e = \sqrt{\frac{1}{2}\left[\left(\sigma_\theta - \sigma_z\right)^2 + \sigma_\theta^{\,2} + \sigma_z^{\,2} \right]} \,, \tag{6}$$

where σ_θ and σ_z are circumferential stress and axial stress, respectively.

The cladding material failure parameter $\omega(\tau)$ is entered into the analysis:

$$\omega(\tau) = A(\tau) \,/\, A_0 \,, \tag{7}$$

where A_0 is SDE at the moment of cladding material failure beginning, known for the given material either from experiment, or from calculation, J/m^3 (Sosnin and Gorev, 1986); $\omega = 0$ – for intact material, $\omega = 1$ – for damaged material.

The proposed method enables us to carry out quantitative assessment of accumulated $\omega(\tau)$ for different NR loading modes, taking into account a real NR load history (Pelykh et al., 2008). The condition of cladding material failure is derived from (4), (5) and (7):

$$\omega(\tau) = \int_0^{\tau} \frac{\sigma_e \cdot \dot{p}_e}{A_0} \cdot d\tau = 1 \tag{8}$$

The CET-method of light-water reactor (LWR) FE cladding operation life estimation can be considered as advancement of the method developed for FE cladding failure moment estimation at loss-of-coolant severe accidents (LOCA) (Semishkin, 2007). The equations of creep and cladding damage accumulation for zirconium alloys are given in (Semishkin, 2007) as

$$\dot{p}_e = f(k_i, T, \sigma_e, \omega(\tau)), \tag{9}$$

$$\dot{\omega}(\tau) = \frac{\sigma_e \cdot \dot{p}_e}{A_0}, \tag{10}$$

where k_i are material parameters defined from experiments with micromodels cut out along the FE cladding orthotropy directions; T is absolute temperature, K.

According to (Semishkin, 2007), for LOCA-accidents only, using the failure condition $\omega(\tau) = 1$, the SDE value A_0 accumulated by the moment of cladding failure and supposed to be temperature-dependent only, is determined from the equations (9)-(10). At the same time, the assumption that the value of A_0 at high-temperature creep and cladding failure analysis is loading history independent, is accepted for LOCA-accidents as an experimentally proved matter.

In contrast to the experimental technique for determining A_0 developed in (Semishkin, 2007), the calculation method proposed in (Pelykh et al., 2008) means that A_0 can be found by any of two ways:

1. As the SDE value at the moment τ_0 of cladding stability loss, which is determined by condition $\sigma_e^{\max}(\tau_0) = \sigma_0^{\max}(\tau_0)$, when equivalent stress $\sigma_e^{\max}(\tau)$ becomes equal to yield stress $\sigma_0^{\max}(\tau)$ for the point of the cladding having the maximum temperature (according to the calculation model, a fuel rod is divided into axial and radial segments).
2. As the SDE value at the rapid creep start moment for the cladding point having the maximum temperature. This way is the most conservative approach, and it is not obvious that such level of conservatism is really necessary when estimating A_0.

The equivalent stress σ_e and the rate of equivalent creep strain $\dot{p}_e$ are calculated by the LWR fuel analysis code FEMAXI (Suzuki, 2000). Though cladding creep test data must have been used to develop and validate the constitutive models used in the finite element code FEMAXI to calculate the equivalent creep strains under cyclic loading, difficulty of this problem is explained by the fact that cladding material creep modeling under the conditions corresponding to real operational variable load modes is inconvenient or impossible as such tests can last for years. As a rule, the real FE operational conditions can be simulated in such tests very approximately only, not taking into account all the variety of possible exploitation situations (Semishkin, 2007).

The code FEMAXI analyzes changes in the thermal, mechanical and chemical state of a single fuel rod and interaction of its components in a given NR power history and coolant conditions. The code analytical scope covers normal operation conditions and transient conditions such as load-following and rapid power increase in a high-burnup region of over 50 MWd/kg-U.

In the creep model used in the code, irradiation creep effects are taken into consideration and rate of equivalent cladding creep strain $\dot{p}_e$ is expressed with a function of cladding stress, temperature and fast neutron flux (MATPRO-09, 1976):

$$\dot{p}_e = K \cdot \Phi(\sigma_\theta + B \cdot \exp(C \cdot \sigma_\theta))\exp(-Q / R \cdot T)\tau^{-0.5}, \quad (11)$$

where $\dot{p}_e$ is biaxial creep strain rate, s^{-1}; K, B, C are known constants characterizing the cladding material properties; Φ is fast neutron flux (E > 1.0 MeV), $1/m^2 \cdot s$; σ_θ is circumferential stress, Pa; $Q = 10^4$ J/mol; R = 1.987 cal/mol·K; T is cladding temperature, K; τ is time, s.

According to (11), creep strain increases as fast neutron flux, cladding temperature, stress and irradiation time increase.

For creep under uniaxial stress, cladding and pellet creep equations can be represented as (Suzuki, 2010):

$$\dot{p}_e = f\left(\sigma_e, \varepsilon^H, T, \Phi, \dot{F}\right), \quad (12)$$

where $\dot{p}_e$ is equivalent creep strain rate, c^{-1}; σ_e is equivalent stress, Pa; ε^H is creep hardening parameter; $\dot{F}$ is fission rate, $1/m^3 \cdot s$.

When equation (12) is generalized for a multi-axial stress state, the creep strain rate vector $\{ \dot{p} \}$ is expressed as a vector function $\{ \beta \}$ of stress and creep hardening parameter:

$$\{ \dot{p} \} = \left\{ \beta\left(\{\sigma\}, \varepsilon^H\right) \right\}, \quad (13)$$

where T, Φ and $\dot{F}$ are omitted because they can be dealt with as known parameters.

When a calculation at time t_n is finished and a calculation in the next time increment Δt_{n+1} is being performed, the creep strain increment vector is represented as

$$\left\{\Delta p_{n+1}\right\} = \Delta t_{n+1}\left\{\dot{p}_{n+\theta}\right\} = \left\{ \beta\{\sigma_{n+\theta}\}, \varepsilon_{n+\theta}^H \right\}, \quad (14)$$

where $\{\sigma_{n+\theta}\} = (1-\theta)\cdot\{\sigma_n\} + \theta\cdot\{\sigma_{n+1}\}$; $\varepsilon_{n+\theta}^H = (1-\theta)\cdot\varepsilon_n^H + \theta\cdot\varepsilon_{n+1}^H$; $0 \le \theta \le 1$.

In order to stress importance of numerical solution stability, θ = 1 is set.

Then, when the $(i+1)$-th iteration by the Newton-Raphson method is being performed after completion of the (i)-th iteration, the creep strain rate vector is expressed (Suzuki, 2010).

As shown in Fig. 1, the analysis model includes a 2-dimensional axisymmetrical system in which the entire length of a fuel rod is divided into AS, and each AS is further divided into concentric ring elements in the radial direction.

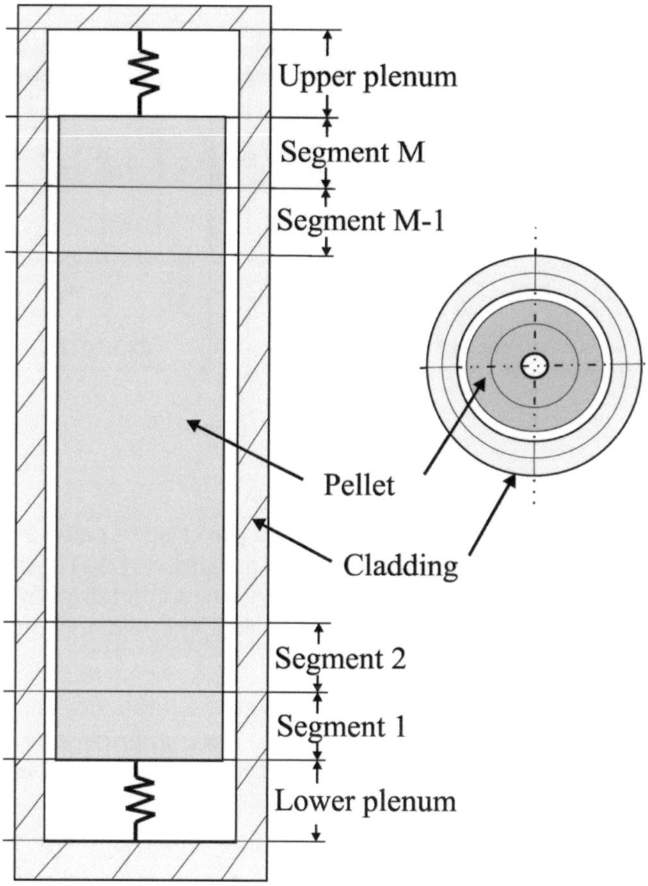

Fig. 1. Analysis model.

In this system, stress/strain analysis is performed using FEM with quadrangular elements having four degrees of freedom, as is shown in Fig. 2.

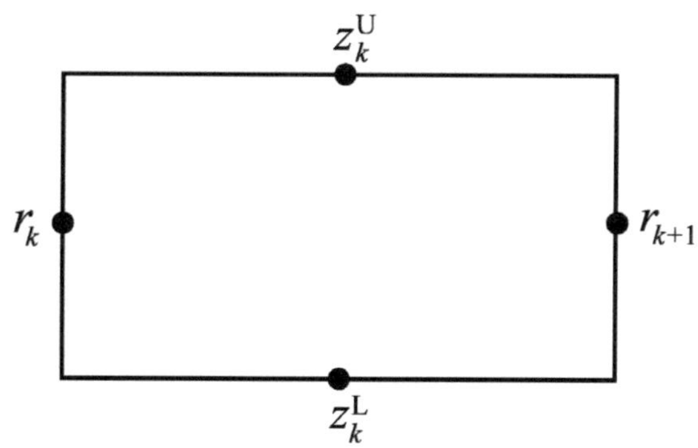

Fig. 2. Quadrangular model element with four degrees of freedom.

Fig. 3 shows relationship between mesh division and degree of freedom for each node in an AS.

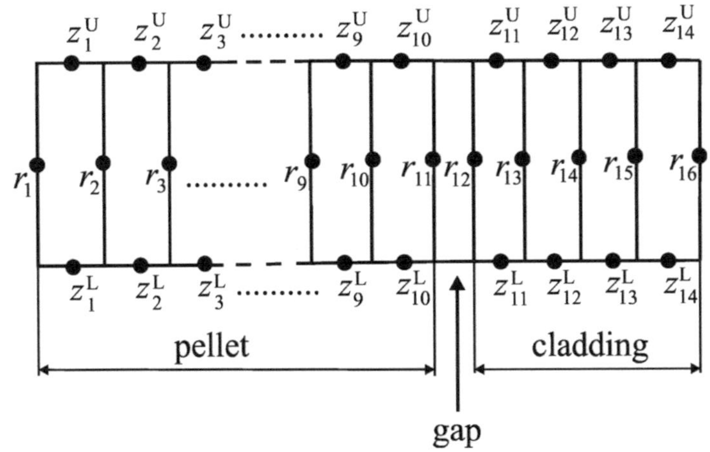

Fig. 3. Mesh division of FEM (for one AS).

In Fig. 3, the number of mesh divisions in the radial direction of pellet and cladding is fixed at 10 and 4, respectively. The inner two meshes of a cladding (11, 12) are metal phase, and the outer two meshes (13, 14) are oxide layer (ZrO_2). The model used in the code takes into account that the oxide layer mesh and metal mesh are re-meshed and change their thickness with the progress of corrosion.

The fuel temperature calculation was carried out with the difference between the numerical solution and analytical solution not exceeding 0.1 %. The numerical error arising in the form of residue from iterative creep calculation on each time step, was not estimated as in most cases this error is exceeded by other uncertainties, first of all by thermal conductivity model error (Suzuki, 2010).

Denoting the number of daily load NR power cycles as N, using the CET-model, the dependence A (N), as well as the borders of characteristic creep stages (unsteady, steady and rapid creep) for zircaloy cladding were obtained for the WWER daily load cycle (16 h – 100 %; 6 h – $k{\cdot}100$ %), where k = 1; 0.75; 0.5; 0.25. Hence the number of daily cycles $N_{e,0}$ that the cladding can withstand prior to the rapid creep stage beginning could be calculated. The conclusion was made that the calculated value of A_0 is not constant for a given material and depends on the operating mode of multiple cyclic power changes (Pelykh, 2008).

It was found, that the calculated equivalent creep strain p_e for zircaloy cladding, for all daily load modes, gradually increases and a hysteresis decrease of p_e can be seen at the last creep stage beginning. Then, after the hysteresis decrease, p_e starts to grow fast and achieves considerable values from cladding reliability point of view. At the rapid creep beginning, the equivalent stress σ_e decrease trend changes into the σ_e increase trend, at the same time p_e decreases a little, that is there is a "hysteresis loop", when the p_e increase has got a phase delay in comparison with the σ_e increase. It should be noted, that the cause of the p_e hysteresis decrease effect must be additionally studied as p_e is expected to continuously increase unless the cladding is subjected to significant compressive creep stresses during the cycle and that this had been properly included in the creep material model.

The following new NR power daily maneuver algorithm was proposed in (Maksimov et al., 2009). It is considered that a nuclear unit is working at the nominal power level (100 %), unwanted xenon oscillations are suppressed by the NR control group movement. At first, boric acid solution is injected so that the NR capacity decreases to 90 %, while the NR inlet coolant temperature is maintained constant at the expense of the Main Steam Line (MSL) pressure rise. To guarantee suppression of xenon oscillations, the optimal instantaneous Axial Offset (AO) is maintained due to the NR control group movement. Further the NR power is lowered at the expense of poisoning. The NR capacity will reach the 80% level in 2–3 h and the capacity will be stabilized by intake of the "pure distillate". The NR capacity will be partly restored at the expense of depoisoning starting after the maximal iodine poisoning. To restore the nominal NR power level, the "pure distillate" is injected into the NR circuit and the MSL pressure is lowered, while the NR coolant inlet temperature is maintained constant. The optimal instantaneous AO to be maintained, the control rod group is extracted from the active core. The automatic controller maintains the capacity and xenon oscillations are suppressed by the control group movement after the NR has reached the nominal power level.

The proposed algorithm advantages: lowering of switching number; lowering of "pure distillate" and boric acid solution rate; lowering of unbalanced water flow; improvement of fuel operation conditions. Also, the proposed NR capacity program meaning the NR inlet coolant temperature stability, while the MSL pressure lies within the limits of 5.8–6.0 MPa and the NR capacity changes within the limits of 100–80 %, has the advantages of the well known capacity program with the first circuit coolant average temperature constancy.

The capacity program with the first circuit coolant average temperature constancy is widely used at Russian nuclear power units with WWER-reactors due to the main advantage of this program consisting of the possibility to change the unit power level when the reactor control rods stay at almost constant position. At the same time, as the MSL pressure lies within the procedural limits, the proposed algorithm is free of the constant first circuit temperature program main disadvantage consisting of the wide range of MSL pressure change. Two WWER-1000 daily maneuver algorithms were compared in the interests of efficiency (Maksimov et al., 2009):

1. The algorithm tested at KhNPP ("Tested") on April 18, 2006: power lowering to 80 % within 1 h - operation at the 80 % power level within 7 h - power rising to 100 % within 2 h.
2. The proposed algorithm ("Proposed"): power lowering to 90 % by boric acid solution injection within 0.5 h – further power lowering to 80 % at the expense of NR poisoning within 2.5 h – operation at the 80 % power level within 4 h – power rising to 100 % within 2 h.

Comparison of the above mentioned daily maneuver algorithms was done with the help of the "Reactor Simulator" (RS) code (Philimonov and Mamichev, 1998). To determine axial power irregularity, AO is calculated as

$$\mathrm{AO} = \frac{N_u - N_l}{N},$$

where N_u, N_l, N are the core upper half power, lower half power and whole power, respectively.

The instantaneous AO corresponds to the current xenon distribution, while the equilibrium AO corresponds to the equilibrium xenon distribution. Having used the proposed method of cladding failure estimation for zircaloy cladding and WWER-type NR, dependence of the irreversible creep deformation accumulated energy from the number of daily load cycles is calculated for the "Tested" and "Proposed" algorithms, and efficiency comparison is fulfilled - see Table 3.

Algorithm	Easy of NR power field stabilization		CF	The number of daily cycles $N_{e,0}$ that cladding can withstand prior to the rapid creep beginning, eff. days
	Divergency of instantaneous and equilibrium AO diagrams	Amplitude of AO change during the maneuver		
"Tested"	considerable divergency	considerable amplitude	0.929	705
"Proposed"	slight divergency	amplitude is more than 10 times less	0.942	706

Table 3. Efficiency comparison for two daily maneuvering algorithms.

For the "Proposed" algorithm, taking into account the lower switching number necessary to enter "pure distillate" and boric acid solution during the maneuver, slight divergency of the instantaneous and equilibrium AO diagrams, the lower amplitude of AO change during the maneuver, the higher turbo-generator efficiency corresponding to the higher CF, as well as in consideration of practically equal cladding operation times for both the algorithms, it was concluded that the "Proposed" algorithm was preferable (Maksimov et al., 2009).

Using this approach, the complex criterion of power maneuvering algorithm efficiency for WWER-1000 operating in the mode of variable loading, taking into account FE cladding damage level, active core power stability, NR capacity factor, as well as control system reliability, has been worked out (Pelykh et al., 2009). Also the Compromise-combined WWER-1000 power control method capable of maximum variable loading operation efficiency, has been proposed and grounded (Maksimov and Pelykh, 2010).

3. Factors influencing durability of WWER FE cladding under normal conditions

Using the CET cladding durability estimation method, an analysis of the cladding (stress relieved zircaloy) durability estimation sensitivity to the WWER-1000 main regime and design initial data uncertainty, under variable loading conditions, has been done. The WWER-1000 main regime and design parameters have been devided into two groups: the parameters that influence the cladding failure conditions slightly and the parameters that determine the cladding failure conditions. The second group includes such initial parameters that any one of them gives a change of τ_0 estimation near 2 % (or greater) if the initial parameter has been specified at the value assignment interval of 3 %. This group consists of outer cladding diameter, pellet diameter, pellet hole diameter, cladding thickness, pellet effective density, maximum FE linear heat rate, coolant inlet temperature, coolant inlet pressure, coolant velocity, initial He pressure, FE grid spacing, etc. (Maksimov and Pelykh, 2009). For example, dependence of cladding SDE on the number of effective days *N*, for pellet centre hole diameter d_{hole} = 0.140 cm, 0.112 cm and 0.168 cm, is shown in Fig. 4.

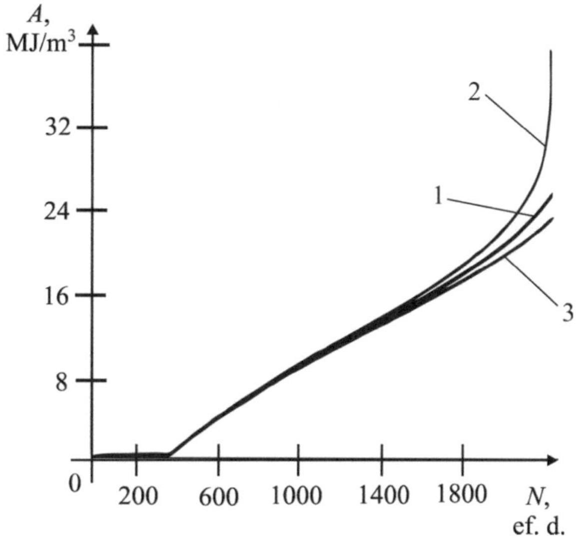

Fig. 4. Dependence of SDE on N for d_{hole} : 0.140 cm (1); 0.112 cm (2); 0.168 cm (3).

Dependence of cladding equivalent stress $\sigma_e^{\max}(\tau)$ and yield stress $\sigma_0^{\max}(\tau)$, for the cladding point having the maximum temperature, on the number of effective days *N*, for d_{hole} = 0.112 cm and 0.168 cm, is shown in Fig. 5.

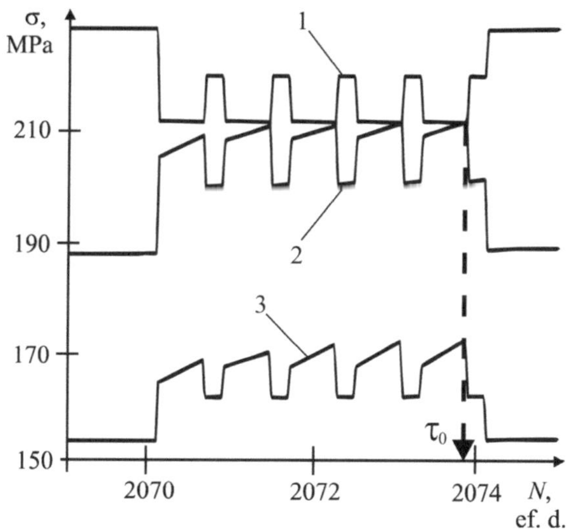

Fig. 5. Dependence of cladding yield stress (1) and equivalent stress (2; 3) on *N* for d_{hole}: 0.112 cm (2); 0.168 cm (3). Determination of τ_0 for d_{hole} = 0.112 cm.

Using the value of τ_0 and the calculated dependence of SDE on *N*, the value of A_0 is found – see Fig. 6.

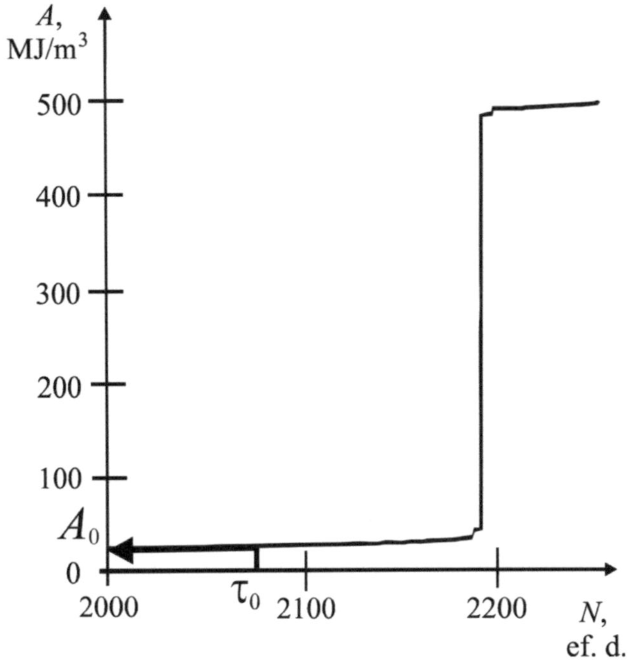

Fig. 6. Calculation of A_0.

For the combined variable load cycle, dependence of cladding SDE on the number of effective days *N* for a medium-loading FE of UTVS , TVS-A and TVS-W, is shown in Fig. 7.

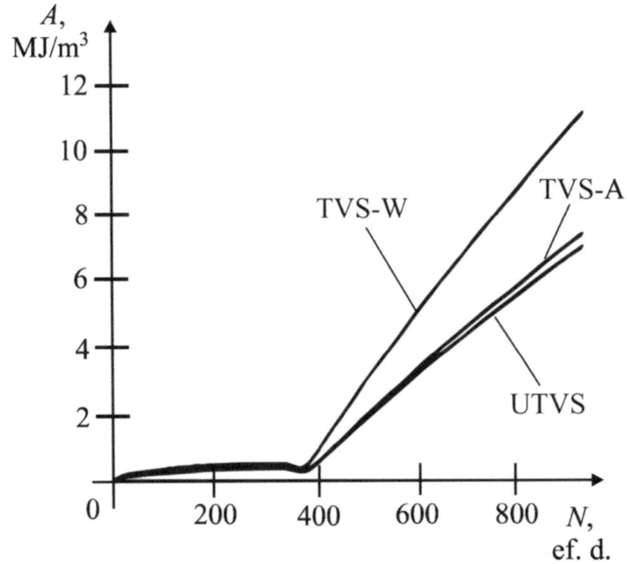

Fig. 7. Dependence of SDE on *N* for UTVS, TVS-A and TVS-W.

For the combined cycle, the maximum SDE value was obtained for a medium-loading FE of the FA produced by WESTINGHOUSE, which has no pellet centre hole (see Table 1). The same result was obtained for the stationary regime of WWER-1000 (Maksimov and Pelykh, 2010).

It has been found that cladding running time, expressed in cycles, for the WWER-1000 combined load cycle decreases from 1925 to 1351 cycles, when FE maximum LHR $q_{l,\max}$ increases from 248 W/cm to 298 W/cm (Maksimov and Pelykh, 2010). Having done estimation of cladding material failure parameter ω after 1576 ef. days, it was found that the WWER-1000 combined load cycle has an advantage in comparison with stationary operation at 100 % power level when $q_{l,\max} \leq 273$ W/cm – see Table 4.

According to FEM, a FE length is divided into n equal length AS. In the first publications devoted to the CET-method it was supposed that the central AS is most strained and shortest-lived. However, this assumption does not consider that segments differ in LHR jump value. In addition, it was assumed that a FA stays in the same place over the whole fuel operating period (Maksimov and Pelykh, 2009).

Parameter	FE maximum LHR, W/cm				
	248	258	263	273	298
	Average fast neutron flux density, $cm^{-2}\cdot s^{-1}$				
	$1\cdot 10^{14}$	$1.04\cdot 10^{14}$	$1.06\cdot 10^{14}$	$1.1\cdot 10^{14}$	$1.2\cdot 10^{14}$
	Stationary loading				
τ_0, ef. d.	2211	2078	2016	1904	1631
A_0, MJ/m³	33.37	35.66	36.87	39.74	47.64
ω, %	60	65	68	74	94
Combined variable loading					
τ_0, ef. d.	2246	2102	2032	1903	1576
A_0, MJ/m³	27.36	29.14	30.05	32.10	37.69
ω, %	57	64	67	74	100

Table 4. Cladding damage parameter for stationary loading and the combined variable loading of WWER-1000.

At last, influence of cladding corrosion rate on cladding durability at variable loading was not taken into account. Thus it is necessary to estimate influence of varying duty on all AS, to take account of a real FA transposition algorithm as well as to consider influence of cladding corrosion rate on its durability.

4. Method to determine the most strained cladding axial segment

The amplitude of LHR jumps in AS occurring when the NR thermal power capacity N increases from 80% to 100% level, was estimated by the instrumentality of the RS code, which is a verified tool of the WWER-1000 calculation modelling (Philimonov and Mamichev, 1998). Using the RS code, the WWER-1000 core neutron-physical calculation numerical algorithms are based on consideration of simultaneous two-group diffusion equations, which are solved for a three-dimensional object (the reactor core) composed of a limited number of meshes.

The amplitude of LHR jumps was calculated for the following daily power maneuvering method: lowering of N from N_1=100% to N_2=90% by injection of boric acid solution within 0.5 h – further lowering of N to N_3=80% due to reactor poisoning within 2.5 h – operation at N_3=80% within 4 h – rising of N to the nominal capacity level N_1=100% within 2 h (Maksimov et al., 2009). According to this maneuvering method, the inlet coolant temperature is kept constant while the NR capacity changes in the range N=100–80%, and the initial steam pressure of the secondary coolant circuit changes within the standard range of 58–60 bar. It was supposed that the only group of regulating units being used at NR power maneuvering was the tenth one, while the control rods of all the other groups of regulating units were completely removed from the active core. The next assumption was that the Advanced power control algorithm (A-algorithm) was used. The WWER-1000 core contains ten groups of regulating units in case of the A-algorithm – see Fig. 8.

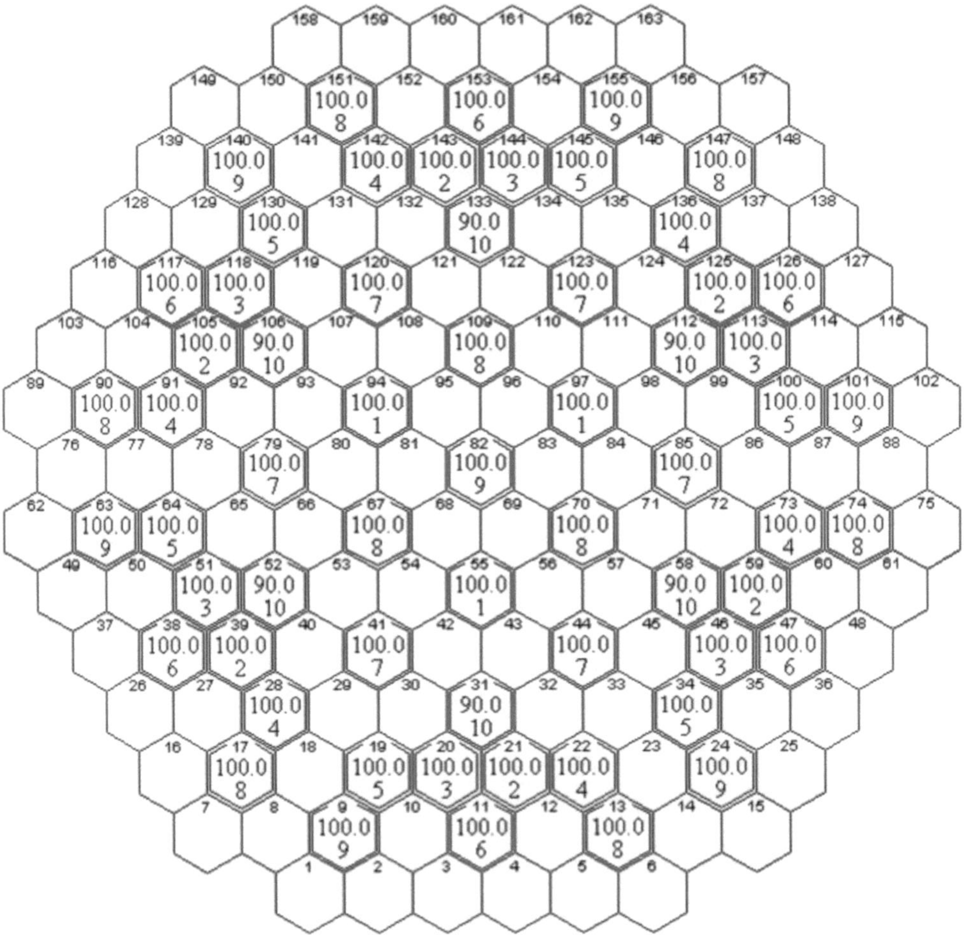

Fig. 8. Disposition of the WWER-1000 regulating units in case of the A-algorithm: (upper figure) the FA number; (middle figure) the lowest control rod axial coordinate (at 100% NR power level) measured from the core bottom, %; (lower figure) the regulating unit group number.

The lowest control rod axial coordinates for N_1=100% and N_3=80% were designated H_1=90% and H_3=84%, respectively. That is when N changes from N_1=100% to N_3=80%, the lowest control rod axial coordinate measured from the core bottom changes from H_1=90% to H_3=84%.

It has been found using the RS code that the WWER-1000 fuel assemblies can be classified into three groups by the FA power growth amplitude occurring when the NR capacity increases from 80% to 100% level - see Table 5 (Pelykh et al., 2010).

FA group	The number of fuel assemblies	FA power growth, %	FA numbers (according to the core cartogram)
1	6	28	31, 52, 58, 106, 112, 133
2	37	26	20, 42, 43, 46, 51, 53…57, 66…71, 80…84, 93…98, 107…111, 113, 118, 121, 122, 144
3	120	≤ 25	all other fuel assemblies

Table 5. Three groups of the WWER-1000 fuel assemblies.

When the eighth, ninth and tenth regulating groups are simultaneously used, the central FA (No. 82) as well as fresh fuel assemblies are regulated by control rods. But when using the A-algorithm, the tenth regulating group is used only. In this case, such a four-year FA transposition algorithm can be considered as an example: a FA stays in the 55-th FA (FE maximum LHR $q_l^{\max}$ = 236.8 W/cm, FA group 2) position for the first year - then the FA stays in the 31-st FA ($q_l^{\max}$ = 250.3 W/cm, group 1) position for the second year - further the FA stays in the 69-th FA ($q_l^{\max}$ = 171.9 W/cm, group 2) position for the third year - at last, the FA stays in the central 82-d FA ($q_l^{\max}$ = 119.6 W/cm, group 2) position for the fourth year (the algorithm 55–31–69–82).

The average LHR for i-segment and j-FA is denoted as $< q_{l,i,j} >$. For all segments (n = 8) of the 55-th, 31-st, 69-th and 82-nd fuel assemblies, the values of $< q_{l,i,j} >$ have been calculated at power levels of N_3=80% and N_1=100% using the RS code. The $< q_{l,i,j} >(100\%)/< q_{l,i,j} >(80\%)$ ratio values are listed in Table 6.

AS	FA number			
	55	31	69	82
8	1.341	1.517	1.328	1.340
7	1.308	1.426	1.297	1.309
6	1.250	1.241	1.263	1.268
5	1.229	1.213	1.238	1.250
4	1.224	1.217	1.232	1.242
3	1.241	1.229	1.243	1.259
2	1.255	1.251	1.271	1.270
1	1.278	1.275	1.288	1.302

Table 6. The $< q_{l,i,j} >(100\%)/< q_{l,i,j} >(80\%)$ ratio values for fuel assemblies 55, 31, 69, 82.

Though the Nb-containing zirconium alloy E-110 (Zr + 1% Nb) has been used for many years in FE of WWER-1000, there is no public data on E-110 cladding corrosion and creep rates for all possible loading conditions of WWER-1000. In order to apply the cladding durability estimation method based on the corrosion and creep models developed for Zircaloy-4 to another cladding alloy used in WWER-1000, it is enough to prove that using these models under the WWER-1000 active core conditions ensures conservatism of the E-110 cladding durability estimation. Nevertheless, the main results of the present analysis will not be changed by including models developed for another cladding alloy.

The modified cladding failure criterion at NR variable loading is given as (Pelykh and Maksimov, 2011):

$$\omega(\tau) = A(\tau) / A_0 = 1; \; A(\tau) = \int_0^{\tau} \sigma_e^{\max}(\tau)\, \dot{p}_e^{\max}(\tau)\, d\tau; \;\; A_0 \text{ at } \sigma_e^{\max}(\tau_0) = \eta\, \sigma_0^{\max}(\tau_0), \qquad (15)$$

where $\omega(\tau)$ is cladding material failure parameter; τ is time, s; $A(\tau)$ is SDE, J/m³; A_0 is SDE at the moment τ_0 of cladding material failure beginning, when $\sigma_e^{\max}(\tau_0) = \eta\, \sigma_0^{\max}(\tau_0)$; $\sigma_e^{\max}(\tau)$ and $\dot{p}_e^{\max}(\tau)$ are equivalent stress (Pa) and rate of equivalent creep strain (s^{-1}) for the cladding point of an AS having the maximum temperature, respectively; $\sigma_0^{\max}(\tau)$ is yield stress for the cladding point of an AS having the maximum temperature, Pa; η is some factor, $\eta \le 1$.

Assuming the 55–31–69–82 four-year FA transposition algorithm and $\eta = 0.6$, the $\omega(\tau)$ values have been calculated by Eq. (15) using the following procedure: calculating $\sigma_e^{\max}(\tau)$, $\dot{p}_e^{\max}(\tau)$ and $\sigma_0^{\max}(\tau)$ by the instrumentality of FEMAXI-V code (Suzuki, 2000); calculating $A(\tau)$; determining the moment τ_0 according to the condition $\sigma_e^{\max}(\tau_0) = \eta\, \sigma_0^{\max}(\tau_0)$; determining $A_0 \equiv A(\tau_0)$; calculating $\omega(\tau)$ - see Table 7 (Pelykh and Maksimov, 2011).

τ, days	AS			
	4	5	6	7
360	0.063	0.151	0.190	0.175
720	0.598	0.645	0.647	0.547
1080	0.733	0.783	0.790	0.707
1440	0.788	0.838	0.848	0.779

Table 7. Cladding failure parameters $\omega(\tau)$ for the axial segments 4–7.

For the other axial segments No. 1–3 and 8, on condition that a FA was transposed in concordance with the 55–31–69–82 four-year algorithm, the $\omega(\tau)$ value was less than 1.0, i.e. there was no cladding collapse up to τ = 2495 days. For τ > 2495 days calculations were not carried out. For all the axial segments, on condition that a FA was transposed in concordance with the 55–31–69–82 four-year algorithm, it has been found that there was no cladding collapse up to τ = 2495 days with $\omega(\tau)$ = 1. At the same time, for all the axial segments, on condition that a FA stayed in the 55-th FA position for all fuel operation period, as well as on condition that a FA stayed in the 55-th FA position for the first year, then it stayed in the 31-st FA position for the remaining fuel operation period, the $\omega(\tau)$ value reached 1.0 and the cladding collapse was predicted at $\tau < 2495$ days with $\eta = 1$.

The prediction shown in Table 7 that the largest value of $\omega(\tau)$ exists at the fifth (central) axial segment and above it the value drops in the sixth segment situated between the axial coordinates z = 2.19 and 2.63 m reflects the fact that the most considerable LHR jumps take place at the core upper region (see Table 6). Thus, taking account of the 55–31–69–82 four-year FA transposition algorithm as well as considering the regulating unit disposition, on condition that the FE length is divided into eight equal-length axial segments, the sixth (counting from the core bottom) AS cladding durability limits the WWER-1000 operation time at daily cycle power maneuvering.

Growth of the water-side oxide layer of cladding can cause overshoot of permissible limits for the layer outer surface temperature prior to the cladding collapse moment. The corrosion models of EPRI (MATPRO-09, 1976) and MATPRO-A (SCDAP/RELAP5/MOD2, 1990) have been used for zircaloy cladding corrosion rate estimation. According to the EPRI model, the cladding corrosion rate for a bubble flow is estimated as

$$dS / dt = (\mathrm{A} / S^2) \exp(-\mathrm{Q}_1 / \mathrm{R}\, T_b)(1 + \mathrm{COR}), \tag{16}$$

where dS / dt is the oxide growth rate, μm/day; A = 6.3×10^9 μm^3/day; S is the oxide layer thickness, μm; Q_1=32289 cal/mol; R=1.987 cal/(mol K); T_b is the temperature at the oxide layer-metal phase boundary, K; COR is an adjusting factor which is added in the FEMAXI code (Suzuki, 2010).

According to the MATPRO-A model, the oxide layer thickness for a nucleate boiling flow is estimated as

$$S = (4.976 \times 10^{-3} \mathrm{A}\, t \exp(-15660 / T_b) + S_0^{\,3})^{1/3} (1 + \mathrm{COR}), \tag{17}$$

where S is the oxide layer thickness, m; A = 1.5 (PWR); t is time, days; T_b is the temperature at the oxide layer-metal phase boundary, K; S_0 is the initial oxide layer thickness, m.

The cladding failure parameter values listed in Table 7 have been obtained using the MATPRO-A corrosion model at COR = 1. If COR is the same in both the models, the MATPRO model estimation of cladding corrosion rate is more conservative than the EPRI-model estimation, under the WWER-1000 conditions. Regardless of the model we use, the factor COR must be determined so that the calculated oxide layer thickness fits to experimental data. The oxide layer thickness calculation has been carried out for the described method of daily power maneuvering, assuming that a FA was transposed in concordance with the 55–31–69–82 four-year algorithm. The calculations assumed that, the Piling-Bedworth ratio was 1.56, the initial oxide layer thickness was 0.1 μm, the maximum oxide layer thickness was restricted by 100 μm, the radial portion of cladding corrosion volume expansion ratio was 80%. It has been found that the calculated cladding oxide layer thickness, for the WWER-1000 conditions and burnup $Bu = 52.5$ MW day / kg, conforms to the generalized experimental data obtained for PWR in-pile conditions (Bull, 2005), when using the EPRI model at COR = - 0.431 – see Fig. 9.

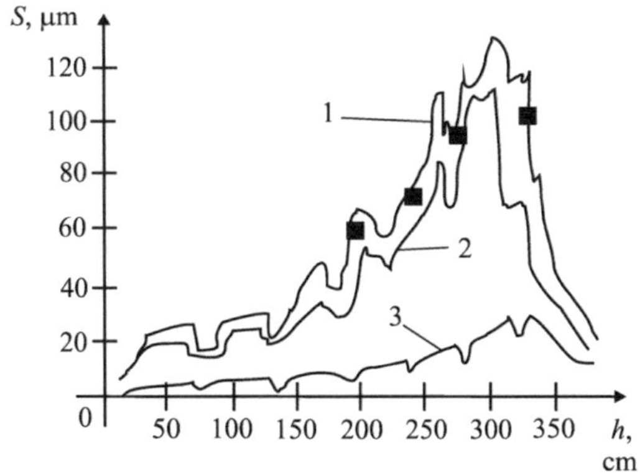

Fig. 9. Cladding oxide layer thickness S subject to height h: (■) calculated using the EPRI model at COR = - 0.431; In accordance with (Bull, 2005): (1) zircaloy-4; (2) improved zircaloy-4; (3) ZIRLO.

The EPRI model at COR = - 0.431 also gives the calculated cladding oxide layer thickness values which were in compliance with the generalized experimental data for zircaloy-4 (Kesterson and Yueh, 2006). For the segments 5–8, assuming that a FA was transposed in concordance with the 55–31–69–82 four-year algorithm, the maximum oxide layer outer surface temperature $T^{max}_{ox,out}$ during the four-year fuel life-time has been calculated (EPRI, COR = - 0.431) – see Table 8. Also, for the segments 5–8, the calculated oxide layer thickness S and oxide layer outer surface temperature $T_{ox,out}$ subject to time τ are listed in Table 8.

The maximum oxide layer outer surface temperature during the four-year fuel life-time does not exceed the permissible limit temperature $T^{lim}_{ox,out}$ =352 °C (Shmelev et al., 2004).

i	$T^{max}_{ox,out}$, °C	S, μm ($T_{ox,out}$, °C)			
		360 days	720 days	1080 days	1440 days
5	345.1	11.3 (342.3)	40.6 (344.8)	58.1 (328.2)	69.8 (316.7)
6	349.6	16.1 (347.6)	49.8 (349.4)	69.3 (332.6)	82.5 (320.1)
7	351.2	18.1 (350.0)	52.7 (351.0)	74.1 (336.1)	88.5 (323.0)
8	348.0	14.2 (347.9)	38.3 (346.9)	58.0 (335.6)	71.2 (323.3)

Table 8. The maximum oxide layer outer surface temperature.

The same result has been obtained for the EPRI model at COR = 0; 1; 2 as well as for the MATPRO-A model at COR = - 0.431; 0; 1; 2. Hence the oxide layer outer surface temperature should not be considered as the limiting factor prior to the cladding collapse moment determined in accordance with the criterion (15). Though influence of the outer oxide layer thickness on the inner cladding surface temperature must be studied.

Having calculated the SDE by the instrumentality of FEMAXI (Suzuki, 2010), assuming that a FA was transposed in concordance with the 55–31–69–82 four-year algorithm, it has been found for the sixth axial segment that the number of calendar daily cycles prior to the beginning of the rapid creep stage was essentially different at COR = - 0.431; 0; 1; and 2. As a result, the rapid creep stage is degenerated for both the corrosion models at COR = - 0.431 (Fig. 10).

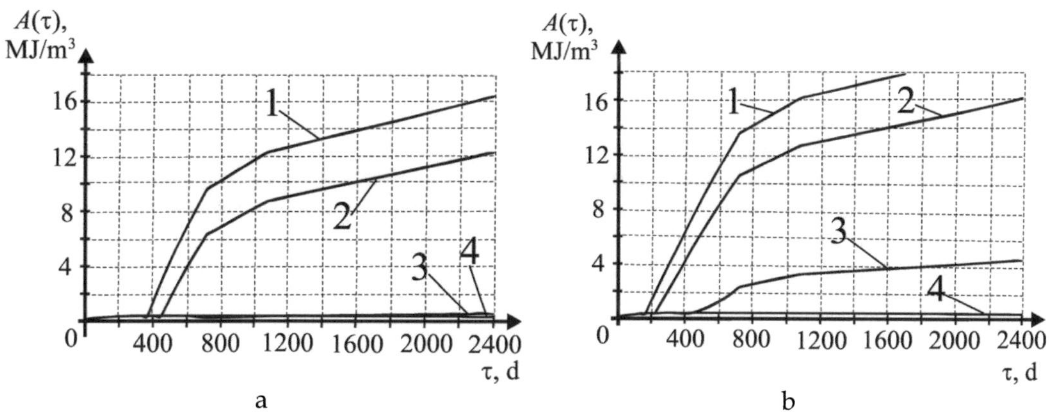

Fig. 10. The SDE as a function of time for the sixth axial segment:(1, 2, 3, 4) at COR = 2, 1, 0, -0.431, respectively; (a) the EPRI model corrosion; (b) the MATPRO-A model corrosion.

Let us introduce a dimensionless parameter I

$$I = \frac{10^{-6}}{{}^\circ\mathrm{C}\cdot \mathrm{day}} \int_0^T T_{clad,in} \cdot dt, \qquad (18)$$

where $T_{clad,in}$ is the cladding inner surface temperature for an axial segment, $^\circ$C; and t is time, days.

Having analysed the described method of daily power maneuvering, the maximum cladding oxide layer outer surface temperature $T_{ox,out}^{\max}$ during the period of 2400 days, as well as I(2400 days) and the 2400 days period averaged cladding inner surface temperature $< T_{clad,in} >$ have been calculated for the sixth segment, using the EPRI corrosion model – see Table 9.

COR	$T_{ox,out}^{\max}$, °C	I(2400 days)	$< T_{clad,in} >$, °C
2	349.2	0.951	396.2
1	349.5	0.947	394.5
0	349.6	0.938	390.7
-0.431	349.6	0.916	381.8

Table 9. Cladding temperatures subject to COR for the sixth segment, the EPRI model corrosion.

This shows that the effect of cladding outer surface corrosion rate (with COR) on the cladding SDE increase rate (see Fig. 10) is induced by the thermal resistance of oxide thickness and the increase in $T_{clad,in}$ (see Table 9).

It should be noticed that the metal wall thickness decrease due to oxidation is considered in the calculation of the SDE, as effect of the cladding waterside corrosion on heat transfer and mechanical behavior of the cladding is taken into account in the FEMAXI code. Since temperature and deformation distributions physically depend on each other, simultaneous equations of thermal conduction and mechanical deformation are solved (Suzuki, 2000).

It is obvious that the cladding temperature at the central point of an AS increases when the outer oxide layer thickness increases. At the same time, according to the creep model (MATPRO-09, 1976) used in the code, the rate of equivalent creep strain $\dot{p}_e^{\max}(\tau)$ for the central point of an axial segment increases when the corresponding cladding temperature increases. Hence the waterside corrosion of cladding is associated with the evaluation of SDE through the creep rate depending on the thickness of metal wall (Pelykh and Maksimov, 2011).

It should be noted, that neutron irradiation has a great influence on the zircaloy corrosion behavior. Power maneuvering will alter neutron flux to give a feedback to the corrosion behavior, either positive or negative. But in this paper, the EPRI model and MATPRO code are used in the corrosion model, where irradiation term is not evidently shown. Although either temperature or reactivity coefficient is introduced in applying the model, it does not fully represent such situation.

For the studied conditions, the maximum cladding hoop stress, plastic strain and oxide layer outer surface temperature do not limit cladding durability according to the known restrictions $\sigma_\theta^{\max} \le 250\ MPa$, $\varepsilon_{\theta,pl}^{\max} \le 0.5\%$ (Novikov et al., 2005) and $T_{ox,out}^{\max} \le 352\ ^{\circ}\mathrm{C}$ (Shmelev et al., 2004), respectively. A similar result has been obtained for the corrosion model MATPRO-A.

Setting COR = 0 and COR = 1 (MATPRO-A), the SDE values for the algorithms 55-31-55-55 and 55-31-69-82 have been calculated. Then the numbers of calendar daily cycles prior to the beginning of rapid creep stage for Zircaloy-4 (Pelykh and Maksimov, 2011) and rapid $\omega(\tau)$ stage for E-110 alloy (Novikov et al., 2005) have been compared under WWER-1000 conditions - see Fig. 11.

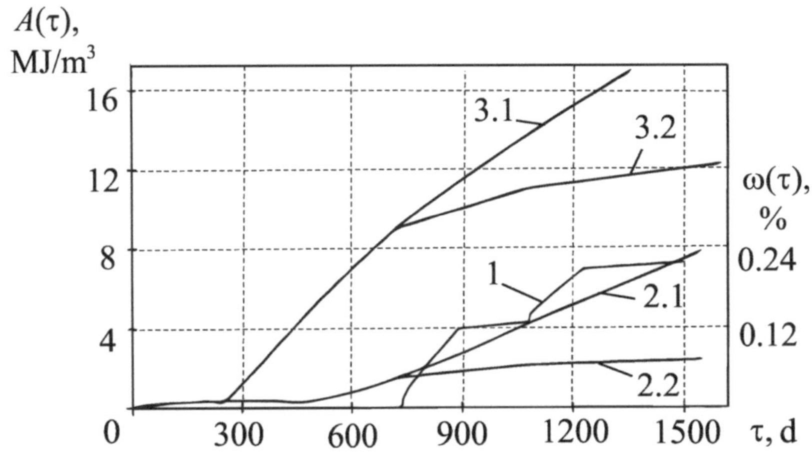

Fig. 11. Cladding damage parameter (E-110) and SDE (Zircaloy-4) as functions of time: (1) $\omega(\tau)$ according to equation (2); (2.1, 2.2) $A(\tau)$ at COR = 0 for the algorithms 55-31-55-55 and 55-31-69-82, respectively; (3.1, 3.2) $A(\tau)$ at COR = 1 for the algorithms 55-31-55-55 and 55-31-69-82, respectively.

It is necessary to notice that line 1 in Fig. 11 was calculated using separate consideration of steady-state operation and varying duty. When using equation (2), the fatigue component has an overwhelming size in comparison with the static one (Novikov et al., 2005).

Use of the MATPRO-A corrosion model under the WWER-1000 core conditions ensures conservatism of the E-110 cladding durability estimation (see Fig. 11). Growth rate of $A(\tau)$ depends significantly on the FA transposition algorithm. The number of daily cycles prior to the beginning of rapid creep stage decreases significantly when COR (cladding outer surface corrosion rate) increases.

Setting the WWER-1000 regime and FA constructional parameters, a calculation study of Zircaloy-4 cladding fatigue factor at variable load frequency $\nu \ll 1$ Hz, under variable loading, was carried out. The investigated WWER-1000 fuel cladding had an outer diameter and thickness of 9.1 mm and 0.69 mm, respectively. The microstructure of Zircaloy-4 was a stress-relieved state. Using the cladding corrosion model EPRI (Suzuki, 2000), AS 6 of a medium-load FE in FA 55 (maximum LHR $q_l^{\max}$ =229.2 W/cm at N=100 %) has been analysed (COR = 1, inlet coolant temperature T_{in}=const=287 °C). The variable loading cycle 100–80–100 % was studied for $\Delta\tau$ =11; 5; 2 h (reactor capacity factor CF=0.9): N lowering from 100 to 80 % for 1 h → exploitation at N = 80 % for $\Delta\tau$ h → N rising to N_{nom}=100 % for 1 h → exploitation at N = 100 % for $\Delta\tau$ h, corresponding to ν =1; 2; 4 cycle/day, respectively($\nu \ll 1$ Hz).

Calculation of the cladding failure beginning moment τ_0 depending on ν showed that if $\nu \ll 1$ Hz and CF=idem, then there was no decrease of τ_0 after ν had increased 4 times, in comparison with the case ν =1 cycle/day, taking into account the estimated error < 0.4 % (η=0.4, AS 6). At the same time, when N=100 % =const (CF=1), the calculated τ_0 decreases significantly - see Table 10.

Hence, the WWER-1000 FE cladding durability estimation based on the CET model corresponds to the experimental results (Kim et al., 2007) in principle.

CF	0.9			1
ν, cycle/day	1	2	4	
τ_0, day	547.6	547.0	549.0	436.6

Table 10. Change of cladding failure time depending on ν and CF.

In the creep model used in the FEMAXI code (Suzuki, 2000), irradiation creep effects are taken into consideration and cladding creep strain rate $\dot{p}_e(\tau)$ is expressed with a function of fast neutron flux, cladding temperature and hoop stress (MATPRO-09, 1976). Thus creep strain increases as fast neutron flux, irradiation time, cladding temperature and stress increase. Fast neutron flux is predominant in cladding creep rate, whereas thermal neutron distribution is a determining factor for reactivity and thermal power (temperature of cladding) in core. It can be seen that both types of neutron flux are important for the cladding life.

One of main tasks at power maneuvering is non-admission of axial power flux xenon waves in the active core. Therefore, for a power-cycling WWER-1000 nuclear unit, it is interesting to consider a cladding rupture life control method on the basis of stabilization of neutron flux axial distribution. The well-known WWER-1000 power control method based on keeping the average coolant temperature constant has such advantages as most favorable conditions for the primary coolant circuit equipment operation, as well as possibility of stable NR power regulation due to the temperature coefficient of reactivity. However, this method has such defect as an essential raise of the secondary circuit steam pressure at power lowering, which requires designing of steam generators able to work at an increased pressure.

Following from this, it is an actual task to develop advanced power maneuvering methods for the ENERGOATOM WWER-1000 units which have such features as neutron field axial distribution stability, favorable operation conditions for the primary circuit equipment, especially for FE claddings, as well as avoidance of a high pressure steam generator design. The described daily power maneuvering method with a constant inlet coolant temperature allows to keep the secondary circuit initial steam pressure within the standard range of 58-60 bar (N=100-80%).

The nonstationary reactor poisoning adds a positive feedback to any neutron flux deviation. Therefore, as influence of the coolant temperature coefficient of reactivity is a fast effect, while poisoning is a slow effect having the same sign as the neutron flux deviation due to this reactivity effect, and strengthening it due to the positive feedback, it can be expected that a correct selection of the coolant temperature regime ensures the neutron flux density axial distribution stability at power maneuvering. The neutron flux axial stability is characterized by AO (Philipchuk et al., 1981):

$$\mathrm{AO} = \frac{N_u - N_l}{N}, \tag{19}$$

where N_u, N_l, N are the core upper half power, lower half power and whole power, respectively.

The variables AO, N_u, N_l, N are represented as

$$\mathrm{AO} = \mathrm{AO}_0 + \delta\mathrm{AO}\,;\; N_u = N_{u,0} + \delta N_u;\; N_l = N_{l,0} + \delta N_l;\; N = N_0 + \delta N\,, \tag{20}$$

where $\mathrm{AO}_0, N_{u,0}, N_{l,0}, N_0$ are the stationary values of AO, N_u, N_l, N, respectively; $\delta\mathrm{AO}, \delta N_u, \delta N_l, \delta N$ are the sufficiently small deviations from $\mathrm{AO}_0, N_{u,0}, N_{l,0}, N_0$, respectively.

The small deviations of N_u and N_l caused by the relevant average coolant temperature deviations $\delta < T_u >$ and $\delta < T_l >$ are expressed as

$$\delta N_u = \frac{\delta N}{\delta < T >} \cdot \delta < T_u >;\; \delta N_l = \frac{\delta N}{\delta < T >} \cdot \delta < T_l >, \tag{21}$$

where δN_u and δN_l are the small deviations of N_u and N_l, respectively; $\delta < T >$ is the average coolant temperature small deviation for the whole core; $\delta < T_u >$ and $\delta < T_l >$ are the average coolant temperature small deviations for the upper half-core and for the lower half-core, respectively.

The term $\delta N / \delta < T >$ is expressed as

$$\frac{\delta N}{\delta < T >} = \frac{\delta\rho / \delta < T >}{\delta\rho / \delta N} \equiv \frac{k_T}{k_N}, \tag{22}$$

where ρ is reactivity; k_T and k_N are the coolant temperature coefficient of reactivity and the power coefficient of reactivity, respectively.

Having substituted equations (20)−(22) in (19), the following equation for a small deviation of AO caused by a small deviation of N is derived after linearization:

$$\delta\text{AO} = \frac{k_T}{k_N} \cdot N_0^{-1} \cdot [(1 - \text{AO}_0) \cdot \delta < T_u > -(1 + \text{AO}_0) \cdot \delta < T_l >] \tag{23}$$

In case of the assumption

$$\text{AO}_0 << 1 \tag{24}$$

equation (23) is simplified:

$$\delta\text{AO} = \frac{k_T}{k_N} \cdot N_0^{-1} \cdot [\delta < T_u > -\delta < T_l >] \tag{25}$$

The criterion of AO stabilization due to the coolant temperature coefficient of reactivity (the coolant temperature regime effectiveness criterion) is obtained from (25):

$$\min \left| \sum_{i=1}^{m} [\delta < T_u > -\delta < T_l >] \right| , \tag{26}$$

where i is the power step number; m is the total number of power steps in some direction at reactor power maneuvering.

Use of the criterion (26) allows us to select a coolant temperature regime giving the maximum LHR axial distribution stability at power maneuvering. Let us study the following three WWER-1000 power maneuvering methods: M-1 is the method with a constant inlet coolant temperature T_{in}=const; M-2 is the method with a constant average coolant temperature $<T>$=const; and M-3 is the intermediate method having T_{in} increased by 1 °C only, when N lowers from 100% to 80%. Comparison of these power maneuvering methods has been made using the RS code. Distribution of long-lived and stable fission products causing reactor slagging was specified for the KhNPP Unit 2 fifth campaign start, thus the first core state having an equilibrium xenon distribution was calculated at this moment. The non-equilibrium xenon and samarium distributions were calculated for subsequent states taking into account the fuel burnup. The coolant inlet pressure and coolant flow rate were specified constant and equal to 16 MPa and $84 \cdot 10^3$ m^3/h, respectively. When using M-1, the coolant inlet temperature was specified at T_{in}=287 °C . When using M-

2, the coolant inlet temperature was specified according to Table 11 (T_{out} is the coolant outlet temperature).

N, %	T_{in}, °C	T_{out}, °C	<T>, °C
100	287	317	302
90	288	316	302
80	290	314	302

Table 11. Change of the coolant temperature at <T>=const in the M-2 method.

Denoting change of the lowest control rod axial coordinate (%) measured from the core bottom during a power maneuvering as ΔH, the first (M-2-a) and second (M-2-b) variants of M-2 had the regulating group movement amplitudes ΔH_{2a} =4% and ΔH_{2b} =6%, respectively. The reactor power change subject to time was set according to the same time profile for all the methods (Fig. 12).

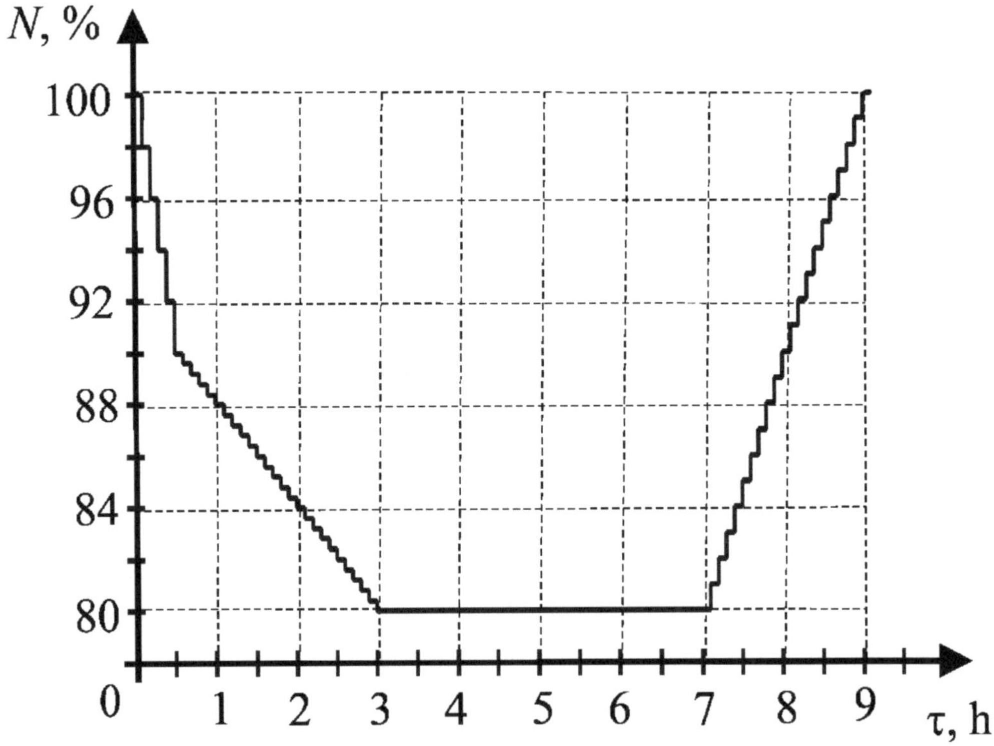

Fig. 12. Change of the reactor power subject to time.

For all the methods, N lowered from N_1=100 % to N_2=90 % within 0.5 h, under the linear law $dN_{1\text{-}2}/d\tau$=-2%/6 min, at the expense of boric acid entering. Also for all the methods, N

lowered from N_2=90% to N_3=80% within 2.5 h, under the law $dN_{2\text{-}3}/d\tau$= -0.4%/6 min, at the expense of reactor poisoning. The coolant concentration of boric acid was the criticality parameter when N stayed constant during 4 h. The NR power increased from N_3=80% to N_1=100% within 2 h, under the law $dN_{3\text{-}1}/d\tau$=1.0%/ 6 min, at the expense of pure distillate water entering and synchronous return of the regulating group to the scheduled position. When N increased from N_3=80% to N_1=100%, change of the regulating group position H subject to time was set under the linear law (Fig. 13).

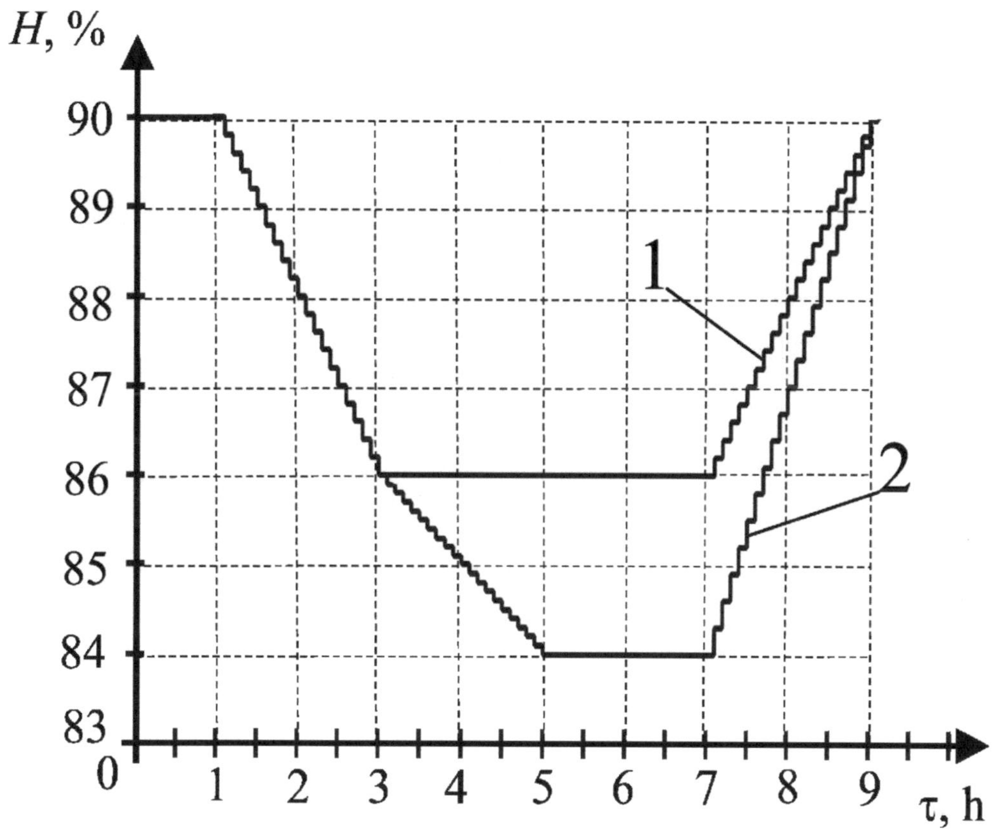

Fig. 13. Change of the regulating group position subject to time: (1) the methods M-1, M-2-a and M-3; (2) the method M-2-b.

Thus, modelling of the non-equilibrium WWER-1000 control was made by assignment of the following control parameters: criticality parameter; $T_{in,0}$; dT_{in}/dN; N_1; N_2; N_3; H_0; ΔH; $dN/d\tau$. Setting the WWER-1000 operation parameters in accordance with the Shmelev's method (Shmelev et al., 2004), for the methods M-1, M-2-a, M-2-b and M-3, when N changed from 100% to 80%, the change of core average LHR distribution was calculated by the RS code. Let us enter the simplifying representation

$$\Delta\delta T \equiv \delta < T_u > - \delta < T_l > \tag{27}$$

Using the obtained LHR distribution, by the FEMAXI code (Suzuki, 2010), the average coolant temperatures of the upper $<T_u>$ and lower $<T_l>$ half-cores were calculated for M-1, M-2-a and M-3 (at time step 0.5 h). Then, on the basis of $<T_u>$ and $<T_l>$, $\left|\sum_{i=1}^{6} \Delta\delta T\right|$ was found for M-1, M-2-a and M-3 having the same ΔH (Table 12).

Method	τ, h	N, %	$<T_u>$	$<T_l>$	$\delta <T_u>$	$\delta <T_l>$	$\Delta\delta T$	$\left\|\sum_{i=1}^{6} \Delta\delta T\right\|$
M-1; M-2-a; M-3	0.1	100	318.3	296.825	0	0	0	
M-1	0.6	90	317.975	296.575	-0.325	-0.25	-0.075	2.65
	1.1	88	316.375	296	-1.6	-0.575	-1.025	
	1.6	86	315.725	295.725	-0.65	-0.275	-0.375	
	2.1	84	315.1	295.525	-0.625	-0.2	-0.425	
	2.6	82	314.5	295.3	-0.6	-0.225	-0.375	
	3.1	80	313.9	295.075	-0.6	-0.225	-0.375	
M-2-a	0.6	90	319.25	298.025	0.95	1.2	-0.25	2.85
	1.1	88	317.875	297.575	-1.375	-0.45	-0.925	
	1.6	86	317.45	297.575	-0.425	0	-0.425	
	2.1	84	316.925	297.575	-0.525	0	-0.525	
	2.6	82	316.65	297.625	-0.275	0.05	-0.325	
	3.1	80	316.35	297.725	-0.3	0.1	-0.4	
M-3	0.6	90	318.4	297.075	0.1	0.25	-0.15	2.70
	1.1	88	316.9	296.55	-1.5	-0.525	-0.975	
	1.6	86	316.35	296.4	-0.55	-0.15	-0.4	
	2.1	84	315.7	296.2	-0.65	-0.2	-0.45	
	2.6	82	315.225	296.125	-0.475	-0.075	-0.4	
	3.1	80	314.775	296	-0.45	-0.125	-0.325	

Table 12. Change of the average coolant temperatures for M-1, M-2-a, M-3.

Having used the criterion (26), the conclusion follows that the coolant temperature regime M-1 ensures the most stable AO, while the regime M-2-a is least favorable – see Table 12. In order to check this conclusion, it is useful to compare AO stabilization for the discussed methods, calculating the divergence ΔAO between the instant and equilibrium axial offsets (Philimonov and Mamichev, 1998) – see Fig. 14.

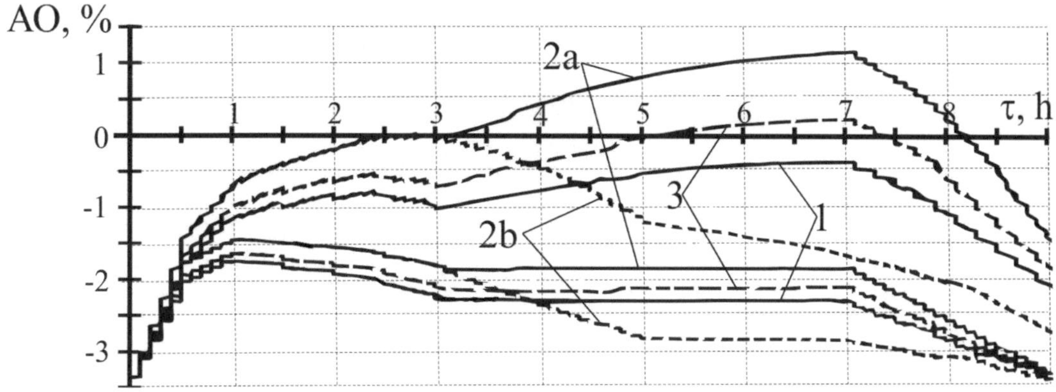

Fig. 14. Equilibrium and instant axial offsets (1, 2a, 2b, 3) subject to time for M-1, M-2-a, M-2-b and M-3, respectively: (lower line) the equilibrium AO; (upper line) the instant AO.

The regulating group movement amplitude is the same (4%) for M-1, M-2-a and M-3, but the maximum divergence between the instant and equilibrium offsets are $\Delta AO_1^{max} \approx 1.9\%$ (M-1), $\Delta AO_{2a}^{max} \approx 3\%$ (M-2-a) and $\Delta AO_3^{max} \approx 2.3\%$ (M-3). This result confirms the conclusion made on the basis of the criterion (26). If the regulating group movement amplitude, at power maneuvering according to the method with a constant average coolant temperature, is increased from 4 to 6%, then the maximum AO divergence lowers from 3% to 1.9% (see Fig. 14). Therefore, when using the method with $<T>$=const, a greater regulating group movement amplitude is needed to guarantee the LHR axial stability, than when using the method with T_{in} = const, on the assumption that all other conditions for both the methods are identical.

Having used the RS code, the core average LHR axial distribution change has been calculated for the methods M-1, M-2-a, M-2-b and M-3, for the following daily power maneuvering cycle: lowering of N from N_1=100% to N_2=90% during 0.5 h by injection of boric acid solution – further lowering of N to N_3=80% during 2.5 h due to reactor poisoning – operation at N_3=80% during 4 h – rising of N to the nominal capacity level N_1=100% during 2 h at the expense of pure distillate water entering and synchronous return of the regulating group to the scheduled position – operation at N_1=100% during 15 h.

When using the criterion (15) for comparative analysis of cladding durability subject to the FA transposition algorithm, the position of an axial segment and the power maneuvering method, the value of η should be set taking into account the necessity of determining the moment τ_0, when the condition $\sigma_e^{max}(\tau_0) = \eta\, \sigma_0^{max}(\tau_0)$ is satisfied. In addition, as the maximum number of power history points is limited by n_p^{lim} =10,000 in the FEMAXI code, the choice of η depends on the analysed time period τ^{max} and the complexity of a power maneuvering method, because a greater time period as well as a more complicated power maneuvering method are described by a greater number of history points n_p. Therefore, the value of η should be specified on the basis of simultaneous conditions $\sigma_e^{max}(\tau_0) = \eta\, \sigma_0^{max}(\tau_0)$; $n_p < n_p^{lim}$; $\tau_0 \le \tau^{max}$. Though the cladding failure parameter values listed in Table 6 were obtained assuming η =0.6 (the MATPRO-A corrosion model, COR = 1), comparison of cladding failure parameters for different power maneuvering methods can be made using the cladding collapse criterion (15), for instance, at η =0.4. Assuming η =0.4, on the basis of the obtained LHR distributions, the cladding failure parameters have been calculated by the instrumentality of the FEMAXI code (Suzuki, 2000) for the methods M-1, M-2-a, M-2-b and M-3, for the axial segments six and seven (the MATPRO-A corrosion model, COR = 1) - see Table 13.

Method			M-1	M-2-a	M-2-b	M-3
Axial Segment	6	τ_0, days	504.4	497.4	496.0	501.4
		A_0, MJ / m^3	1.061	1.094	1.080	1.068
		ω(500 days)	0.957	1.027 (+7.3%)	1.040 (+ 8.7%)	0.988 (+3.2%)
	7	τ_0, days	530.0	519.4	519.0	525.0
		A_0, MJ / m^3	1.044	1.055	1.019	1.043
		ω(500 days)	0.766	0.848 (+10.7%)	0.848 (+10.7%)	0.804 (+5.0%)

Table 13. Cladding failure parameters for the methods M-1, M-2-a, M-2-b and M-3.

Among the regimes with the regulating group movement amplitude ΔH =4%, the coolant temperature regime M-1 ensuring the most stable AO is also characterized by the least calculated cladding failure parameter ω(500 days), while the regime M-2-a having the least stable AO is also characterized by the greatest ω(500 days) - see Table 12, Fig. 14 and Table 13. The intermediate method M-3 having T_{in} increased by 1 °C only, when N lowers from 100% to 80%, is also characterized by the intermediate values of AO stability and ω(500 days).

In addition, the second variant of M-2 (M-2-b) having the regulating group movement amplitude ΔH_{2b} =6% is characterized by a more stable AO in comparison with the method M-2-a (see Fig. 14) and, for the most strained axial segment six, by a greater value of ω(500 days) - see Table 13.

It should be stressed that the proposed cladding rupture life control methods are not limited only in WWER-1000. Using the FEMAXI code, these methods can be extended into other reactor types (like PWR or BWR). At the same time, taking into account a real disposition of regulating units, a real coolant temperature regime as well as a real FA transposition algorithm, in order to estimate the amplitude of LHR jumps at FE axial segments occurring when the NR (PWR or BWR) capacity periodically increases, it is necessary to use another code instead of the RS code, which was developed for the WWER-1000 reactors.

The FA transposition algorithm 55-31-69-82 is characterized by a lower fuel cladding equivalent creep strain than the algorithm 55-31-55-55. At the same time, it has a lower fuel burnup than the algorithm 55-31-55-55 (see Table 14).

FA transposition algorithm	55-31-69-82		55-31-55-55	
COR	0	1	0	1
B_U, MW·day/kg	57.4		71.4	
$\sigma_e^{\max}$, MPa (% of σ_0)	69.9 (33)	127.4 (61)	107.2 (51)	146.7 (70)
p_e, %	4.22	11.22	9.36	16.02

Table 14. Fuel burnup and cladding equivalent creep strain for AS 6 (after 1500 d).

Thus, an optimal FA transposition algorithm must be set on the basis of cladding durability-fuel burnup compromise.

5. Methods of fuel cladding durability control at NPP with WWER

As is shown, the operating reactor power history as well as the WWER–1000 main regime and design parameters included into the second conditional group (pellet hole diameter, cladding thickness, pellet effective density, maximum FE linear heat rate, etc.) influence significantly on fuel cladding durability. At normal operation conditions, the WWER-1000 cladding corrosion rate is determined by design constraints for cladding and coolant, and depends slightly on the regime of variable loading. Also the WWER-1000 FE cladding rupture life, at normal variable loading operation conditions, depends greatly on the coolant temperature regime and the FA transposition algorithm. In addition, choice of the group of regulating units being used at NR power maneuvering influences greatly on the offset stabilization efficiency (Philimonov and Mamichev, 1998).

Hence, under normal operation conditions, the following methods of fuel cladding durability control at NPP with WWER can be considered as main ones:

- choice of the group of regulating units being used at power maneuvering.
- balance of stationary and variable loading regimes;
- choice of FE consrtuction and fuel physical properties, e.g., for the most strained AS, making the fuel pellets with centre holes;
- assignment of the coolant temperature regime;
- assignment of the FA transposition algorithm;

To create a computer-based fuel life control system at NPP with WWER, it is necessary to calculate the nominal and maximum permissible values of pick-off signals on the basis of calculated FA normal operation probability (Philipchuk et al., 1981). Though a computer-based control system SAKOR-M has already been developed for NPP with WWER at the OKB "Gidropress" (Bogachev et al., 2007), this system does not control the remaining life of fuel assemblies.

As the described CET-method can be applied to any type of LWR including prospective thorium reactors, the future fuel life control system for NPP with LWR can be created using this physically based method.

6. Conclusions

Taking into account the WWER-1000 fuel assembly four-year operating period transposition algorithm, as well as considering the disposition of control rods, it has been obtained that the axial segment, located between z = 2.19 m and z = 2.63 m, is most strained and limits the fuel cladding operation time at day cycle power maneuvering.

For the WWER-1000 conditions, the rapid creep stage is degenerated when using the Zircaloy-4 cladding corrosion models MATPRO-A and EPRI, at the correcting factor COR = - 0.431. This phenomenon proves that it is possible, for four years at least, to stay at the steady creep stage, where the cladding equivalent creep strain and radial total strain do not exceed 1-2%, on condition that the corrosion rate is sufficiently small.

The WWER-1000 thermal neutron flux axial distribution can be significantly stabilized, at power maneuvering, by means of a proper coolant temperature regime assignment. Assuming the maximum divergence between the instant and equilibrium axial offsets equal to 2%, the regulating unit movement amplitude at constant average coolant temperature is 6%, while the same at constant inlet coolant temperature is 4%. Therefore, when using the method with $\langle T \rangle$=const, a greater regulating unit movement amplitude is needed to guarantee the linear heat rate axial stability, than when using the method with T_{in} = const, on the assumption that all other conditions for both the methods are identical.

The WWER-1000 average cladding failure parameter after 500 day cycles, for the most strained sixth axial segment, at power maneuvering according to the method with $\langle T \rangle$=const, is 8.7% greater than the same for the method with T_{in} = const, on the assumption that the thermal neutron flux axial distribution stability is identical for both the methods.

The physically based methods of WWER-1000 fuel cladding durability control include: optimal choice of the group of regulating units being used at reactor power maneuvering, balance of stationary and variable loading regimes, choice of fuel element consrtuction and fuel physical properties considering the most strained fuel element axial segment, assignment of the coolant temperature regime and the fuel assembly transposition algorithm.

7. References

Bogachev, A.V. et al., 2005. Operating experience of system of the automated control of a residual cyclic resource for RP with VVER-1000. In: Proc. 18-th Int. Conf. on Structural Mechanics in Reactor Technology, Beijing, China.

Bull, A., 2005. The future of nuclear power, Materials challenges, Birmingham, 21 pp.

Kesterson, R. L. and Yueh, H. K., 2006. Cladding optimization for enhanced performance margins. In: Proc. Int. Conf. TopFuel, Salamanca.

Kim, J.H. et al., 2007. Deformation behavior of Zircaloy-4 cladding under cyclic pressurization. Journal of Nuclear Science and Technology 44, 1275–1280.

Maksimov, M.V. et al., 2009. Model of cladding failure estimation for a cycling nuclear unit. Nuclear Engineering and Design 239, 3021–3026.

Maksimov, M.V. and Pelykh, S.N., 2009. Comparison of fuel-element cladding durability for a WWER-1000 reactor operating in the mode of variable loadings. Odes'kyi Natsional'nyi Politechnichnyi Universytet. Pratsi 1, 49–53 (in Russian).

Maksimov, M.V. and Pelykh, S.N., 2010. Method for evaluating the service life of VVER-1000 fuel-element cladding in different loading regimes. Atomic Energy. 5, 357-363.

MATPRO-09, 1976. A Handbook of Materials Properties for Use in the Analysis of Light Water Reactor Fuel Rod Behavior, USNRC TREE NUREG-1005.

Nemirovskiy, Y., 2001. About an estimation of construction safe operation time, Proc. Int. Conf. RDAMM-2001, Novosibirsk, 328 - 333 (in Russian).

Novikov, V. V. et al., 2005. Nuclear fuel operability assurance in maneuver regimes. In: Proc. Ukrainian-Russian Conf. on Experience of the new WWER fuel exploitation, Khmelnitskiy, p. 22 (in Russian).

Pelykh, S.N. et al., 2008. Model of cladding failure estimation under multiple cyclic reactor power changes. In: Proc. of the 2-nd Int. Conf. on Current Problems of Nuclear Physics and Atomic Energy, Kiev, Ukraine.

Pelykh, S.N. et al., 2009. A complex power maneuvering algorithm efficiency criterion for a WWER-1000 reactor working in the mode of variable loadings. Odes'kyi Natsional'nyi Politechnichnyi Universytet. Pratsi 2, 53–58 (in Russian).

Pelykh, S.N. et al., 2010. Estimation of local linear heat rate jump values in the variable loading mode. In: Proc. of the 3-rd Int. Conf. on Current Problems of Nuclear Physics and Atomic Energy, Kiev, Ukraine.

Pelykh, S.N. and Maksimov, M.V., 2011. Cladding rupture life control methods for a power-cycling WWER-1000 nuclear unit. Nuclear Engineering and Design 241, 2956–2963.

Philimonov, P.E. and Mamichev, V.V., 1998. The "reactor simulator" code for modelling of maneuvering WWER-1000 regimes. Atomnaya Energiya 6, 560–563 (in Russian).

Philipchuk, E. V. et al., 1981. Control of the nuclear reactor neutron field, Energoatomizdat, Moscow, 280 pp. (in Russian).

SCDAP/RELAP5/MOD2, 1990. Code manual, Vol. 4. MATPRO-A: A Library of Materials Properties for Light Water Reactors Accident Analysis, NUREG/CR-5273.

Semishkin, V.P., 2007. Calculation-experimental methods to ground the WWER fuel element and fuel assembly behaviour under the LOCA emergency conditions. Author's abstract of dissertation for a degree of Doctor of Technical Science, Moscow, 48 p (in Russian).

Semishkin, V.P. et al., 2009. Standard durability and reliability requirements for WWER reactor unit elements, and safety problems. In: Abs. of the 6-th Int. Conf. on Safety Assurance of NPP with WWER, Podolsk, Russia, p. 119 (in Russian).

Shmelev, V.D. et al., 2004. The WWER active cores for nuclear stations, Akademkniga, Moscow, 220 pp. (in Russian).

Sosnin, O., Gorev, B.V., 1986. Energy Variant of the Theory of Creep, The Siberian Branch of the Russian Academy of Sciences, Novosibirsk, 95 pp. (in Russian).

Suzuki, M., 2000. Light Water Reactor Fuel Analysis Code FEMAXI-V (Ver.1). JAEA Report, Japan Atomic Energy Research Institute, 285 pp.

Suzuki, M., 2010. Modelling of light-water reactor fuel element behaviour in different loading regimes, Astroprint, Odessa, 248 pp. (in Russian).

11

Improving the Performance of the Power Monitoring Channel

M. Hashemi-Tilehnoee[1,*] and F. Javidkia[2]
[1]*Department of Engineering, Aliabad Katoul Branch, Islamic Azad University, Aliabad Katoul*
[2]*School of Mechanical Engineering, Shiraz University, Shiraz*
Iran

1. Introduction

In this chapter, different methods for monitoring and controlling power in nuclear reactors are reviewed. At first, some primary concepts like neutron flux and reactor power are introduced. Then, some new researches about improvements on power-monitoring channels, which are instrument channels important to reactor safety and control, are reviewed. Furthermore, some new research trends and developed design in relation with power monitoring channel are discussed. Power monitoring channels are employed widely in fuel management techniques, optimization of fuel arrangement and reduction in consumption and depletion of fuel in reactor core. Power reactors are equipped with neutron flux detectors, as well as a number of other sensors (e.g. thermocouples, pressure and flow sensors, ex-vessel accelerometers). The main purpose of in-core flux detectors is to measure the neutron flux distribution and reactor power. The detectors are used for flux mapping for in-core fuel management purposes, for control actions and for initiating reactor protection functions in the case of an abnormal event (IAEA, 2008). Thus, optimization on power monitoring channel will result in a better reactor control and increase the safety parameters of reactor during operation.

2. Neutron flux

It is convenient to consider the number of neutrons existing in one cubic centimeter at any one instant and the total distance they travel each second while in that cubic centimeter. The number of neutrons existing in a cm^3 of material at any instant is called neutron density and is represented by the symbol n with units of $neutrons/cm^3$. The total distance these neutrons can travel each second will be determined by their velocity.

A good way of defining neutron flux (ϕ) is to consider it to be the total path length covered by all neutrons in one cubic centimeter during one second. Mathematically, this is the equation below.

$$\phi = n\, v \tag{1}$$

* Corresponding Author

where: ϕ = neutron flux (neutrons cm^{-2} s^{-1}), n = neutron density (neutrons cm^{-3}), and v = neutron velocity (cm s^{-1}). The term neutron flux in some applications (for example, cross section measurement) is used as parallel beams of neutrons travelling in a single direction. The intensity of a neutron beam is the product of the neutron density times the average neutron velocity. The directional beam intensity is equal to the number of neutrons per unit area and time (neutrons cm^{-2} s^{-1}) falling on a surface perpendicular to the direction of the beam. One can think of the neutron flux in a reactor as being comprised of many neutron beams travelling in various directions. Then, the neutron flux becomes the scalar sum of these directional flux intensities. Macroscopic cross sections for neutron reactions with materials determine the probability of one neutron undergoing a specific reaction per centimeter of travel through that material. If one wants to determine how many reactions will actually occur, it is necessary to know how many neutrons are travelling through the material and how many centimeters they travel each second. Since the atoms in a reactor do not interact preferentially with neutrons from any particular direction, all of these directional beams contribute to the total rate of reaction. In reality, at a given point within a reactor, neutrons will be travelling in all directions (DOE, 1993).

3. Power monitoring in nuclear reactors

In order to ensure predictable temperatures and uniform depletion of the fuel installed in a reactor, numerous measures are taken to provide an even distribution of flux throughout the power producing section of the reactor. This shaping, or flattening, of the neutron flux is normally achieved through the use of reflectors that affect the flux profile across the core, or by the installation of poisons to suppress the neutron flux where desired. The last method, although effective at shaping the flux, is the least desirable since it reduces the neutron economy by absorbing the neutrons (DOE, 1993).

In recent years, power monitoring systems are under developing in research centers. Sakai et al. (Sakai et al., 2010) invented a power monitoring system for boiling water reactors (BWRs). In the BWR, the output power alternately falls and rises due to the generation and disappearance of voids, respectively, which may possibly generate power oscillation whereby the output power of the nuclear reactor oscillates and is amplified. The power monitoring system has a local power range monitor (LPRM) unit that has a plurality of local power channels to obtain local neutron distribution in a nuclear reactor core; an averaged power range monitor (APRM) unit that receives power output signals from the LPRM unit and obtains averaged output power signal of the reactor core as a whole; and an oscillation power range monitor (OPRM) unit that receives the power output signals from the LPRM unit and monitors power oscillation of the reactor core. The output signals from the LPRM unit to the APRM unit and the output signals from the LPRM unit to the OPRM unit are independent. A new flux mapping system (FMS) in Korea Electric Power Research Institute (KEPRI) was installed in Kori's unit 1 nuclear power plant. An in-core neutron FMS in a pressurized water reactor (PWR) yields information on the neutron flux distribution in the reactor core at selected core locations by means of movable detectors. The FMS having movable neutron detectors is equipped with detector cable drive units and path selectors located inside the reactor containment vessel. The drive units push and pull their detector cables, which run through guide tubes, and the path selectors route the detector cables into the predetermined guide tubes. Typically, 36–58 guide tubes (thimbles) are allocated in the reactor depending on the number of fuel assemblies. A control system of FMS is located at the main control room to control the detector drive system and measure the flux signal sensed by the detectors. The flux mapping data are used to verify the reactor core design parameters, and to determine the fission power distribution in the core. The new designed path selector for a guide the neutron detectors through the reactor core are shown in Figure 1.

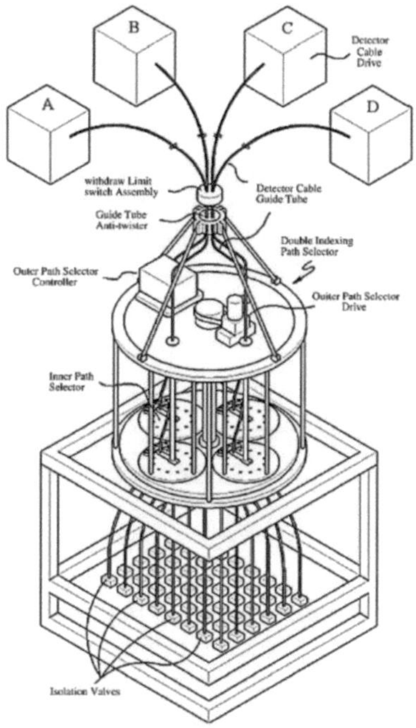

Fig. 1. The new designed path selector for KEPRI unit 1 reactor (Cho et al., 2006)

The path selector system is composed of four inner path selectors and an outer path selector. With the benefit of the double indexing path selector mechanism, the reliability of the detector drive system has been improved five times higher than that of a conventional system. Currently, the developed in-core flux mapping systems have been deployed at the Kori nuclear units 1–4.

4. Neutron flux monitoring and measurement in nuclear reactors

In thermal nuclear reactors, most of the power is generated through fission induced by slow neutrons. Therefore, nuclear sensors those are to be part of reactor control or safety systems are generally based on detectors that respond primarily to slow neutrons. In principle, many of detector types can be adapted for application to reactor measurements. However, the extreme conditions associated with reactor operation often lead to substantial design changes, and a category of slow neutron detectors designed specifically for this application has gradually evolved.

4.1 Neutron detectors and instruments

It is conventional to subdivide reactor instruments into two categories: in-core and out-of-core. In-core sensors are those that are located within narrow coolant channels in the reactor core and are used to provide detailed knowledge of the flux shape within the core. These sensors can be either fixed in one location or provided with a movable drive and must obviously be of rather small size (typically on the order of 10mm diameter). Out-of-core detectors are located some distance from the core and thus respond to properties of the neutron flux integrated over the entire core. The detectors may be placed either inside or outside the pressure vessel and normally will be located in a much less severe environment compared with in-core detectors. Size restrictions are also less of a factor in their design. The majority of neutron sensors for reactor use are of the gas-filled type. Their advantages in this application include the inherent gamma-ray discrimination properties found in any gas detector, their wide dynamic range and long-term stability, and their resistance to radiation damage. Detectors based on scintillation processes are less suitable because of the enhanced gamma-ray sensitivity of solid or liquid scintillators, and the radiation induced spurious events that occur in photomultiplier tubes. Semiconductor detectors are very sensitive to radiation damage and are never used in reactor environments.

4.2 In-core neutron detectors

There is often a need to place neutron sensors within the core of a nuclear reactor to provide information on the spatial variation of the neutron flux. Because of the small size (1-7 cm) of the channel in which these instruments must be located, emphasis is placed on compactness and miniaturization in their design. They may either be left in a fixed position or provided with a motorized drive to allow traverses through the reactor core. Miniaturized fission chambers can be tailored for in-core use over any of the power ranges likely to be encountered in reactor operation. Walls of the chamber are usually lined with highly enriched uranium to enhance the ionization current. These small ion chambers are typically made using stainless steel walls and electrodes, and operating voltage varies from about 50 to 300 V. Argon is a common choice for the chamber fill gas and is used at a pressure of several atmospheres. The elevated pressure ensures that the range of fission fragments within the gas does not exceed the small dimensions of the detector. The gradual burn up of neutron-sensitive material is a serious problem for the long term operation of in-core detectors. Although the change in current-voltage characteristics with increased neutron flux may be greater for in-core detectors than out of core detectors, a similar effect is observed in both the compensated and uncompensated ion chambers used in pressurized water reactors (Knoll, 2000).

4.3 Self-powered detectors

A unique type of neutron detector that is widely applied for in-core use is the self-powered detector (SPD). These devices incorporate a material chosen for its relatively high cross section for neutron capture leading to subsequent beta or gamma decay. In its simplest form, the detector operates on the basis of directly measuring the beta decay current following capture of the neutrons. This current should then be proportional to the rate at which neutrons are captured in the detector. Because the beta decay current is measured directly, no external bias voltage need be applied to the detector, hence the name self-powered. Another form of the self-powered detector makes use of the gamma rays emitted following neutron capture. Some fraction of these gamma rays will interact to form secondary electrons through the Compton, photoelectric, and pair production mechanisms. The current of the secondary electrons can then be used as the basic detector signal. Nonetheless, the self powered neutron detector (SPND) remains the most common term applied to this family of devices. Compared with other neutron sensors, self-powered detectors have the advantages of small size, low cost, and the relatively simple electronics required in conjunction with their use. Disadvantages stem from the low level of output current produced by the devices, a relatively severe sensitivity of the output current to changes in the neutron energy spectrum, and, for many types, a rather slow response time. Because the signal from a single neutron interaction is at best a single electron, pulse mode operation is impractical and self-powered detectors are always operated in current mode.

Figure 2 shows a sketch of a typical SPD based on beta decay.

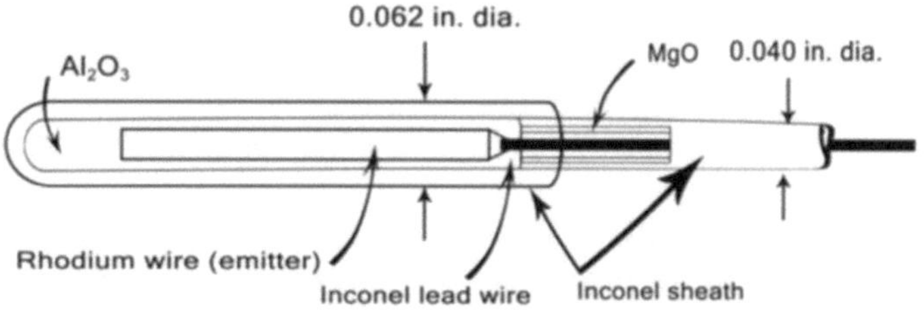

Fig. 2. Cross sectional view of a specific SPD design (Knoll, 2000)

The heart of the device is the emitter, which is made from a material chosen for its relatively high cross section for neutron capture leading to a beta-active radioisotope. Ideally, the remainder of the detector does not interact strongly with the neutrons, and construction materials are chosen from those with relatively low neutron cross sections.

4.4 Neutron instruments (NI) and detectors in pressurized water reactors

Detectors for the routine monitoring of reactor power in a PWR are located outside the reactor pressure vessel and are characterized by the following typical environmental conditions: neutron flux up to 10^{11} n cm^{-2} s^{-1}, gamma irradiation rates up to 10^6 R h^{-1}, and temperatures of approximately 100 °C. Out-of-core sensors are the usual basis of reactor control and safety channels in a PWR. In choosing specific detector types, consideration must be given to the expected neutron signal level compared with noise sources, the speed of response of the detector, and the ability to discriminate against gamma-induced signals.

Each of these criteria assumes different importance over various ranges of reactor power, and as a result multiple detector systems are usually provided, each designed to cover a specific subset of the power range (Knoll, 2000). Figure 3 illustrates a typical scheme for a PWR in which three sets of sensors with overlapping operating ranges are used to cover the entire power range of the reactor.

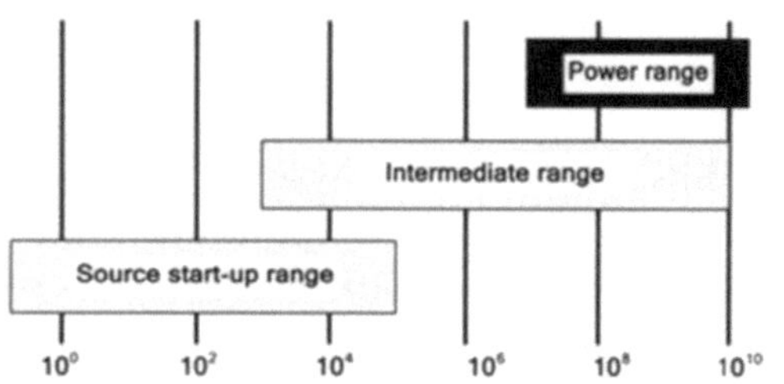

Fig. 3. Typical ranges covered by out-of-core neutron detectors in a PWR (Knoll, 2000)

The lowest range, usually called the source start-up range, is encountered first when bringing up reactor power from shut-down conditions. This range is characterized by conditions in which the gamma flux from the fission product inventory in the core may be large compared with the small neutron flux at these low power levels. Under these conditions, good discrimination against gamma rays is at a premium. Also, the expected neutron interaction rates will be relatively low in this range. Pulse mode operation of either fission chambers or BF_3 proportional counters is therefore possible, and the required gamma-ray discrimination can be accomplished by accepting only the much larger amplitude neutron pulses. As the power level is increased, an intermediate range is encountered in which pulse mode operation is no longer possible because of the excessive neutron interaction rate. In this region the gamma-ray-induced events are still significant compared with the neutron flux, and therefore simple current mode operation is not suitable. The MSV mode of operation can reduce the importance of the gamma-ray signal in this range, but a more common method used in PWRs is to employ direct gamma-ray compensation using a compensated ionization chamber (CIC). A third range of operation corresponds to the region near the full operating power of the reactor. The neutron flux here is usually so large that gamma-ray-induced currents in ion chambers are no longer significant, and simple uncompensated ion chambers are commonly used as the principal neutron sensor. Because these instruments are often part of the reactor safety system, there is a premium on simplicity that also favors uncompensated ion chamber construction.

4.5 Neutron instruments and detectors in boiling water reactors

The BWR NI system, like the PWR system, has three overlapping ranges as illustrated in Figure 4.

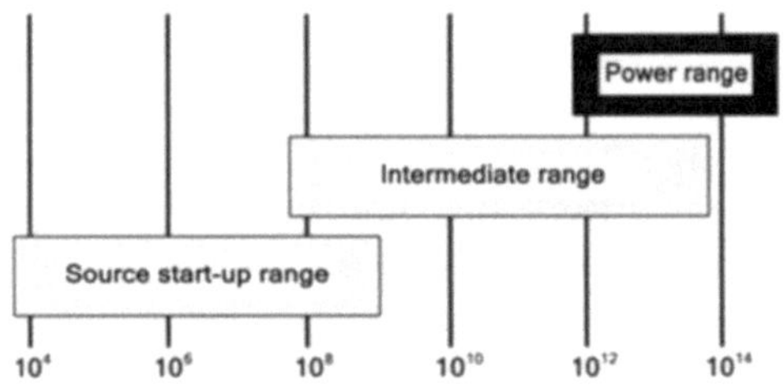

Fig. 4. Typical ranges covered by in-core neutron detectors in a BWR (Knoll, 2000)

The three systems are called source, intermediate, and power range monitors. Unlike the PWR, which uses out-of-core neutron detectors, the neutron detectors are all located in-core. There are also many more detectors used in the BWR NI system than in the PWR system.

The source range monitoring system typically consists of four in-core fission chambers operating in pulse mode. Pulse mode operation provides good discrimination against gamma rays, which is necessary when measuring a relatively low neutron flux in the presence of a high gamma flux. A typical intermediate range monitoring system has eight in-core fission chambers operating in the mean square voltage (MSV) mode. The MSV mode promotes the enhanced neutron to gamma response required to provide a proper measure of neutron flux in the presence of gamma rays for both control and safety requirements. The power range monitoring system typically consists of 144-164 fission ion chambers distributed throughout the core. The fission chambers operate in current mode and are called local power range monitors (LPRM). Current mode operation provides satisfactory neutron response at the high flux levels encountered between 2 and 150% full power. In a typical system, approximately 20 LPRMs are summed to provide input to one of the seven or eight average power range monitoring (APRM) systems. The APRM system provides input for both control and reactor protection systems. In-core flux detectors are used at high power levels (above 10% of full power) because they provide spatial information needed, at high power, to control xenon-induced flux tilts and to achieve the optimum flux distribution for maximum power output. The control system flux detectors are of two types. One type has an inconel emitter and is used for the zone control system. The other type has a vanadium emitter, and is used for the flux mapping system. For power mapping validation, channel temperature differentials are used with measured flows (instrumented channels) or predicted flows (other channels) to determine the estimated channel powers, which are then compared with the powers calculated from the flux mapping readings; this provides an ongoing validation of the accuracy of the flux mapping channel powers.

4.6 Neutron instruments and detectors in CANDU reactors

In CANDU reactors, three instrumentation systems are provided to measure reactor thermal neutron flux over the full power range of the reactor (Knoll, 2000). Start-up instrumentation covers the eight-decade range from 10^{-14} to 10^{-6} of full power; the ion chamber system extends from 10^{-7} to 1.5 of full power, and the in-core flux detector system provides accurate spatial measurement in the uppermost decade of power (10% to 120% of full power). The fuel channel temperature monitoring system is provided for channel flow verification and for power mapping validation. The self-powered in-core flux detectors are installed in flux detector assemblies to measure local flux in the regions associated with the liquid zone controllers. The flux mapping system uses vanadium detectors distributed throughout the core to provide point measurements of the flux. The fast, approximate estimate of reactor power is obtained by either taking the median ion chamber signal (at powers below 5% of full power) or the average of the in-core inconel flux detectors (above 15% of full power) or a mixture of both (5% to 15% of full power).

5. Several advanced power measuring and monitoring systems

The power range channels of nuclear reactors are linear, which cover only one decade, so they do not show any response during the startup and intermediate range of the reactor operation. So, there is no prior indication of the channels during startup and intermediate operating ranges in case of failure of the detectors or any other electronic fault in the channel. Some new reliable instrument channels for power measurement will be studied in this section.

5.1 A wide-range reactor power measuring channel

The power range channels of nuclear reactors are linear, which cover only one decade, so they do not show any response during the startup and intermediate range of the reactor operation. So, there is no prior indication of the channels during startup and intermediate operating ranges in case of failure of the detectors or any other electronic fault in the channel. A new reliable instrument channel for power measurement will be studied in this section. The device could be programmed to work in the logarithmic, linear, and log-linear modes during different operation time of the reactor life cycle. A new reliable nuclear channel has been developed for reactor power measurement, which can be programmed to work in the logarithmic mode during startup and intermediate range of operation, and as the reactor enters into the power range, the channel automatically switches to the linear mode of operation. The log-linear mode operation of the channel provides wide-range monitoring, which improves the self-monitoring capabilities and the availability of the reactor. The channel can be programmed for logarithmic, linear, or log-linear mode of operation. In the log-linear mode, the channel operates partially in log mode and automatically switches to linear mode at any preset point. The channel was tested at Pakistan Research Reactor-1 (PARR-1), and the results were found in very good agreement with the designed specifications. A wide range nuclear channel is designed to measure the reactor power in the full operating range from the startup region to 150% of full power. In the new channel, the status of the channels may be monitored before their actual operating range. The channel provides both logarithmic and linear mode of operation by automatic operating mode selection. The channel can be programmed for operation in any mode, log, linear or log-linear, in any range. In the log-linear mode, the logarithmic mode of operation is used for monitoring the operational status of the channel from reactor startup to little kilowatt reactor power where the mode of operation is automatically changed to linear mode for measurement of the reactor power. At the low power operation, the channel will provide monitoring of the proper functioning of the channel, which includes connection of the electronics with the chamber and functioning of the chamber, amplifier, high-voltage supply of the chamber, and auxiliary power supply of the channel. The channel has been developed using reliable components, and design has been verified under recommended reliability test procedures. The channel consists of different electronic circuits in modular form including programmable log-linear amplifier, isolation amplifier, alarm unit, fault monitor, high-voltage supply, dc-dc converter, and indicator. The channel is tested at PARR-1 from reactor startup to full reactor power. Before testing at the reactor, the channel was calibrated and tested in the lab by using a standard current source. The channel has been designed and developed for use in PARR-1 for reactor power measurement. The response of the channel was continuously compared with ^{16}N channel of PARR-1, and the test channel was calibrated according to the ^{16}N channel at 1 MW. After calibration, it was noticed that the test channel gave the same output as the ^{16}N channel. The channel response with Reactor Power is shown in Figure 5.

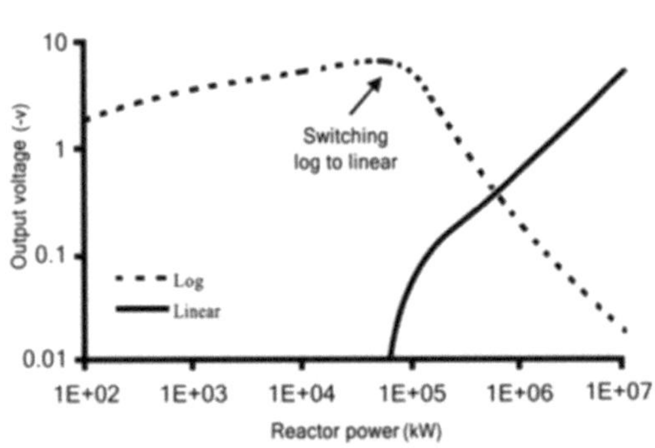

Fig. 5. Response of the channel at different reactor power at PARR-1 (Tahir Khaleeq et al. 2003)

The channel shows an excellent linearity. A very important check was the response of the test channel at the operating mode switching level, and it was found that the channel smoothly switched from log to linear operating mode. The designed channel has shown good performance throughout the operation and on applying different tests. The self-monitoring capabilities of the channel will improve the availability of the system.

5.2 A new developed monitoring channel using ^{16}N detector

^{16}N is one of the radioactive isotopes of nitrogen, which is produced in reactor coolant (water) emitting a Gamma ray with energy about 6 MeV and is detectable by out-core instruments. In this section, a ^{16}N instrument channel in relation to reactor power measurement will be studied. The reactor power and the rate of production of ^{16}N have a linear relation with good approximation. A research type of ^{16}N power monitoring channel subjected to use in Tehran Research Reactor (TRR). Tehran Research Reactor is a 5 MW pool-type reactor which use a 20% enriched MTR plate type fuel. When a reactor is operating, a fission neutron interacts with oxygen atom (^{16}O) present in the water around the reactor core, and convert the oxygen atom into radioactive isotope ^{16}N according to the following (n, p) reaction. Also another possible reaction is production of ^{19}O by $^{18}_{8}O\,(n,\gamma)\ ^{19}_{8}O$ reaction.

Of course, water has to be rich of $^{18}_{8}O$ for at least 22% to have a significant role in ^{19}O producing, but ^{18}O is exist naturally (0.2%)

$$^{1}_{0}n + {}^{18}_{8}O \rightarrow {}^{19}_{8}O + \gamma(2.8\ \text{MeV}) \tag{2}$$

$$^{1}_{0}n + {}^{16}_{8}O \rightarrow {}^{1}_{1}P + {}^{16}_{7}N^{*} \tag{3}$$

$^{16}N^{*}$ is produced and radiate gamma rays (6MeV) and β particles during its decay chain.

$$^{16}_{7}N^{*} \xrightarrow{7.2\ s} {}^{16}_{8}O + {}^{0}_{-1}\beta + \gamma(6.13\ \text{MeV}) \tag{4}$$

In addition to $^{16}_{8}O$ (99.76%) and $^{18}_{8}O$ (0.2%), other isotope of oxygen is also exist naturally in water, including $^{17}_{8}O$ (0.04%). $^{17}_{7}N$ (0.037%) produced from $^{17}_{8}O$ by the (n, p) reaction which will decayed through beta emission.

$$^{17}_{8}O + {}^{1}_{0}n \rightarrow {}^{17}_{7}N + {}^{1}_{1}H \tag{5}$$

$$^{17}_{7}N \xrightarrow{4.14\ s} {}^{16}_{8}O + {}^{0}_{-1}\beta + {}^{1}_{0}n \tag{6}$$

Since activity ratio of ^{16}N to ^{17}N is 257/1, thus activity of ^{17}N does not count much and is negligible. Primary water containing this radioactive ^{16}N is passed through the hold-up tank (with capacity of 384.8 m^3, maximum amount of water that can pour to the hold-up tank is 172 m^3 and reactor core flow is 500 $m^3\ h^{-1}$), which is placed under the reactor core and water flow from core down to this tank by gravity force. The hold-up tank delays the water for about 20.7 min. During this period activity of the short lived ^{16}N ($T_{1/2}$ = 7.4 s) decays down to low level. The decay tank and the piping connection to the reactor pool are covered with heavy concrete shielding in order to attenuate the energy of gamma emitted by the ^{16}N nuclei. To investigate the amount of ^{16}N in Tehran Research Reactor by direct measurements of gamma radiation and examine the changes with reactor power, the existing detectors in the reactor control room used and experiment was performed. To assess gamma spectrum for the evaluation of ^{16}N in reactor pool a portable gamma spectroscopy system which includes a sodium-iodide detector is used. The sodium-iodide (NaI) detector which is installed at reactor outlet water side is used for counting Gamma rays due to decay of ^{16}N which depends directly on the amount of ^{16}N. Some advantages of the power measurement using ^{16}N system:

- Power measurement by ^{16}N system uses the gamma from decay of ^{16}N isotope only, so other gammas from impurities do not intervened the measurements.
- Since ^{16}N system installed far from the core, fission products and its gamma rays would not have any effects on the measurements.
- Energy dissipation of heat exchanged with surroundings would not intervene, because water temperature would not use in this system for reactor power measurements.

It is expected that the amount of ^{16}N which is produced in reactor water has linear relation with the reactor power. Comparison of theory and experience is shown in Figure 6.

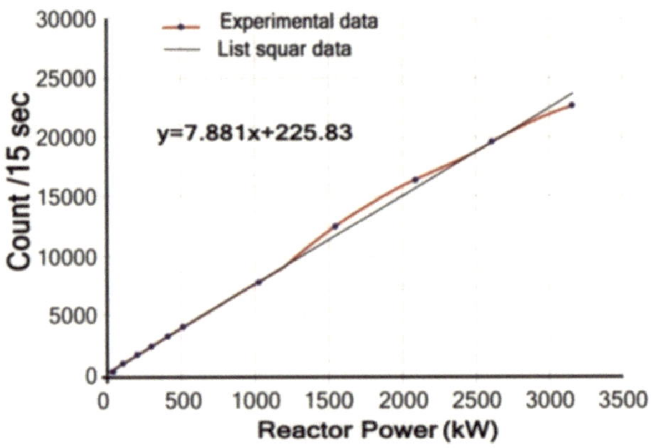

Fig. 6. Comparison of theory and experimental data from ^{16}N channel (Sadeghi, 2010)

Based on graph which resulted from experimental data and the straight line equation using least squire fit, it is appear that the experimental line deviated from what it expected; it means that the line is not completely straight. It seems this small deviation is due to the increasing water temperature around the core in higher power, density reduction and outlet water flow reduction which cause ^{16}O reduction and so ^{16}N. At the same time the amount of ^{16}N production decreases and thus decreasing gamma radiations, this will reduce the number of counting, but on the other hand, since the number of fast neutron production in reactor can increase according to reactor power and moderator density became less, the possibility of neutron interaction with water would increased. During past years, linearity of the curve as the experimental condition and the measurements were improved. Now that this linearity is achieved, by referring to the graph, it could conclude that ^{16}N system is suitable to measure the reactor power. Safety object of the new channel is evaluated by the radiation risk of ^{16}N, dose measurement performed in the area close to the hold-up tank for gamma and beta radiations. The dose received in these areas (except near the hold-up tank charcoal filter box which is shielded) are below the recommended dose limits for the radiation workers (0.05 Sv/year), therefore it can be seen that the radiation risk of ^{16}N is reduced due to design of the piping system and hold-up tank which is distanced from the core to overlap the decay time. Thus, ^{16}N decay through the piping and hold-up tank is reduced to a safe working level. It could be seen that ^{16}N system is able to measure the reactor power enough accurately to be used as a channel of information. For the pool type research reactor which has only one shut down system also could be used to increase the reactor safety (Sadeghi, 2010).

6. Power monitoring by some developed detectors and new methods

In this section, several neutron detectors and power monitoring systems are reviewed.

Application of a micro-pocket fission detector for in-core flux measurements is described in section 6.1. SIC neutron monitoring system is examined experimentally and theoretically. Development of an inconel self-powered neutron detector (SPND) for in-core power monitoring will be reviewed in section 6.3. Furthermore, a prototype cubic meter antineutrino detector which is used as a new device for measuring the thermal power as an out-core detection system, will be discussed. Finally, two passive approaches for power measurement are discussed.

6.1 Micro-pocket fission detectors (MPFD) for in-core neutron flux monitoring

There is a need for neutron radiation detectors capable of withstanding intense radiation fields, capable of performing "in-core" reactor measurements, capable of pulse mode and current mode operation, capable of discriminating neutron signals from background gamma ray signals, and that are tiny enough to be inserted directly into a nuclear reactor without significantly perturbing the neutron flux. A device that has the above features is the subject of a Nuclear Engineering Research Initiative (NERI) research project, in which miniaturized fission chambers are being developed and deployed in the Kansas State University (K-State) TRIGA Mark-II research reactor (McGregor, 2005). The unique miniaturized neutron detectors are to be used for three specific purposes (1) as reactor power-level monitors, (2) power transient monitors, and (3) real-time monitoring of the thermal and fast neutron flux profiles in the core. The third application has the unique benefit of providing information that, with mathematical inversion techniques, can be used to infer the three-dimensional (3D) distribution of fission neutron production in the core. Micro-pocket fission detectors (MPFD) are capable of performing near-core and in-core reactor power measurements. The basic design utilizes neutron reactive material confined within a micro-sized gas pocket, thus forming a miniature fission chamber. The housing of the chamber is fabricated from inexpensive ceramic materials, the detectors can be placed throughout the core to enable the 3D mapping of the neutron flux profile in "real-time". Initial tests have shown these devices to be radiation hard and potentially capable of operating in a neutron fluence exceeding 10^{19} cm^{-2} without noticeable degradation. Figure 7 shows a cutaway view of the basic detector concept. It consists of a small ceramic structure, within which is a miniature gas-filled pocket.

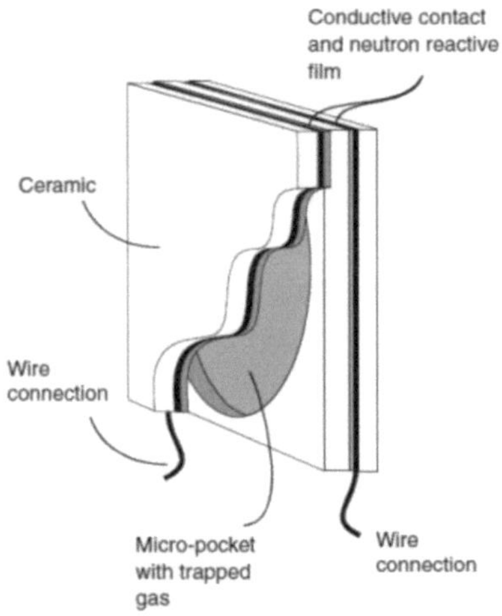

Fig. 7. Cut away view of MPFD (McGregor, 2005)

A conductive layer is deposited on opposing sides of the device, but not the perimeter. Neutron reactive material, such as ^{235}U, ^{232}Th, ^{10}B, or some material containing ^{6}Li, is applied over the conductive contact(s). Although both sides may be coated with neutron reactive material, only one side needs to be coated for the device to work. The ceramic pieces must be insulators and must not be composed of neutron-absorbing material. For instance, aluminum oxide or oxidized silicon may be used. Connecting wires must be sealed well so that no gas leaks out. Additionally, the ceramic pieces must be sealed with high temperature cement such that the seal integrity is secure within the hostile environment of a reactor core.

By in-core evaluation the device demonstrated excellent count-rate linearity with reactor power. Further, the small size and minute amount of uranium used permitted pulse mode operation without appreciable deadtime distortions or problems. MPFDs have, thus far, shown exceptional radiation hardness to neutrons, gamma rays, and charged-particle reaction products, while showing no performance degradation for devices exposed to neutron fluences exceeding 10^{19} cm^{-2}. Further, pulse mode operated devices have shown a linear relation to reactor power for neutron fluxes up to 10^{12} cm^{-2} s^{-1}, and smaller MPFDs are expected to operate in pulse mode in even higher neutron fluxes. The next generation of MPFDs will be composed of a triad of detectors on a single substrate, one with a ^{232}Th coating, one with a ^{235}U coating, and one with no coating. Such a triad permits monitoring of the fast neutron flux, the thermal neutron flux, and the gamma ray background, all at the same time. Further, the devices behave as point detectors, which greatly simplify data interpretation. Data from such a MPFD array can be converted into a power density map of the reactor core for real-time analysis. Mathematical models are under development that can relate the power density profiles in the reactor's fuel rods to the flux densities at the detector locations. Key to this formulation is the construction of an appropriate response function that gives the flux at any position in the core to the fast neutrons born at an arbitrary axial depth in any of the core fuel rods. Response functions have been derived and used to illustrate the analysis methods. Thus far, modeled results using predicted sensitivities of the MPFDs indicated that the power density in the fuel can be determined provided that appropriate boundary conditions regarding device placement are met. Good matching to power density profiles can be achieved with as few as five detector triads per detector string.

6.2 Experimental and computational evaluation of the response of a SiC neutron monitoring system in a thermal neutron field

Silicon carbide (SiC) is an interesting material for nuclear-reactor power monitor detectors. It has a wide band-gap, small volume and high break down electric field. In addition, SiC is chemically and neutronically inactive. Using SiC power monitors as in-core detectors provides the ability for high counting rate that may help to increase the safety margins of nuclear reactors. To observe the triton response in the SiC p-n diode, a detector with a 1.56 μm LiF converter (with 95% enriched ^{6}LiF) was used. ^{6}Li atoms in the LiF converter may absorb thermal neutrons and generate 2.05 MeV alpha and 2.73 MeV triton particles ($^{6}Li(n,^{3}H)\alpha$ reaction). An 8 μm Al layer was used to minimize damage in the SiC by blocking all alpha particles. However, most tritons have enough energy to pass through this layer and reach the 4.8 μm SiC active layer. The diameter of the LiF converter is 0.508 cm and the SiC diode area is 1.1 mm x 1.1 mm (diode is a square). The active area of the diode is approximately 0.965 mm^{2}. Upon irradiation in the thermal column (TC) facility, one can observe the triton peak in the recorded detector pulse-height spectra and the concomitant triton induced radiation damage on the detector. A schematic of the detector is shown in Figure 8.

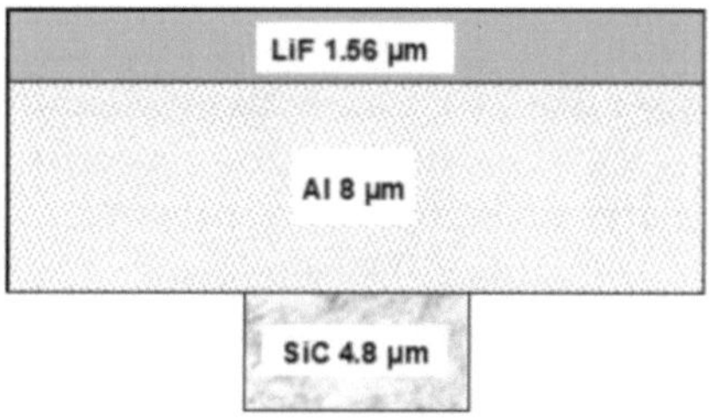

Fig. 8. Schematic of side view of SiC detector. The diameter of the LiF converter is 0.508 cm and the SiC diode active area is 0.965 mm^{2} (diode is a square). Only the active region of SiC is shown (Blue and Miller, 2008).

The SiC detector package was connected to a pulse processing system consisting of a preamplifier (ORTEK 142 B) and a digital spectrum analyzer (Canberra DSA 2000). An oscilloscope (Hewlett Packard 54601B, 100 MHz) was used to study the shape of the signal from the amplifier. Bias voltage was provided by the DSA to the detector through the preamplifier. A power monitoring program was used to verify the reactor power that was displayed in the control room. In addition, the degradation of the SiC detectors in the TC's thermal neutron environment was evaluated in terms of dose and dose rate effects. After irradiating the detector at 455 kW, the count rate per kW decreased by a factor of 2 after 11 hr. The I-V characteristics recorded during pre-irradiation and post-irradiation, confirm degradation of the detectors. A theoretical model of the SiC schottky diode detectors was constructed based on MCNP and TRIM computer codes to study the damage induced by tritons for a given diode detector package configuration in the TC's thermal neutron environment. The predicted count rate was compared with the experimental results that were obtained in the TC irradiation field using a charge sensitive preamplifier. The experimental results are in agreement with the predicted response to within a factor of three. I-V measurements show some annealing effects occurring at room temperature. Maintaining the detectors at a higher temperature during irradiation may cause more annealing to occur, thus reducing degradation of the detector. Experiments are necessary to test the degradation of the detector at elevated temperatures, to determine if the effects of annealing are sufficiently great so that the detectors may be useful for neutron power monitoring at high count rates.

6.3 Development of an inconel self powered neutron detector for in-core reactor monitoring

An inconel600 self-powered neutron detector has been developed and tested for in-core neutron monitoring (Alex, 2007). The sensing material in a self-powered detector is an emitter from which electrons are emitted when exposed to radiation. These electrons penetrate the thin insulation around the emitter and reach the outer sheath without polarising voltage. Some electrons are emitted from the insulator and sheath also. The net flow of electrons from the emitter gives rise to a DC signal in an external circuit between the emitter and sheath, which is proportional to the incident neutron flux. Rh and V SPDs work on the basis of (n, β) reaction and are used for flux mapping while Co and Pt SPDs work on the basis of (n, γ − e) prompt reaction and are used for reactor control and safety. However, the build-up of the ^{60}Co and ^{61}Co gives rise to background signal in the cobalt detector thereby reducing the useful life. In the case of the platinum detector, the detector responds to both reactor neutrons via (n, γ, e) interaction and reactor gamma rays via (γ, e) interaction. Since the neutron sensitivity varies with irradiation as a result of burn up while the gamma sensitivity remains the same, the dynamic response of a mixed response detector varies with time. This mixed and time-dependent response of platinum SPD gives rise to anomalous behaviour in some situations. Development of SPDs with inconel emitters as alternative to Co and Pt prompt SPDs has been reported in literatures. The detector (Figure 9) consists of a 2 mm diameter × 21 cm long inconel 600 emitter wire surrounded by a high purity alumina ceramic tube (2.2 mm ID × 2.8 mm OD). The assembly is enclosed in a 3 mm ID × 3.5 mm OD inconel600 tube.

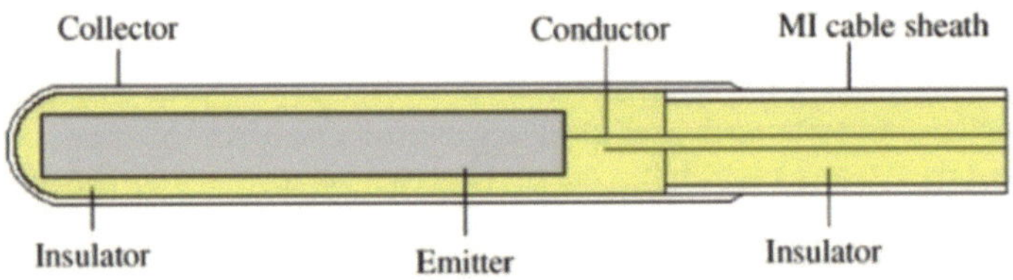

Fig. 9. Schematic diagram of self powered neutron detector

One end of the emitter is coupled to the conductor of a 2 mm diameter × 12 m long twin core mineral insulated (MI) cable while the detector sheath is laser welded to the MI cable sheath. The detector is integrally coupled to the MI cable and the cold end of the cable is sealed by a twin core ceramic-to-metal seal over which a Lemo connector is fitted.

The gamma sensitivity of the detectors was measured in pure gamma field using ^{60}Co source facility. The detectors were placed at a distance of 1m from the source for better source to detector geometry and 1m above the ground to minimize background from scattered rays. To estimate the gamma field at the detector location, a miniature gamma ion chamber (6 mm diameter and 25 mm long) was used. The calculated gamma sensitivity, 24.8 (fA R^{-1} h) was used to determine the gamma field at the self-powered neutron detector location. The three SPNDs (inconel600, cobalt, platinum) and were tested together with the miniature gamma chamber in a 200 kCi ^{60}Co source facility. The results showed that the gamma response of the inconel600 and Co detector was found to be similar. However, it was observed that unlike the platinum detector, which has positive response, the Co and inconel detectors showed negative response. The gamma sensitivity of the inconel600 detector is about 7.7 times lower than Pt detector. This low gamma response of the inconel600 detector improves the neutron to gamma ratio and makes it desirable for reactor safety and control applications. In addition to gamma sensitivity, the neutron sensitivity of SPNDs was tested in dry tube (55 mm diameter ×8.4 m long) in-core location of the Pool type reactor. The neutron sensitivity and the total sensitivity of the inconel600 detector were found to be lower than the Co detector. The total sensitivity of the inconel SPD is about 20–25% of the sensitivity of cobalt and about 35% of the sensitivity of platinum detectors of similar dimensions; however, it is proposed to improve the sensitivity by helically winding the detector with a short axial length. Finally by comparison, the performance of the inconel detector with cobalt and platinum detectors of similar dimensions, it is obvious that inconel SPD is a useful alternative to Co and Pt SPDs.

6.4 Monitoring the thermal power of nuclear reactors with a prototype cubic meter antineutrino detector

A new power monitoring method applied to a pressurized water reactors designed by combustion engineering. The method estimate quickly and precisely a reactor's operational status and thermal power can be monitored over hour to month time scales, using the antineutrino rate as measured by a cubic meter scale detector. Antineutrino emission in nuclear reactors arises from the beta decay of neutron-rich fragments produced by heavy element fissions, and is thereby linked to the fissile isotope production and consumption processes of interest for reactor safeguards. On average, fission is followed by the production of approximately six antineutrinos. The antineutrinos emerge from the core isotropically, and effectively without attenuation. Over the few MeV energy range within which, reactor antineutrinos are typically detected, the average number of antineutrinos produced per fission is significantly different for the two major fissile elements, ^{235}U and ^{239}Pu. Hence, as the core evolves and the relative mass fractions and fission rates of these two elements change, the measured antineutrino flux in this energy range will also change. It is useful to express the relation between fuel isotopic and the antineutrino count rate explicitly in terms of the reactor thermal power, P_{th}. The thermal power is defined as

$$P_{th} = \sum_i N_i^f . E_i^f \quad (7)$$

where N_i^f is the number of fissions per unit time for isotope i, and E_i^f is the thermal energy released per fission for this isotope. The sum runs over all fissioning isotopes, with ^{235}U, ^{238}U, ^{239}Pu, and ^{241}Pu accounting for more than 99% of all fissions. The antineutrino emission rate $n_{\bar{\nu}}(t)$ can then be expressed in terms of the power fractions and the total thermal power as:

$$n_{\bar{\nu}}(t) = P_{th}(t) \sum_i \frac{f_i(t)}{E_i^f} \int \varphi_i (E_{\bar{\nu}})\, dE_{\bar{\nu}} \quad (8)$$

where the explicit time dependence of the fission fractions and, possibly, the thermal power are noted. $\varphi(E_{\bar{\nu}})$, is the energy dependent antineutrino number density per MeV and fission for the ith isotope. $\varphi(E_{\bar{\nu}})$ has been measured and tabulated. Equation 7 defines the burn-up effect. The fission rates $N_i^f(t)$ and power fractions $f_i(t)$ change by several tens of percent throughout a typical reactor cycle as ^{235}U is consumed and ^{239}Pu produced and consumed in the core. These changes directly affect the antineutrino emission rate $n_{\bar{\nu}}(t)$. Reactor antineutrinos are normally detected via the inverse beta decay process on quasi-free protons in hydrogenous scintillator. In this charged current interaction, the antineutrino $\bar{\nu}$ converts the proton into a neutron and a positron: $\bar{\nu} + p \rightarrow e^+ + n$. For this process, the cross section σ is small, with a numerical value of only $\sim 10^{-43} \mathrm{cm}^2$. The small cross section can be compensated for with an intense source such as a nuclear reactor. For example, cubic meter scale hydrogenous scintillator detectors, containing $\sim 10^{28}$ target protons N_p, will register thousands of interactions per day at standoff distances of 10-50 meters from typical commercial nuclear reactors. In a measurement time T, the number of antineutrinos detected via the inverse beta decay process is:

$$N_{\bar{\nu}}(t) = \left(\frac{TN_p}{4\pi D^2}\right) P_{th}(t) \sum_i \frac{f_i(t)}{E_i^f} \int \sigma\, \varphi_i\, \epsilon\, dE_{\bar{\nu}} \quad (9)$$

In the above equation, σ is the energy dependent cross section for the inverse beta decay interaction, N_p is the number of target protons in the active volume of the detector, and D is the distance from the detector to the center of the reactor core. ϵ is the intrinsic detection efficiency, which may depend on both energy and time. The antineutrino energy density and the detection efficiency are folded with the cross section σ, integrated over all antineutrino energies, and summed over all isotopes i to yield the antineutrino detection rate. The SONGS1 detector consists of three subsystems; a central detector, a passive shield, and a muon veto system. Figure 10 shows a cut away diagram of the SONGS1 detector. Further information can be found in (Bowden, 2007) and (Bernstein et al., 2007).

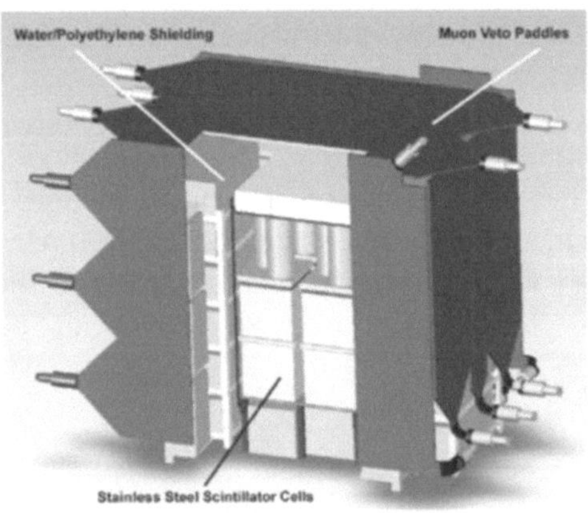

Fig. 10. A cut away diagram of the SONGS1 detector (showing the major subsystems).

This prototype that is operated at 25 meter standoff from a reactor core, can detect a prompt reactor shutdown within five hours, and monitor relative thermal power to 3.5% within 7 days. Monitoring of short-term power changes in this way may be useful in the context of International Atomic Energy Agency's (IAEA) Reactor Safeguards Regime, or other cooperative monitoring regimes.

6.5 Application of Cherenkov radiation and a designed detector for power monitoring

Cherenkov radiation is a process that could be used as an excess channel for power measurement to enhance redundancy and diversity of a reactor. This is especially easy to establish in a pool type research reactor (the TRR). A simple photo diode array is used in Tehran Research Reactor to measure and display power in parallel with the existing conventional detectors (Arkani and Gharib, 2009). Experimental measurements on this channel showed that a good linearity exists above 100 kW range. The system has been in use for more than a year and has shown reliability and precision. Nevertheless, the system is subject to further modifications, in particular for application to lower power ranges. TRR is originally equipped with four channels, namely, a fission chamber (FC), a compensated ionization chamber (CIC), and two uncompensated ionization chambers (UIC). However, in order to improve the power measuring system, two more channels have also been considered for implementation in recent years. One of these channels is based on 16O (n,p) ^{16}N reaction which is very attractive due to the short half life of ^{16}N (about 7 s). The other channel, at the center of our attention in this work, is based on measurement of Cherenkov radiation produced within and around the core. This channel has a fast response to power change and has been in operation since early 2007. It has been established that the movement of a fast charged particle in a transparent medium results in a characteristic radiation known as Cherenkov radiation. The bulk of radiation seen in and around a nuclear reactor core is mainly due to Beta and Gamma particles either from fission products or directly emanating from the fission process (prompt fission gamma rays). As it will be explained more thoroughly in the following section, Cherenkov radiation is produced through a number of ways when: (a) beta particles emitted by fission products travel with speeds greater than the speed of light in water and (b) indirect ionization by Gamma radiation produces electrons due to photo electric effect, Compton effect and pair production effect. Among these electrons, Compton electrons are the main contributors to Cherenkov radiation. It is established that Cherenkov light is produced by charged particles which pass through a transparent medium faster than the phase velocity of light in that medium. Considering the fact that speed of light in water is 220,000 km/s, the corresponding electron energy that is required to produce Cherenkov light is 0.26 MeV. This is the threshold energy for electrons that are energetic enough to produce Cherenkov light. It is the principal basis of Cherenkov light production in pool type research reactors in which the light is readily visible. For prompt Gamma rays, in general, it makes it possible to assume that Cherenkov light intensity is a linear function of reactor power. It is clear that neutron intensity, fission rate, power density, and total power itself are all inter-related by a linear relationship. In other words, Cherenkov light intensity is also directly proportional to the fission rate. This leads us to the fact that the measured Cherenkov light intensity at any point in a reactor is linearly proportional to the instantaneous power. As long as the measurement point is fixed, the total power could easily be derived from the light intensity with proper calibration. It should be noted here that, as mentioned before, Cherenkov light is also emitted by the electrons produced by the indirect ionization of fission products by Gamma rays, which are confined in fuel elements. For this reason, a linear relationship between reactor power and Cherenkov light intensity would only hold at the higher power range where fission power is dominant in comparison with residual power. Cherenkov light emanating from core is collected by a collimator right above the core and reflected by a mirror onto a sensitive part of the PDA. Figure 11 shows the integrated system at work, overlooking the core.

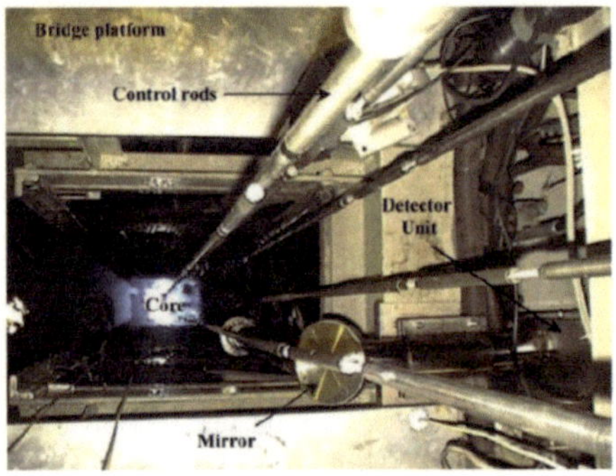

Fig. 11. Power measuring channel at work in TRR while receiving Cherenkov light (Arkani and Gharib, 2009).

An important factor to be checked is the system fidelity. This means that the response of the system must be the same when the reactor power is raised or lowered. There is a good fidelity within the linearity range by comparison of the Cherenkov system with the output of CIC power monitoring channel. Moreover, there has been no drift observed in the system in the long run as the system functioned properly for almost 2 years since it was installed. Finally, it is necessary to examine whether the reading from the Cherenkov detector is consistent with other channels. Finally, it is necessary to examine whether the reading from the Cherenkov detector is consistent with other channels. Figure 12 shows its good consistency with other conventional channels (only the fission chamber is shown for the sake of simplicity) within a typical shift operation.

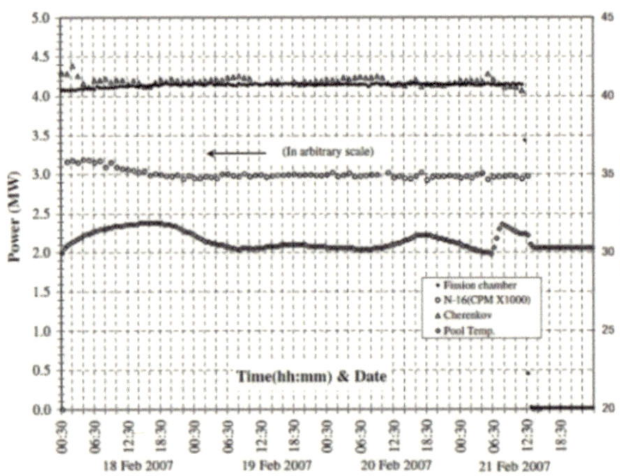

Fig. 12. Comparison of Cherenkov detector output with other regular channels within a typical operation shift of TRR (Arkani and Gharib, 2009).

It is observed that the steadiness and stability of the Cherenkov detector is as good as other existing channels. The ^{16}N counts and pool average temperature are also included as further confirmation of the general behavior of the reactor during the operation. Reasonable stability is observed in the hourly readings of all the channels. Based on statistics, the output value of the present PDA system is valid within ±1% at its nominal power. It is concluded that, at least for the case of research reactors, one can simply increase redundancy and diversity of medium-range reactors by employing the Cherenkov detector as an auxiliary tool for monitoring purposes. It is seen that such a system can provide a stable and reliable tool for the major part of power range, and it can assist in the reactor operation with additional safety interlocks to issue appropriate signals. The advantage of the present detector system over conventional ones is that it is far from the radiation source and thus easily accessible for maintenance and fine tuning. It contains no consumable materials to degrade in long term, and it is relatively inexpensive and simple. Nevertheless, a drawback of the Cherenkov system, which is also true about uncompensated ionization chambers, is its lack of linearity in the low power range.

6.6 A digital reactivity meter related to reactor power measuring process

Reactivity is a physical characteristic of the core (based on composition, geometry, temperature, pressure, and the ability of the core to produce fission neutrons) and may be either constant or changing with time. In reactor operation or experiments, signals indicating reactor power (or neutron flux) and reactor period are generally used for direct information on the state of the reactor. However, the most important time dependent parameter is reactivity and continuous information on its value from instant to instant should be highly useful. Since reactivity measurement is one of the challenges of monitoring, control and investigation of a nuclear reactor and is in relation with reactor power measuring. Thus, design and construction of a digital reactivity meter as a continuous monitoring of the reactivity will be reviewed in a research reactor. The device receives amplified output of the fission chamber, which is in *mA* range, as the input. Using amplifier circuits, this current is converted to voltage and then digitalized with a microcontroller to be sent to serial port of computer. The device itself consists of software, which is a MATLAB real time programming for the computation of reactivity by the solution of neutron kinetic equations. After data processing the reactivity is calculated and presented using LCD. Tehran research reactor is selected to test the reactivity meter device. The results of applying this reactivity meter in TRR are compared with the experimental data of control rod worth, void coefficient of reactivity and reactivity changes during approach to full power. Three experiments for system verification for TRR are; determination of control rod worth, void coefficient experiment, and measuring of reactivity during approach to full power (Khalafi and Mosavi, 2011). For investigating the results of reactivity meter, the reactor power and reactivity plots during the step-wise approach to full power of a particular run of TRR reactor are shown in Figure 13. In this experiment the reactor power was initially stable and critical at 100 kW and a positive reactivity insertion was introduced in the core by changes in control rods positions.

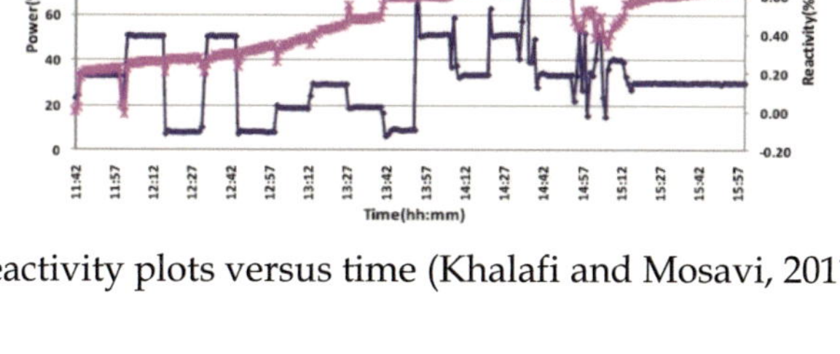

Fig. 13. Power and reactivity plots versus time (Khalafi and Mosavi, 2011).

The maximum relative error in three experiments is 13.3%. This error is caused by discrete signal that is transferred to the reactivity meter device. A great portion of the data is lost in the discrete signal and some others in the sampling process. As described in this section, the system of a digital reactivity meter developed on a PIC microcontroller and the personal computer is proved to function satisfactorily in the nuclear research reactor and the utilization of the plant instrument signals makes the system simple and economical. Besides, this device can be used to determine the positive reactivity worth of the fresh fuel and the reflector elements added to the core, effectively. According to the above experiments, the relative error of the digital reactivity meter can be reduced by increasing the sampling frequency of the device. Also by using digital signal processing (DSP) utilities, the rate and accuracy of the reactivity meter can be improved. Because derivative circuits are not used in this device, the error due to the noise that is observed in analog circuits decreases extremely.

7. Application of computational codes in simulation, modeling and development of the power monitoring tools

Some developed codes and simulators for improving the power monitoring will be reviewed in this section. For example, MCNP (monte-carlo n-particle transport code) is developed for neutron detector design, or modeling a fission chamber to optimize its performance

7.1 Computational tools to conduct experimental optimization

Research reactors need a handy computational tool to predict spatial flux changes and following power distribution due to experimental requirements. Therefore it is important to get accurate and precise information ahead of any modifications. To meet this demand, flux measurements were conducted in case that a typical flux trap inside the core to be allocated. In TRR, one of standard fuel boxes, in position D6 in core configuration of the year 1999, was taken out of the core and a water trap was formed in its place. With the aid of miniature neutron detector (MND) using standard procedure, thermal neutron flux is measured inside the water trap. To calculate the flux and power theoretically, two different computational approaches such as diffusion and Monte Carlo methods were chosen. Combination of cell calculation transport code, WIMS-D5, and three-dimensional core calculation diffusion code, such as CITATION, were used to calculate neutron flux inside the whole core either in two or five energy groups. However, MCNP-4, as a Monte Carlo code, was used to calculate neutron flux again inside the whole core as well as inside the trap (Khalafi and Gharib, 1999). Figure 14 shows axial thermal flux distribution along the D6 position by measurement and computation.

It is obvious from the figure; the both calculation codes are satisfactory and a good agreement exists between detector measurements and code computations. However, diffusion method is a rational choice especially for survey calculation where the Monte Carlo approach is more time demanding. For some consideration, in order to measure spectrum, a fixed point on the midplane along D6 axis was chosen. A variety of foils of different material was selected as measuring windows to determine differential fluxes at specified energy bins. Metal foils such as Ti, Se, Mg, Ni, Al, Co, Au, In, and Fe were selected as energy windows. These foils are sensitive to a part of neutron energy spectrum starting form high energies and ending to thermal energies. Induced activity of each foil is measured based on gamma spectroscopy using high purity Germanium (HPGe) detector. By providing raw counts to SAND-II computer code, neutron energy spectrum was calculated. The measured and calculated spectrum using neutron detector, MCNP and WIMS codes is shown in Figure 15.

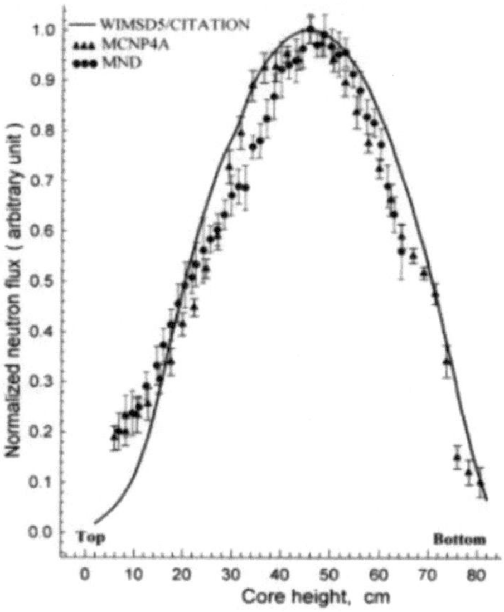

Fig. 14. Axial thermal neutron flux distribution in trap at D6 position (Khalafi and Gharib, 1999)

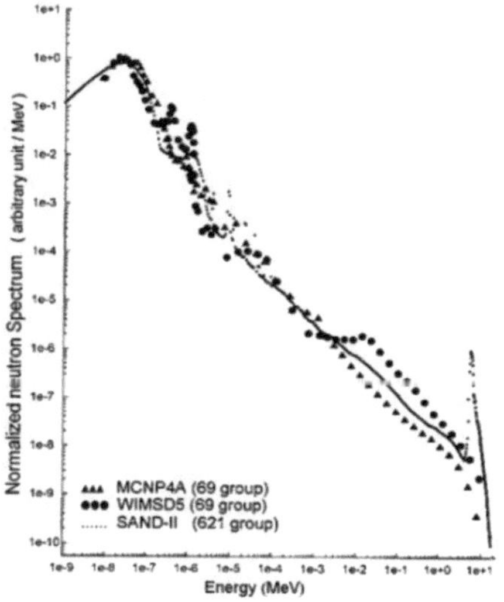

Fig. 15. Detector measured and MCNP and WIMS code calculated neutron spectrum(Khalafi and Gharib, 1999)

Spectrum calculations were also checked against measurements. Monte Carlo shows a better prediction while WIMS provides a fair result. It is notable that combination of WIMS/CITATION would be sufficient for neutron flux calculations while Monte Carlo technique should be reserved for the final stages of simulation. A good choice of computational tools would save time a lot in this respect and one is encouraged to perform a comprehensive simulation ahead of design and construction of irradiation facility.

7.2 Optimizing the performance of a neutron detector in the power monitoring channel of TRR

A fission chamber was utilized for neutron detection in TRR. It was a valuable instrument for in-core/out-core information and the core status monitoring during normal and transient operations. A general theoretical model is presented to calculate the current-voltage characteristics and associated sensitivity for a fission chamber. The chamber was used in the research nuclear reactor, TRR, and a flux-mapping experiment was performed. The experimental current measurement in certain locations of the reactor was compared with the theoretical model results. The characteristic curves were obtained as a function of fission rate, chamber geometry, and chamber gas pressure. An important part of the calculation was related to the operation of the fission chamber in the ionization zone and the applied voltages affecting two phenomena, recombination and avalanche. In developing the theoretical model, the MCNP code was used to compute the fission rate and the SRIM program for ion-pairs computations. In modeling the source for MCNP, the chamber was placed in a volume surrounded by standard air. Figure 16 illustrates the geometrical details of the MCNP simulation (Hashemi-Tilehnoee and Hadad, 2009).

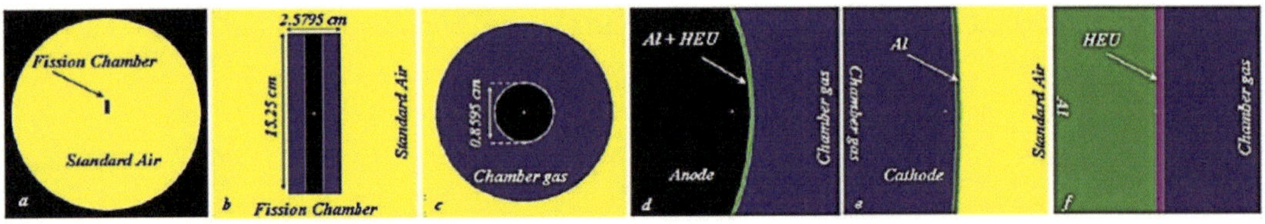

Fig. 16. The geometrical details of the MCNP simulation. (a) The chamber is placed in a volume surrounded by standard air, (b) the chamber geometrical details, width = 2.5795cm and length = 15.25 cm, (c) the cross section of the fission chamber, (d) the anode, (e) the cathode and (f) the fissile element coating.

The theoretical model together with the mentioned codes was used to evaluate the effects of different applicable variations on the chamber's parameters. An effective approach in decreasing the minimum voltage in the plateau zone, and retaining the chamber in the ionization zone, is to reduce the chamber gas pressure. However, by reducing the pressure, we decrease the gas density. This leads to the reduction of ion-pairs generation rate. Reduction of ion-pairs would affect the sensitivity. At high pressures, the plateau zone width would be extended. This extension needs a stronger electric field, which in turn causes the distortion of the electric field due to space charge effect. Thus, pressure is an important parameter in design considerations. Variations in the enrichment of the fissile element resulted in the enhancement of the fission rate and hence the sensitivity while retaining the applied voltage and plateau zone width. However, surface mass increase would require more applied voltage. Sensitivity of detection of the neutron flux would increase by decreasing the inter-electrode gap. In addition, it increases the width of the plateau zone. This extension optimizes the chamber performance and decreases the detection errors. Furthermore, by decreasing the inter-electrode gap, the fission chamber can be used in a low flux neutron surrounding for detection with high resolution. In contrast, by increasing the inter-electrode gap, the fission chamber can be used in a high flux nuclear reactor. Since the pressure variations have significant effects on the sensitivity, the detector components should be designed in accordance with the location, temperature, and neutron flux of the nuclear reactor core. Finally, applying the proper voltage not only enhances the sensitivity and readout, but also increases the longevity of the chamber.

In addition, the chamber is modeled by GEANT4 to evaluate its sensitivity to gamma ray, which exists as background. Figure 17 illustrates geometry of the modeled chamber in GEANT4. The unwanted noises from gamma ray in the core are dispensable, but in laboratory, this sensitivity must be accounted for the experiments as a disturbance signal.

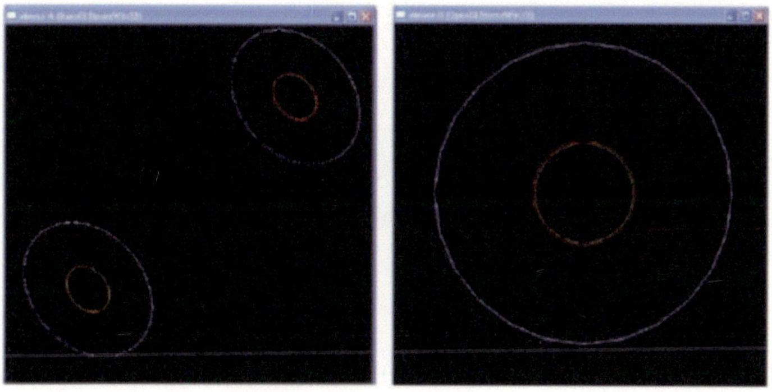

Fig. 17. Geometry of the modeled chamber in GEANT4

8. Thermal methods for power monitoring of nuclear reactor

Power monitoring using thermal power produced by reactor core is a method that is used in many reactors. To explain how the method is used for reactor power measurement, a research reactor is studied in this section. In IPR-R1, a TRIGA Mark I Research Reactor, the power is measured by four nuclear channels. The departure channel consists of a fission counter with a pulse amplifier that a logarithmic count rate circuit. The logarithmic channel consists of a compensated ion chamber, whose signal is the input to a logarithmic amplifier, which gives a logarithmic power indication from less than 0.1 W to full power. The linear channel consists of a compensated ion chamber, whose signal is the input to a sensitive amplifier and recorder with a range switch, which gives accurate power information from source level to full power on a linear recorder. The percent channel consists of an uncompensated ion chamber, whose signal is the input to a power level monitor circuit and meter, which is calibrated in percentage of full power. The ionization chamber neutron detector measures the flux of neutrons thermalized in the vicinity of the detector. In the present research, three new processes for reactor power measurement by thermal ways were developed as a result of the experiments. One method uses the temperature difference between an instrumented fuel element and the pool water below the reactor core. The other two methods consist in the steady-state energy balance of the primary and secondary reactor cooling loops. A stainless steel-clad fuel element is instrumented with three thermocouples along its centerline in order to evaluate the reactor thermal hydraulic performance. These processes make it possible on-line or off-line evaluation of the reactor power and the analysis of its behavior.

8.1 Power measuring channel by fuel and pool temperature

To evaluate the thermal hydraulic performance of the IPR-R1 reactor one instrumented fuel element was put in the core for the experiments. The instrumented fuel is identical to standard fuel elements but it is equipped with three chromel-alumel thermocouples, embedded in the zirconium pin centerline. The sensitive tips of the thermocouples are located one at the center of the fuel section and the other two 25.4 mm above, and 25.4 mm below the center. Figure 18 shows the diagram and design of the instrumented fuel element (Zacarias Mesquita and Cesar Rezende, 2010).

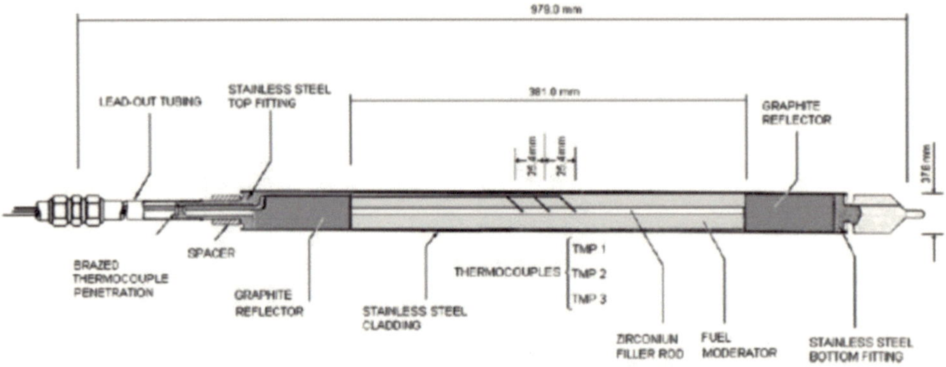

Fig. 18. Diagram of the instrumented fuel element (Zacarias Mesquita and Cesar Rezende, 2010)

The instrumented fuel element which is placed in proper thimble (B6 position) is obvious in Figure 19, a core upper view.

Fig. 19. Core upper view with the instrumented fuel element (Zacarias Mesquita and Cesar Rezende, 2010)

During the experiments it was observed that the temperature difference between fuel element and the pool water below the reactor core (primary loop inlet temperature) do not change for the same power value. Figure 20 compares the reactor power measuring results using the linear neutron channel and the temperature difference channel method (Zacarias Mesquita and Cesar Rezende, 2007).

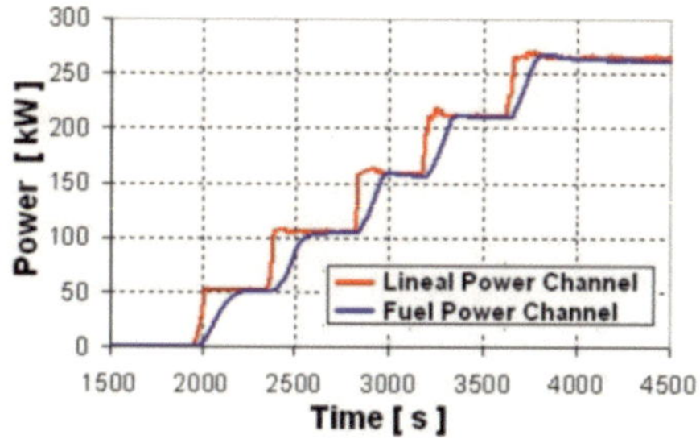

Fig. 20. Reactor power measured by neutron channel and by fuel element temperature (Zacarias Mesquita and Cesar Rezende, 2007).

There is a good agreement between the two results, although the temperature difference method presents a delay in its response, and it is useful for steady-state or very slow transient. It is notable that the thermal balance method presented in this report is now the standard methodology used for the IPR-R1 TRIGA Reactor power calibration. The heat balance and fuel temperature methods are accurate, but impractical methods for monitoring the instantaneous reactor power level, particularly during transients. For transients the power is monitored by the nuclear detectors, which are calibrated by the thermal balance method (Zacarias Mesquita and Cesar Rezende, 2007).

8.2 Power measuring channel by thermal balance

The new developed on-line monitoring method which is based on a temperature difference between an instrumented fuel element and the pool water below a research reactor in practice, as known power measuring by thermal balance is as following. The reactor core is cooled by natural convection of demineralized light water in the reactor pool. Heat is removed from the reactor pool and released into the atmosphere through the primary cooling loop, the secondary cooling loop and the cooling tower. Pool temperature depends on reactor power, as well as external temperature, because the latter affects heat dissipation in the cooling tower. The total power is determined by the thermal balance of cooling water flowing through the primary and secondary loops added to the calculated heat losses. These losses represent a very small fraction of the total power (about 1.5% of total). The inlet and outlet temperatures are measured by four platinum resistance thermometers (PT-100) positioned at the inlet and at the outlet pipes of the primary and secondary cooling loops. The flow rate in the primary loop is measured by an orifice plate and a differential pressure transmitter. The flow in the secondary loop is measured by a flow-meter. The pressure transmitter and the temperature measuring lines were calibrated and an adjusted equation was added to the data acquisition system. The steady-state is reached after some hours of reactor operation, so that the power dissipated in the cooling system added with the losses should be equal to the core power. The thermal power dissipated in the primary and secondary loops were given by:

$$q_{cool} = \dot{m}.c_P.\Delta T \qquad (10)$$

where q_{cool} is the thermal power dissipated in each loop (kW), $\dot{m}$ is the flow rate of the coolant water in the loop (kg. s^{-1}), c_P is the specific heat of the coolant (kJ kg^{-1} °C^{-1}), and ΔT is the difference between the temperatures at loop the inlet and outlet (°C). Figure 21 shows the power evolution in the primary and secondary loops during one reactor operation.

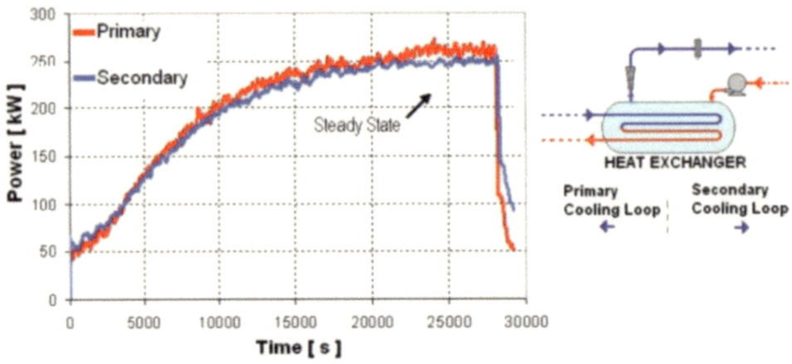

Fig. 21. Thermal power evolution in the cooling system (Zacarias Mesquita and Cesar Rezende, 2007).

9. Acknowledgement

The authors are thankful InTech publication for their scientific effort to develop the sciences. We also, thank Mrs. M. Mohammadi for her effort to preparing the manuscript.

10. Conclusion

Power monitoring channels play a major role in retaining a safe reliable operation of nuclear reactors and nuclear power plants. Accurate power monitoring using advanced developed channels could make nuclear reactors a more reliable energy source and change public mind about this major energy resource. Regarding harsh accidents such as Chernobyl, Three-Mile Island and the recent accidents in Fukushima nuclear power plants and their dangerous effects on the environment and human life, the importance of developing reactor safety system like power monitoring channels are more attended. New generations of nuclear power plants are much safer than their predecessors because of their new accurate safety systems and more reliable monitoring channels. They produce energy from nuclear fission and are the cleanest, safe and environment-friendly source of energy among many investigated power resources (Javidkia et al., 2011).

There is no doubt that nuclear power is the only feasible green and economic solution for today's increasing energy demand. Therefore, studying, researches and more investments on the power monitoring systems and channel in nuclear reactors will help to create an inexhaustible source of safe and clean energy.

11. References

IAEA, (2008). *On-line monitoring for improving performance of nuclear power plants part 2: process and component condition monitoring and diagnostics,* International atomic energy agency (IAEA), Vienna, ISSN 1995–7807; no. NP-T-1.2, STI/PUB/1323, ISBN 978–92–0–101208–1.

Department of energy, (1993). DOE fundamentals handbook, Nuclear physics and reactor theory, Vol.1 and 2, DOE-HDBK-1019/1-93 & 2-93.

Sakai, H.; Oda, N.; Miyazaki, T.; Sato, T. (2010). *Power monitoring system,* US 2010/0254504 A1.

Cho, B.; Shin, C.; Byun, S.; Park, J.; Park, S. (2006). *Development of an innovative neutron flux mapping system,* Nuclear Engineering and Design, Vol. 236, 1201–1209.

Knoll, G.E. (2000). *Radiation detection and measurement,* 3rd edition, ISBN 0-471-07338 , John Wiley & Son.

Tahir Khaleeq, M.; Zaka, I.; Qaiser, Hameed.; Nayyar, B. H.; Ahmad Ghumman, I.; Ali, A. (2003). *A New Wide-Range Reactor Power-Measuring Channel,* IEEE transactions on nuclear science, Vol. 50, No. 6.

Sadeghi, N. (2010). *Estimation of reactor power using N16 production rate and its radiation risk assessment in Tehran Research Reactor (TRR),* Nuclear Engineering and Design, Vol. 240, 3607–3610.

McGregor, D. S.; F. Ohmes, M.; E. Ortiz, R.; Sabbir Ahmed, A.S.M.; Kenneth Shultis, J. (2005). Micro-pocket fission detectors (MPFD) for in-core neutron flux monitoring, Nuclear Instruments and Methods in Physics Research A, Vol. 554, 494–499.

Blue, T. E.; Miller, D.W. (2008). *Nuclear reactor power monitoring using silicon carbide semiconductor radiation detectors,* nuclear energy research initiative (NERI), Project Number DE-FG03-02SF22620.

Alex, M.; Ghodgaonkar, M.D. (2007). *Development of an inconel self powered neutron detector for in-core reactor monitoring.* Nuclear Instruments and Methods in Physics Research A, Vol. 574, 127–132.

Bowden, N.S.; Bernstein, A.; Allen, M.; Brennan, J.S.; Cunningham, M., Estrada, J.K.; Greaves, C.M.R.; Hagmann, C.; Lund, J.; Mengesha, W.; Weinbeck, T.D.; Winant, C.D. (2007). *Experimental results from an antineutrino detector for cooperative monitoring of nuclear reactors,* Nuclear Instruments and Methods in Physics Research A, Vol. 572, 985–998.

Bernstein, A.; Bowden, N. S.; Misner, A.; Palmer, T. (2007). *Monitoring the thermal power of nuclear reactors with a prototype cubic meter antineutrino detector,* Journal of Applied Physics, UCRL-JRNL-233165.

Arkani, M.; Gharib, M. (2009), *Reactor core power measurement using Cherenkov radiation and its application in Tehran Research Reactor,* Annals of Nuclear Energy, Vol. 36 , 896–900.

Khalafi, H.; Mosavi, S.H.; Mirvakili, S.M. (2011), *Design and construction of a digital real time reactivity meter for Tehran research reactor,* Progress in Nuclear Energy, Vol. 53 , 100-105.

Khalafi, H.; Gharib, M. (1999). *Calculational tools to conduct experimental optimization in Tehran Research Reactor,* Annals of Nuclear Energy, Vol. 26 , 1601–1610.

Hashemi-Tilehnoee, M. ; Hadad, K. (2009). *Optimizing the performance of a neutron detector in the power monitoring channel of Tehran Research Reactor (TRR),* Nuclear Engineering and Design, Vol. 239, 1260–1266.

Zacarias Mesquita, A.; Cesar Rezende, H. (2010). *Thermal methods for on-line power monitoring of the IPR-R1 TRIGA Reactor,* Progress in Nuclear Energy, Vol. 52, 268–272.

Zacarias Mesquita, A.; Cesar Rezende, H. (2007). *Power measure channels of the IPR-R1 TRIGA research nuclear reactor by thermal methods*, 19th International Congress of Mechanical Engineering, Proceedings of COBEM 2007, Brazil.

Javidkia, F.; Hashemi-Tilehnoee, M.; Zabihi ,V. (2011) *A Comparison between Fossil and Nuclear Power Plants Pollutions and Their Environmental Effects*, Journal of Energy and Power Engineering, Vol. 5, 811-820.

12

Multiscale Materials Modeling of Structural Materials for Next Generation Nuclear Reactors

Chaitanya Deo
*Nuclear and Radiological Engineering Programs,
George W. Woodruff School of Mechanical Engineering,
Georgia Institute of Technology
USA*

1. Introduction

In the process of energy production via fission – and fusion in the years to come - both fuel components and structural materials within nuclear reactors can sustain substantial radiation damage. Regardless of the type of reactor, this damage initially appears in the form of local intrinsic point defects within the material - vacancies and interstitials. The point defects agglomerate, interact with the underlying microstructure and produce effects such as void swelling and irradiation creep. Vacancies provide a pathway for solutes to segregate to grain boundaries and dislocation leading to chemical inhomogeneities that translate into phase transformations and/or property variations in these materials, rendering them unsuitable for the desired application.

Many deleterious effects of irradiation on material properties—e.g. void swelling, irradiation creep, radiation-induced hardening and embrittlement—can be traced back to the formation of the aforementioned point defect clusters and gas bubbles. These effects include such phenomena as swelling, growth, phase change, segregation, etc. For example (Was 2007), a block of pure nickel, 1cm on a side, irradiated in a reactor (to a fluence of say, 10^{22} n/cm^2) will measure 1.06cm on a side, representing a volume change of 20%. The volume change, or swelling, is isotropic and is due to the formation of voids in the solid. Other examples are irradiation growth which is distortion at constant volume, phase changes under irradiation where new phases form as a consequence of diffusion of supersaturated defect concentrations and radiation induced segregation.

In addition, the transmutation of reactor elements produces extrinsic defects such as hydrogen, deuterium and helium. For example, Zircalloy high-pressure-tubes used in light water reactors are known to absorb deuterium which can cause delayed hydride cracking (Cirimello, G. et al. 2006). Similarly, in Pebble Bed Modular reactors and in other technologies based on inert gas cooling, formation of ionic gas bubbles within both fuel and structural materials is common(Was 2007). This is critical in structural materials as their behaviour depends on their microstructure, which is in turn affected by neutron radiation.

Cladding materials are also exposed to fission product gasses produced in the fuel during operation.

The effect of irradiation on materials is a classic example of an inherently multiscale phenomenon, as schematically illustrated in Figure 1 for the case of deformation and plasticity in irradiated materials. Irradiation changes the properties of cladding and duct materials as follows: increase in the ductile to brittle transition temperature; reduction of fracture toughness from low temperature irradiation (below 400°C); irradiation creep, helium embrittlement (above ~500C) and swelling. Experimental irradiation programs can be conducted to test the irradiation induced mechanical property changes, but these take significant time and the conditions are limited. As a consequence, it is desirable and efficient to develop models with reliable predictive capabilities for both, design of new reactor components, and simulation of the in-service response of existing materials.

Although these phenomena have been known for many years (Olander 1981; de la Rubia, Zbib et al. 2000; Was 2007) , the underlying fundamental mechanisms and their relation to the irradiation field have not been clearly demonstrated. Most models and theories of irradiation induced deformation and mechanical behavior rely on empirical parameters fit to experimental data. With improvements in computational techniques and algorithms, it is now possible to probe structure-property connections through the elucidation of fundamental atomic mechanisms. Often, these mechanisms involve defects of different dimensionality that exist and interact with each other to significantly affect material properties.

For irradiated materials, point defects and clusters affect crystal plasticity. Dislocation-defect interactions may be associated with defects in the matrix, the modification of the local elastic constants due to the presence of defects and defect clusters and the effects of clusters and voids on the stacking fault energy. The formation of self interstitial loop rafts and the decoration of dislocations with self interstitial clusters have become important issues for understanding radiation hardening and embrittlement under cascade damage conditions(Wen, Ghoniem et al. 2005). Dislocation motion is thought to be the main mechanism for deformation, because a fairly high density of network dislocations are generated during irradiation, and the dislocation sink strength for point defects is much higher that the grain boundary sink strength for point defects. Computational studies of dislocation activity can be performed at several different length and time scales(Ghoniem, Busso et al. 2003) that are shown in Fig. 1.

Pertinent processes span more than 10 orders of magnitude in length from the sub-atomic nuclear to structural component level, and span 22 orders of magnitude in time from the sub-picosecond of nuclear collisions to the decade-long component service lifetimes (Odette, Wirth et al. 2001; Wirth, G.R. et al. 2004). Many different variables control the mix of nano/microstructural features formed and the corresponding degradation of physical and mechanical properties in nuclear fuels, cladding and structural materials. The most important variables include the initial material composition and microstructure, the thermo-mechanical loads, and the irradiation history. While the initial material state and thermo-mechanical loading are of concern in all materials performance-limited engineering applications, the added complexity introduced by the effects of radiation is a large concern for materials in advanced nuclear energy systems.

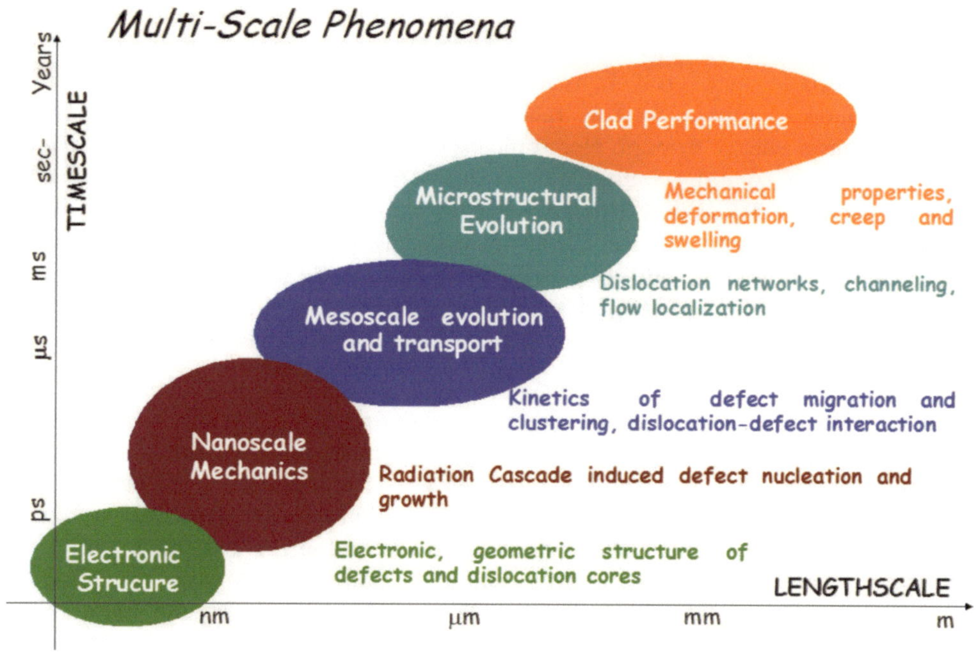

Fig. 1. Multiscale processes that govern deformation processes in irradiated materials

The scientific challenge for next-generation extreme materials - whatever their composition - is to understand their failure modes, and to prolong their useful lifetimes by interrupting or arresting these failures. Damage starts with atomic displacements that create interstitials and vacancies, which then migrate and aggregate to form clusters and ever-larger extended structures. Eventually, the damage reaches macroscopic dimensions, leading to degradation of performance and failure. This problem is massively multiscale, covering nine orders of magnitude in its spatial dimension, and neither experiment nor theory has yet captured this complexity in a single framework.

On the experimental side, *in situ* measurements of neutron irradiation with atomic or nano-scale resolution are needed to observe the initial damage processes, followed by coarser-grained experiments to capture migration, aggregation and ultimately macroscopic failure. The modeling challenge is equally dramatic: kinetic energy from an incident particle is transferred successively to electronic, atomic, vibrational and structural systems, requiring a diverse mix of theoretical formulations appropriate for different spatial scales.

This chapter will review choices for structural materials for these environments, and review the methods and techniques available for simulating these materials at various length and time scales Methods that will be reviewed will include first principles calculations, molecular dynamics calculations, kinetic Monte Carlo methods, and microstructural mechanics methods.

2. Radiation damage in materials

At the smallest scales, radiation damage is continually initiated with the formation of energetic primary knock-on atoms (PKA) primarily through elastic collisions with high-energy neutrons. Concurrently, high concentrations of fission products (in fuels) and transmutants (in cladding and structural materials) are generated, which can cause pronounced effects in the overall chemistry of the material, especially at high burnup. The primary knock-on atoms, as well as recoiling fission products and transmutant nuclei quickly lose kinetic energy through electronic excitations (that are not generally believed to produce atomic defects) and a chain of atomic collision displacements, generating cascade of vacancy and self-interstitial defects. High-energy displacement cascades evolve over very short times, 100 picoseconds or less, and small volumes, with characteristic length scales of 50 nm or less, and are directly amenable to molecular dynamics (MD) simulations if accurate potential functions are available and chemical reactions are not occurring. If change in electronic structure need to be included, then ab initio MD is needed and this is beyond current capabilities.

In order to simulate the appropriate reactor conditions for all models, it is important to connect the parameters of the atomistic models with reactor conditions and the type of irradiation encountered. The radiation damage event is composed of several distinct processes concluded by a displacement cascade (collection of point defect due to the PKA) and by the formation of an interstitial -which occurs when the PKA comes to rest-. In order to simulate the radiation effects, it is important to determine the type of energetic particle interaction we wish to model. In nuclear reactors, neutrons and charged fission product particles are the dominant energetic species produced (Beta and Gamma rays are also produced, but these create less damage than the neutrons and charged particles). The type of reactor determines the nature of the dominant energetic particle interaction. The proposed study will focus on neutrons and He ions. The additional aspect to consider concerns the energy of the PKA, which in the case of D-T fusion reaction can reach the order of ~1MeV. These energies are out of reach of atomistic simulations. Nonetheless cascade event simulations at lower energies -ranging from 5to 45 KeV- can yield significant insight on the evolution of defect type and number as a function of PKA energy. Was (2007) and Olander (1981) have extensively documented how it is possible to determine the primary damage state due to irradiation by energetic particles. The simplest model is one that approximates the event as colliding hard spheres with displacement occurring when the transferred energy is high enough to knock the struck atom off its lattice site.

The physics of primary damage production in high-energy displacement cascades has been extensively studied with MD simulations(Was 2007). The key conclusions from those MD studies of cascade evolution have been that i) intra-cascade recombination of vacancies and self-interstitial atoms (SIAs) results in ~30% of the defect production expected from displacement theory, ii) many-body collision effects produce a spatial correlation (separation) of the vacancy and SIA defects, iii) substantial clustering of the SIAs and to a lesser extent, the vacancies occurs within the cascade volume, and iv) high-energy displacement cascades tend to break up into lobes, or sub-cascades which may also enhance recombination(Calder and Bacon 1993; Calder and Bacon 1994; Bacon, Calder et al. 1995; Phythian, Stoller et al. 1995).

It is the subsequent transport and evolution of the defects produced during displacement cascades, in addition to solutes and transmutant impurities, that ultimately dictate radiation effects in materials, and changes in material microstructure(Odette et al. 2001; Wirth et al. 2004). Spatial correlations associated with the displacement cascades continue to play an important role over much larger scales, as do processes including defect recombination, clustering, migration and gas and solute diffusion and trapping. Evolution of the underlying materials structure is thus governed by the time and temperature kinetics of diffusive and reactive processes, albeit strongly influenced by spatial correlations associated with the sink structure of the microstructure and the continual production of new radiation damage. Extended defects including dislocations, precipitate interfaces and grain boundaries, which exist in the original microstructure and evolve during irradiation, serve as sinks for point defect absorption and for vacancy - self-interstitial recombination and/or point defect - impurity clustering.

The inherently wide range of time scales and the "rare-event" nature of the controlling mechanisms make modeling radiation effects in materials extremely challenging and experimental characterization is often unattainable. Indeed, accurate models of microstructure (point defects, dislocations, and grain boundaries) evolution during service are still lacking. To understand the irradiation effects and microstructure evolution to the extent required for a high fidelity nuclear materials performance model will require a combination of experimental, theoretical, and computational tools.

Furthermore, the kinetic processes controlling defect cluster and microstructure evolution, as well as the materials degradation and failure modes may not entirely be known. Thus, a substantial challenge is to develop knowledge of, and methodologies to determine, the controlling processes so that they can be included within the models. Essentially, this is to avoid the detrimental consequences of in-service surprises. A critical issue that needs to be addressed is not only the reliability of the simulation but also the accuracy of the model for representing the critical physical phenomena.

3. Multiscale materials models

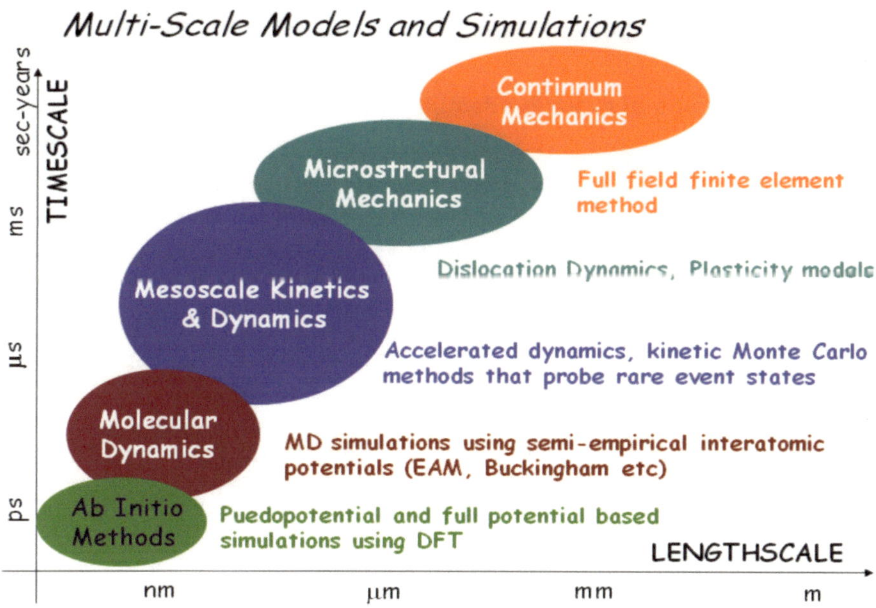

Fig. 2. Illustration of the length and time scales (and inherent feedback) involved in the multiscale processes responsible for microstructural changes in irradiated materials

A hierarchy of models is employed in the theory and simulation of complex systems in materials science and condensed matter physics: macroscale continuum mechanics, macroscale models of defect evolution, molecular scale models based on classical mechanics, and various techniques for representing quantum-mechanical effects. These models are classified according to the spatial and temporal scales that they resolve (Figure 2). In this figure, individual modeling techniques are identified within a series of linked process circles showing the overlap of relevant length and timescales. The modeling methodology includes *ab initio* electronic structure calculations, molecular dynamics (MD), accelerated molecular dynamics, kinetic Monte Carlo (KMC), phase field equations or rate theory simulations with thermodynamics and kinetics by passing information about the controlling physical mechanisms between modeling techniques over the relevant length and time scales. The key objective of such an approach is to track the fate of solutes, impurities and defects during irradiation and thereby, to predict microstructural evolution. Detailed microstructural information serves as a basis for modeling the physical behavior through meso (represented by KMC, dislocation dynamics, and phase field methods) and continuum scale models, which must be incorporated into constitutive models at the continuum finite element modeling scale to predict performance limits on both the test coupons and components.

4. Modeling and simulation examples

To span length and temporal scales, these methods can be linked into a multi-scale simulation. The objective of the lower length scale modeling is to provide constitutive properties to higher length scale, continuum level simulations, whereas these higher length scale simulations can provide boundary conditions to the lower length scale models, as well as input regarding the verification/validity of the predicted constitutive properties. Here some examples are provided of models and simulations of materials under irradiation. The emphasis is on the development of the model, the assumptions and the underlying physics that goes into model development.

4.1 Atomistic simulations of radiation damage cascades

Damage to materials caused by neutron irradiation is an inherently multiscale phenomenon. Macroscopic properties, such as plasticity, hardness, brittleness, and creep behavior, of structural reactor materials may change due to microstructural effects of radiation. Atomistic models can be a useful tool to generate data about the structure and development of defects, on length and time scales that experiments cannot probe. Data from simulations can be fed into larger scale models that predict the long term behavior of materials subject to irradiation.

The amount of energy that an incident particle can transfer to a lattice atom is a function of their masses and the angle at which the collision occurs. Energy can be lost through inelastic collisions, (n, 2n) or (n, γ) reactions, and, most importantly, elastic collisions. Elastic collisions between neutrons and nuclei can be treated within the hard sphere model with the following equations:

$$T = \frac{\gamma}{2} E_i (1 - cos\theta) \tag{1}$$

$$\gamma = \frac{4A}{A+1} \tag{2}$$

where, T is the total energy transferred, E_i is the energy of the incident neutron, θ is the angle of collision, and A is the mass of the lattice nucleus. With the assumption that scattering is isotropic in the center of mass system, the average energy transferred over all angles can be shown to be the average of the minimum and maximum possible transfer energies, i.e., .

For iron, the energy required to displace an atom is about 40 eV, depending on the direction from which it is struck. So, a neutron needs a minimum energy of about 581 eV to displace an iron atom. Neutrons produced from fission of uranium carry around 2 MeV of kinetic energy and so have potential to cause damage as they slow down. Additionally, deuterium-tritium fusion reactions produce neutrons with energy of 14.1 MeV.

The first attempt to create a model for defect production based on PKA energy comes from Kinchin and Pease (1955). In this model, above a certain threshold Ed, energy is lost only to electron excitation, while below it, energy is lost only in hard-sphere elastic scattering. Norgett, Robinson et al. (1975) proposed a revised model, taking into account a more realistic energy transfer cross section, based on binary collision model simulations, . Here, N_D is the number of Frenkel pairs surviving relaxation and the damage energy E_D is the amount of energy available for creating displacements through elastic collisions and is a function of T. Since some of the energy of the cascade is lost to electronic excitation, E_D will be less than T; for the energy range considered in this paper E_D can be estimated as equal to T. This model is frequently used as the standard for estimating DPA, but many molecular dynamics simulations have shown that it tends to strongly overestimate the actual damage efficiency. Bacon, Gao et al. (2000)proposed an empirical relationship between N_D and T,where A and n are weakly temperature dependent constants fit to particular materials, respectively equal to 5.57 and 0.83 for Fe at 100 K, and T is in keV.

The study of irradiation damage cascades has been a popular topic over the last fifteen or so years. A through literature review of the many different of damage cascade simulations, such as binary collision approximation and kinetic Monte Carlo, that have been performed in a variety of materials is beyond the scope of this paper and readers are referred to (Was 2007). The following brief review will concentrate solely on molecular dynamics simulations in α-Fe. A thorough review of results from many papers was written by Malerba (2006).

Malerba (2006) states that the first published MD study in alpha-iron was performed by Calder and co-workers (Calder and Bacon 1993; Calder and Bacon 1994). Eighty cascades with PKAs of up to 5 keV were analyzed for properties such as percent of defects surviving relaxation, channeling properties, temperature dependence, and clustering. The interatomic potential used was developed by (Finnis and Sinclair 1984) and stiffened by Calder and Bacon to treat small interatomic distance properly. This article established a large base of data for future papers to compare with.

Following this initial study, many papers came out which utilized both the modified FS potential mentioned above and competing multi-body potentials including those from Johnson and Oh (1989), Harrison, Voter et al. (1989), and Simonelli, Pasianot et al. (1994). These papers had three main motivations: to generate data from a new potential, to compare data between two or more potentials, or to compare damage in α-Fe with that in another material such as copper. The main difficulties in comparing results from different authors are defining what makes up a cluster of defects and non-reporting of exactly how cascades were generated.

Many authors contributed to generate databases; some papers of note are described here. Stoller, Odette et al. (1997), using the FS potential modified by Calder and Bacon, ran a number of cascades at energies up to 40 keV. They found evidence for vacancy clustering, a feature not seen in previous works. Bacon et al. (2000) performed a study comparing the cascade characteristics of bcc, hcp, and fcc metals. They found that there were no major differences in interstitial and vacancy production, so concluded that any differences observed experimentally must be due to evolution of the microstructure following the primary damage event. Caturla, Soneda et al. (2000) compared bcc Fe with fcc Cu, finding that clustering in Fe was at least an order of magnitude less than in Cu. Terentyev, Lagerstedt et al. (2006) produced a study looking solely at differences between four available potentials by applying the same defect counting criteria to each. They found that the stiffness of a potential, a somewhat arbitrary feature, was the most important factor in determining cascade properties.

In all primary damage cascade simulations, first, the incident radiation has an interaction with an atom in the crystal lattice, transferring enough energy to remove the atom from its site. This atom, the primary knock-on atom (PKA), goes on to interact with other atoms in the crystal, removing them from their sites and generating a displacement cascade in the thermal spike phase. Atoms that are removed from their perfect lattice sites and come to rest between other atoms are known as interstitials; the empty lattice sites they leave are called vacancies. At some time shortly after the PKA is created, some peak number of Frenkel pairs, N_p, will exist in the crystal, where a Frenkel pair is defined as one vacancy plus one interstitial. After this point, the defects will begin to recombine as the energy is dissipated. After a few picoseconds, only a few defects, N_d will remain. This generally results in a core of vacancies surrounded by a shell of interstitials. A profile of the number of defects over time in a typical cascade can be seen in Figure 3.

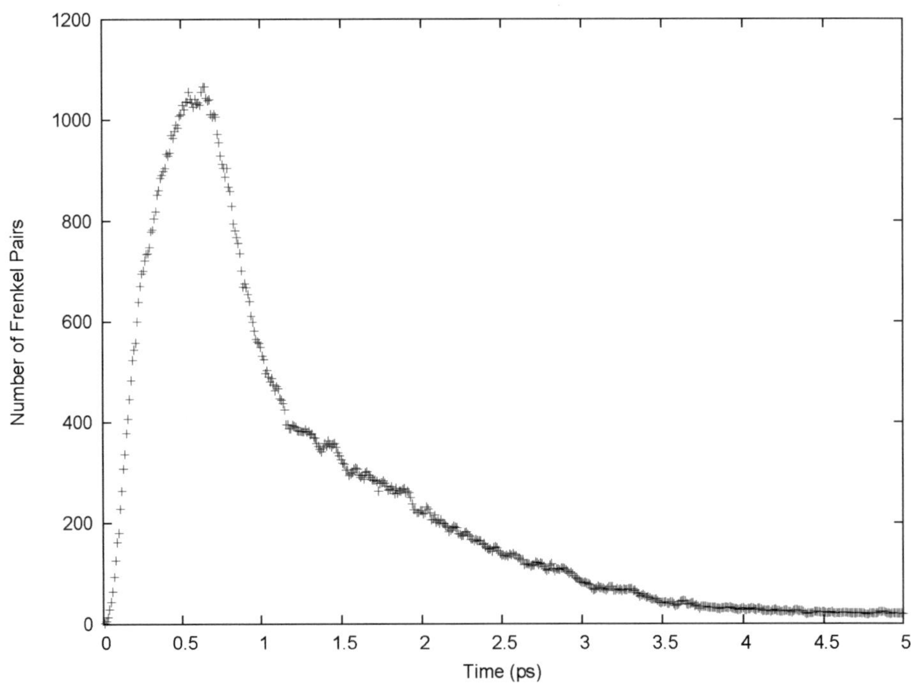

Fig. 3. Initial stages of radiation damage cascade, the number of vacancies and interstitials as a functions of time (from Hayward and Deo 2010)

The atomistic simulations provide information on the number of surviving defects after initial damage and are not able to simulate large time scale or length scales. They provide a good atomistic picture of the unit processes affecting the formation of defects and the evolution of the primary damage state. Experiments cannot yet access this small time and length scale of radiation damage processes, therefore, experimental corroboration is hard to find for such simulations. Results of atomistic simulations of radiation cascades can be used to develop higher length and time scale theories and simulation of radiation effects in materials. Parameters that can be passed to other simulations/theories include the number and spatial distribution of defects created at the conclusion of the radiation cascade phase.

4.2 Molecular dynamics calculations of dislocation defect interactions

While atomistic methods can probe the primary damage state with great detail, they can also be used to probe the interactions of the defects formed with the underlying microstructure. An example is the case of creep due to irradiation in materials. Creep of metals and alloys under irradiation has been the subject of many experimental and theoretical studies for more than 30 years. Although a vast amount of knowledge of irradiation creep has accumulated, the database on irradiation creep comes from many relatively small experiments, and there were often differences in experimental conditions from one study to the next. Theoretical models are based on linear elasticity. Among the many theories that exist to describe the driving force for irradiation creep, the most important are the SIPN, SIPA, and SIPA-AD effects.

Stress Induced Preferential Nucleation of loops (SIPN) is based on the idea that the application of external stress will result in an increased number of dislocation loops nucleating on planes of preferred orientations. Interstitial loops will tend to be oriented perpendicular to the applied tensile stress, while vacancy loops will prefer to be oriented parallel to the stress. The net result is elongation of the solid in the direction of applied stress. While there is some experimental support of this theory, it is thought that it cannot account fully for creep seen in materials.

An alternative theory is Stress Induced Preferential Absorption/Attraction (SIPA). The essential idea behind SIPA is that interstitials are preferentially absorbed by dislocations of particular orientations, resulting in climb; this is described by an elastic interaction between the stress fields of the defect and dislocation. A variant on SIPA that accounts for anistropic diffusion is SIPA-AD. This theory uses the full diffusion equations, derived by Dederichs and Schroeder (1978), to take into account anisotropic stress fields. Savino and Tome developed this theory and found that it generally gives a larger contribution to dislocation climb than the original SIPA (Tome, Cecatto et al.). A thorough review of many dislocation creep models was prepared by Matthews and Finnis (1988).

These models go a long way towards explaining irradiation creep due to dislocations. However, all models based on linear elasticity break down near a dislocation core due to the $1/r$ terms in the stress and strain field expressions. Atomistic calculations do not suffer from this problem, so they can be used to verify the range of validity of theoretical expressions and successfully predict true behavior at the core.

Molecular statics calculations can be performed in order to understand the interactions between vacancies and interstitials and line dislocations in bcc iron. These can be compared to similar results given by dipole tensor calculations based in linear elasticity theory. Results from two methods are used to calculate the interaction energy between a dislocation and a point defects in bcc iron are compared. For vacancies and a variety of self-interstitial dumbbell configurations near both edge and screw dislocation cores, there are significant differences between direct calculations and atomistics. For vacancies some interaction is seen with both edge and screw dislocations where none is predicted. Results for interstitials tended to have a strong dependence on orientation and position about the core. Particularly for the screw, continuum theory misses the tri-fold splitting of the dislocation core which has a large influence on atomistic results.

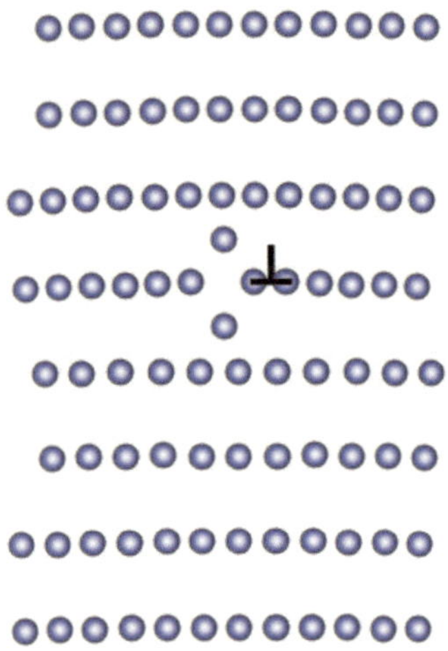

Fig. 4a. Initial position of the interstitial near the dislocation core

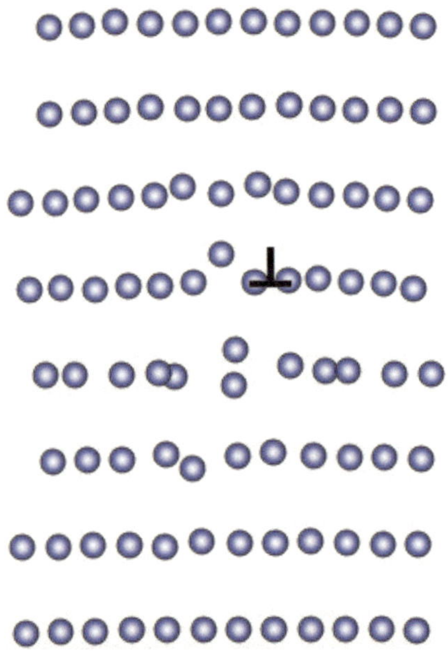

Fig. 4b. Intermediate position of the interstitial as it folds into the dislocation

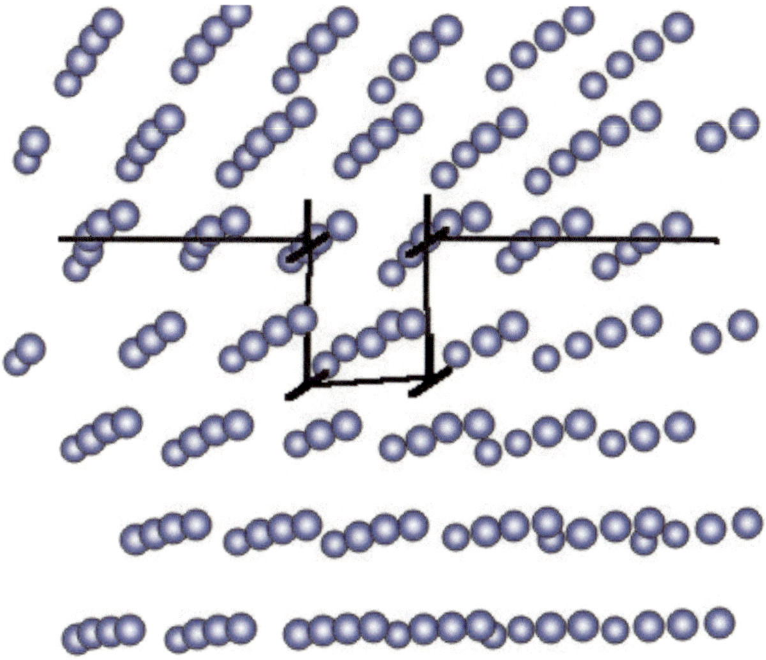

Fig. 4c. Intermediate position of interstitial near dislocation creates a jog in the dislocation

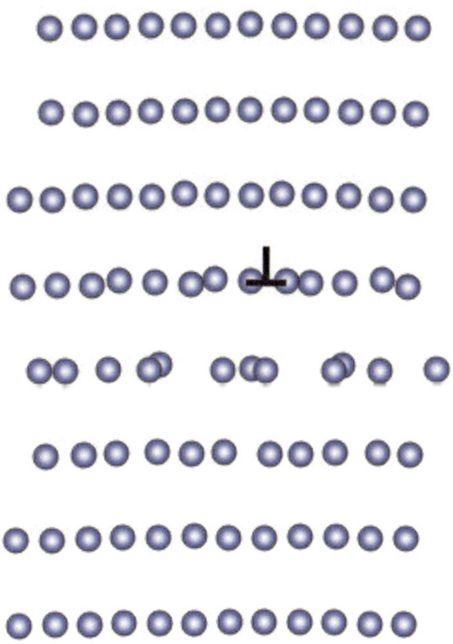

Fig. 4d. Final position of the dislocation with the interstitial absorbed in the core

Figures 4a-d shows the evolution of a defect in the vicinity of the dislocation. In this example an interstitial is positioned near a dislocation core and the energy of the system is minimized. The interstitial moves into the dislocation core and forms an extended jog in the dislocation core structure. Such a process is not captured by linear elasticity calculations which fail to capture the core structure and the core-defect interactions.

The relationship between crystal plasticity and dislocation behavior in materials has motivated a wide range of experimental and computational studies of dislocation behavior (Vitek 1976; Osetsky, Bacon et al. 1999). Computational studies of dislocation activity can be performed at several different length and time scales. In some cases, a multi-scale modeling approach is adopted(Ghoniem et al. 2003). Core properties and atomic mechanisms are simulated using first principles calculations and molecular dynamics simulations. These results can be used to form the rules that govern large-scale Dislocation Dynamics (DD) simulations (Wen et al. 2005) that account for the activity of a large number of dislocation segments. Polycrystalline plasticity models are then developed that utilize the information at the atomistic scale to parameterize partial differential equations of rate dependent viscoplasticity(Deo, Tom et al. 2008). While understanding dislocation creep processes, dislocation climb rates and hence, the interaction of dislocations with point defects is an important quantity to be calculated. Here, we show how the dislocation core affects the interaction energy between the dislocation and the point defect using both linear elasticity as well as atomistic calculations.

4.3 Kinetic Monte Carlo simulations of defect evolution

Kinetic Monte Carlo models and simulations have been employed to study cascade ageing and defect accumulation at low doses using the input from atomistic simulations of cascades and defect migration properties (Barashev, Bacon et al. 1999; Gao, Bacon et al. 1999; Barashev, Bacon et al. 2000; Heinisch, Singh et al. 2000; Heinisch and Singh 2003; Domain, Becquart et al. 2004; Becquart, Domain et al. 2005; Caturla, Soneda et al. 2006). These have mostly focused on intrinsic defect (interstitial and vacancy) migration, clustering and annihilation under irradiation conditions. In this example, we focus on extrinsic gas atoms such as helium or hydrogen migration in irradiated materials.

For the Fe-He system, modeling of helium clusters has been performed by Morishita, Wirth and co-workers (Morishita, Sugano et al. 2003a; Morishita, Sugano et al. 2003b) and Bringa, Wirth et al. (2003)using semi-empirical potentials. In one paper, Morishita, Sugano et al. (2003a) performed molecular dynamics (MD) calculations to evaluate the thermal stability of helium–vacancy clusters in Fe In another paper, Morishita, Sugano et al. (2003b)have looked at dissolution of helium-vacancy clusters as a function of temperature increase using the empirical potentials for the Fe-He system. Wirth and Bringa (Wirth and Bringa 2004) have simulated the motion of one single 2He-3Vac cluster at 1000 K using the same potential system. First principles calculations of helium atoms in interstitial and substitutional sites has been performed by Fu and Willaime 2005 and can be used to provide an input parameter set for kinetic Monte Carlo calculations.

The kinetic Monte Carlo model(Deo, Srivilliputhur et al. 2006; Deo, Okuniewski et al. 2007; Deo, Srinivasan et al. 2007) consists of helium interstitials on the octahedral sublattice and vacancies on the bcc iron lattice. The migration of the free (not clustered) helium is shown in figure 5(a) and that of the vacancies and self interstitial atoms is shown in Figure 5(b). The

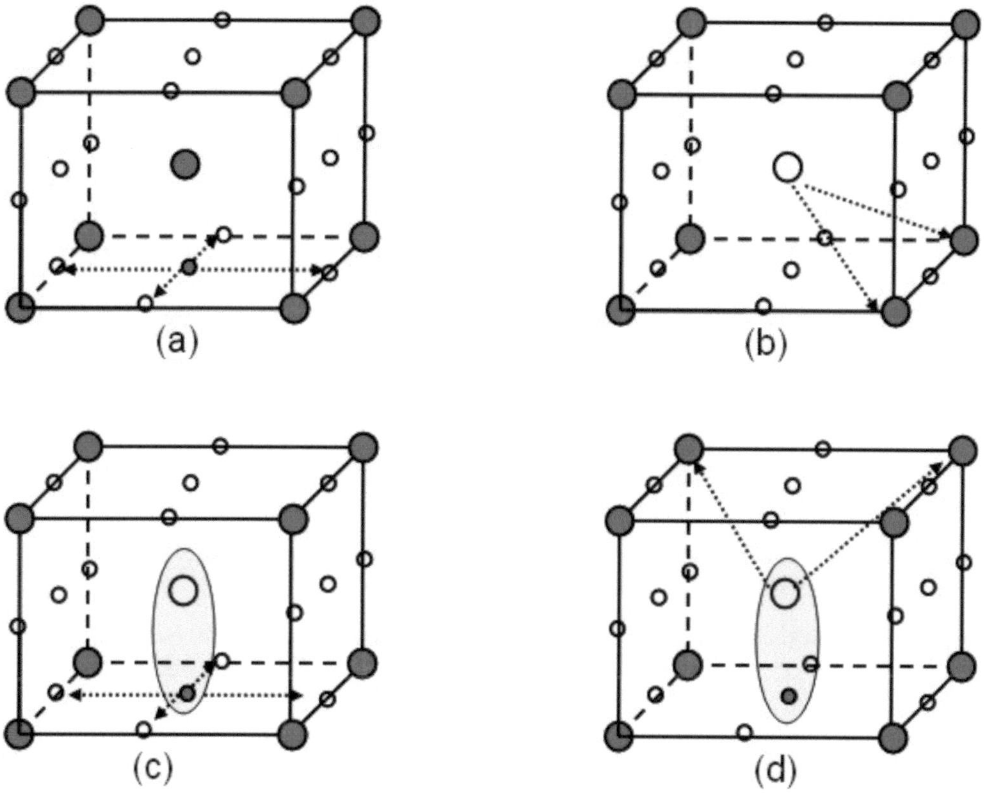

Fig. 5. Shows the basic mechanisms of helium and vacancy activity in single crystal bcc iron. Large filled circles represent iron, large open circles represent vacancies, small filled circles represent helium atoms and small open circles represent the octahedral bcc sites. (a) Helium migration on the octahedral sublattice (b) Vacancy and self interstitial migration in bcc iron (c) Dissociation of helium from an embryonic bubble (d) dissociation of vacancy from an embryonic bubble.

lowest-energy migration path of the SIA corresponds to a nearest-neighbor translation-rotation jump in the <111> direction. The rates of migration of the point defect entities are calculated as

$$r^i_{migration} = \nu^i_{migration} \exp\left(-\frac{E^i_{migration}}{k_B T}\right), \tag{3}$$

where the superscript i refers to the helium, self interstitial atoms and the vacancy point defect entities. The rate of migration of the point defect entity is $r^i_{migration}$, the attempt frequency is $v^i_{migration}$, the migration barrier is $E^i_{migration}$, while k_B and T are the Boltzmann constant and the temperature respectively. Two point defect entities are considered to be in a cluster when the distance between them is less than a_0, which is the lattice constant of bcc iron. Interstitial atoms are not allowed to dissociate from clusters. Dissociation of the helium and the vacancy from the cluster is described in Figures 2(c) and 2(d) respectively. The rate of dissociation of a point defect entity (i = helium or vacancy) from a cluster into the bulk lattice is considered to be thermally activated and is calculated as:

$$r^i_{dissociation} = v^i_{dissociation} \exp\left(-\frac{E^i_{dissociation}}{k_B T}\right) \tag{4}$$

where $r^i_{dissociation}$ is the rate or dissociation, $v^i_{dissocation}$ is the attempt frequencies, $E^i_{dissociation}$ is the energy of dissociation. The dissociation energy $E^i_{dissociation}$ of a point defect from a cluster is taken to be the sum of the energy to bind a point defect entity to the cluster and $E^i_{migration}$. Small bubbles migrate with a Arrhenius migration rate parameterized by Table 1. Larger bubbles migrate by surface diffusion at the bubble-matrix interface (Cottrell 2003; Evans 2004),

$$D_B = D_S\left(\frac{3\Omega^{4/3}}{2\pi r^4}\right), \tag{5}$$

where D_B and D_s are the bubble and surface diffusivities respectively, Ω is the atomic volume and r the radius of the bubble.

Morishita et al. (Morishita et al. 2003) and Fu et al (Fu and Willaime 2005) have calculated the migration energies of helium and vacancies as well as the binding energies of some helium-vacancy clusters. We employ these migration barriers to calculate the rate of migration of the point defect entities. Parameters used to calculate these quantities are described in Table 1. These parameters are used to calculate the rates migration and dissociation events (Equations 1-3) in the system and build the event catalog for the kMC simulation.

The event catalog is generated by calculating the rates of migration or dissociation of the point defect entities using Equations (3-6). The kMC event catalog consists of the migration, clustering and dissociation of the point defect entities, helium, self interstitial atoms and vacancies. The transition probability of each event is proportional to the rate of event occurrence, calculated by the Equations (3-6). We follow the well established kMC simulation algorithm (Bortz, Kalos et al. 1975; Fichthorn and Weinberg 1991) which is a stochastic, atomic-scale method to simulate the time-evolution of defects and nano/microstructural evolution that focuses on individual defects and not on atomic vibrations.

Reaction pathways available in the system are tabulated in table 1. Rates of migration of events are calculated at each kMC step. Parameters are obtained both from literature using first principles calculations (Fu, Willaime et al. 2004; Fu and Willaime 2005) and also from molecular statics calculations using semi-empirical potentials as employed by Morishita et al (Morishita et al. 2003) and Wirth and Bringa (Wirth and Bringa 2004) (using the Ackland Finnis–Sinclair potential, the Wilson–Johnson potential and the Ziegler–Biersack–Littmark–Beck potential for describing the interactions of Fe–Fe, Fe–He and He–He, respectively). Helium atoms are introduced at random positions at the beginning of the simulation.

Self interstitial atoms (SIA) are produced in the simulation along with vacancies as Frenkel pairs as cascade debris. Self interstitial atoms are mobile and cluster. Self interstitial clusters up to size 5 migrate one dimensionally. Self interstitials of higher size are stationary. Mono-

and clustered vacancies are mobile. Vacancies can dissociate from a vacancy cluster as well as a helium-vacancy cluster. Helium atoms migrate on the octahedral sublattice as well as part of helium-vacancy bubbles. A substitutional helium is considered as a 1-1 He_1V_1 and is mobile. If a bubble has a helium-vacancy ratio greater than 5, it emits a self interstitial atom. Small bubbles migrate according to Equation 3 and Table 1 while large bubbles migrate according to Eq. 3. Self interstitial atoms and vacancies annihilate when they meet either as point defects or in a cluster. The boundary of the simulation cell acts as a sink for the point defect entities.

Entity	Event	E (eV)	v_0 (s^{-1})	Remarks
Helium	Migration	0.078	1e14	Helium migrates on the interstitial sublattice
Vacancy	Migration	0.65	6e13	Vacancy migration on the substitutional sublattice
SIA	Migration	0.3	6e13	SIA migration on the substitutional sublattice
Helium	Dissociation from He_nV_m	2.0	1e14	Helium dissociation from the He-V cluster
Vacancy	Dissociation from He_nV_m	2.0	6e13	Vacancy dissociation from the helium-vacancy cluster
Helium	Dissociation from He_n	0.30	1e14	Helium dissociation from the helium-helium cluster
Vacancy	Dissociation from V_m	0.20	6e13	Vacancy dissociation from the vacancy cluster
Interstitial clusters	1D migration	0.1	6e13	Interstitial clusters up to size 4 are considered mobile
He-V clusters	3D migration	1.1	1e14	Clusters containing up to 3 vacancies atoms migrate according to this rate
He-V clusters	3D migration	---	---	Diffusivity of clusters containing more than 3 vacancies calculated by considering surface diffusion mechanisms (Eq. 3)

Table 1. A table of the events included in the kinetic Monte Carlo model. Migration energies and attempt frequencies are provided where applicable.

At each kMC step, the system is monitored to identify a clustering event. When two point defect entities (helium-helium, vacancy-vacancy, helium-vacancy) are in a cluster the simulation creates a mapping between the entities and the cluster such that for each cluster there are at least two entities associated with the cluster. The event catalog is updated with the new rates of event occurrence and the transition probabilities for the next kMC event are calculated using Equations (4-6) using the parameters from Table 1.

Simulations were performed for a damage energy of 100keV. A production rate of randomly distributed cascades is assumed such that the damage is introduced at a rate of 10^{-6} dpa/s. The simulation cell size is $400a_0 x 400a_0 x 400a_0$, where a_0 is the lattice parameter of iron. The number of Frenkel pairs introduced in the simulation cell are calculated by the Norgett-Robinson-Torrens relationship,

$$displacements\ per\ cascade = \frac{0.8E_D}{2E_d} \tag{6}$$

where E_D is the energy of incident particles while E_d is the threshold energy (40eV for iron (1994)). Initial concentration of helium is a parameter in the simulation and is varied from 1 to 25 appm/dpa. Damage measured in dpa is another parameter in the system and determines the length of the simulation. The incident energy is also a simulation control parameter that determines the number of defects introduced in a single cascade. The simulation is performed until sufficient damage is accumulated in the simulation cell.

Simulations are performed for values of the He concentrations varying from 1 to 25 appm/dpa (appm = atomic parts per million). Initial concentrations of both point defect entities are calculated from the NRT formula (Equation 4) using an incident energy of 100keV. Overall damage is introduced at the rate of 10^{-6} dpa/s and the simulation is carried out until the required amount of damage has accumulated in the system.

Figure 3 shows the concentration of bubbles as a function of He concentration (appm/dpa) after damage equal to 0.1 and 1 dpa is introduced. The simulation temperature is $0.3T_m$. The bubble density increases with increasing He/dpa ratio. The bubble density can be described by a power law expression,

$$c_B = K\left(c_{He}\right)^m \text{ , } c_B = K\left(c_{He}\right)^m \tag{7}$$

where is the bubble density, is the helium concentration expressed as appm He/dpa and K and m are constants determined by the kinetic Monte Carlo simulations. We find that the exponent m is approximately 0.5. In Figure 6, the solid lines are fit to the data assuming a square root dependence of the bubble density on helium concentration. Thus the bubble density increases as the square root of the He/dpa ratio. While experimental evidence of this variation is difficult to find, such dependence has been suggested by rate theory calculations (SINGH and TRINKAUS 1992) for the case of cold helium implantation annealing. The square root dependence is also found for equilibrium bubbles containing an ideal gas(Marksworth 1973).

The bubble density increases with damage (expressed as dpa) as damage is accumulated till 1 dpa. At higher dpa ratios more vacancies are produced that may serve as nucleation sites for embryonic bubbles by trapping helium. Thus the bubble density scales directly with increasing damage. That bubble density increases with accumulating dpa is observed in experiments (Sencer, Bond et al. 2001; Sencer, Garner et al. 2002; Zinkle, Hashimoto et al. 2003). These experiments suggest much smaller values of bubble density than those calculated in the present simulations. Embryonic bubbles are submicroscopic and difficult to estimate from experimental observations; while the kMC simulations have a large fraction of nanometer size bubbles; making comparison with experiment difficult.

This example demonstrates the use of the kinetic Monte Carlo method to simulate defect diffusion in irradiated materials. The method depends on the ability to calculate rates of migration events of individual defects. The advantage of the method is that all atoms do not necessarily need to be simulated and large time scales are accessible to the simulation provided the underlying physics remains invariant.

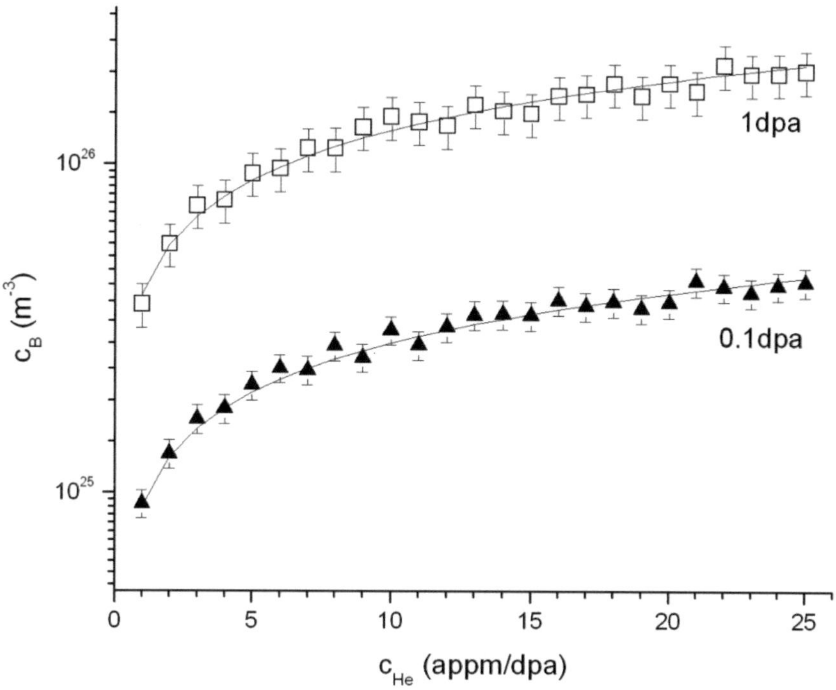

Fig. 6. is a plot of the concentration of bubbles (c_B) as a function of helium concentration (expressed in terms of appm/dpa) for two damage levels, 0.1 dpa (filled triangles) and 1 dpa (open squares). The line is a fit to the data assuming a square root dependence of the bubble density on helium concentration. The simulation temperature is 543K (0.3 T_m, the melting temperature for iron).

4.4 Microstructural mechanics of irradiation hardening

The previous examples have looked at the atomistic and mesoscale of radiation damage and defect formation. This information can be used by plasticity models and microstructural mechanics models of the effect of radiation on materials properties. Here an example is presented where the atomistic calculations are used to parameterize a viscoplasticity treatment of hardening in materials due to irradiation.

Hardening and embrittlement are controlled by interactions between dislocations and irradiation induced defect clusters. Radiation hardening and embrittlement that occurs in metals irradiated at low temperatures (below ~0.3 T_m, where T_m is the melting temperature) is a an important technical challenge for advanced nuclear energy systems(Zinkle and Matsukawa 2004). In this example, the Visco Plastic Self Consistent (VPSC) polycrystalline code (Lebensohn and Tome 1993) is employed in order to model the yield stress dependence in ferritic steels on the irradiation dose. The dispersed barrier hardening model is implemented in the VPSC code by introducing a hardening law, function of the strain, to describe the threshold resolved shear stress required to activate dislocations. The size and number density of the defect clusters varies with the irradiation dose in the model. Such modeling efforts can both reproduce experimental data and also guide future experiments of irradiation hardening.

In order to describe the nature of the yield stress dependence on the irradiation dose, we implemented a new microstructural model at the grain level in the VPSC code. The model assumes that hardening is affected by the presence of the defects and defect clusters produced by irradiation. These defects interact with the pre-existing dislocations in the microstructure leading to an increase in the critical stress necessary to move the dislocations. This leads to an increase in the overall yield stress of the material.

Defects are treated as barriers to the motion of dislocations. Two approximate dislocation barrier models have historically been used to describe radiation hardening in metals (Zinkle and Matsukawa 2004) and are reviewed in (Koppenaal and Arsenault 1971; Kocks 1977). The dispersed barrier model (Seeger, Diehl et al. 1957) is based on straightforward geometrical considerations for obstacles intersecting the dislocation glide plane and it is most appropriate for strong obstacles. An alternative hardening relationship was developed by Friedel–Kroupa–Hirsch (FKH) for weak obstacles (Friedel 1955; Kroupa and Hirsch 1964), where the effective interparticle spacing is increased compared to the planar geometric spacing due to less extensive dislocation bowing prior to obstacle breakaway. Using the simple approximation for dislocation line tension, the functional dependence of polycrystalline yield strength increase on defect cluster size and density for these two limiting cases is given by the following equations:

$$\Delta\sigma = M\alpha\mu b\sqrt{Nd}\,, \tag{8}$$

$$\Delta\sigma = \frac{1}{8}M\mu bN^{\frac{2}{3}}d\,, \tag{9}$$

Equation 8 corresponds to the dispersed barrier hardening model and Equation 9 to the FKH model. In the two equations, $\Delta\sigma$ is the change in the yield stress, M is the Taylor factor (3.06 for non-textured BCC and FCC metals), α is the defect cluster barrier strength, μ is the shear modulus, b is the Burgers vector of the primary glide dislocations, and N and d are the defect cluster density and diameter.

Most radiation hardening studies have used the dispersed barrier model (Equation 8) for data interpretation, and in this work we find that it provides a better representation of our experimental results. However, the FKH model (Equation 9) may be more appropriate for many radiation-induced small defect clusters which are weak obstacles to dislocation motion. According to some early analyses (Kocks_1977), the FKH model is adequate for barrier strengths up to 1/4 of the Orowan (impenetrable obstacle) limit, i.e., α < 0:25, and the dispersed barrier model is more appropriate for barrier strengths of $0.25<\alpha<1$. Typical experimental values of the defect cluster barrier strength for copper and austenitic stainless steel neutron-irradiated and tested near room temperature are α = 0.15–0.2 (Zinkle 1987). The reported barrier strengths for the visible defect clusters in BCC metals (Rice and Zinkle

1998) are $\alpha = 0.4$ or higher. It is possible that hardening from atomic scale voids in the BCC metals might cause one to overestimate the reported barrier strength for the visible defect clusters.

It is possible to introduce a hardening law that is a function of the strain to describe the threshold resolved shear stress required to activate dislocations. In the present application, however, evolution is not simulated and only the initial threshold is required. We assume that the initial critical resolved shear stress (CRSS) in each grain is affected by irradiation according to the dispersed barrier hardening law and follows the Orowan expression,

$$\tau = \tau_0 + \alpha \mu b \sqrt{Nd} \,, \tag{10}$$

where τ is the initial CRSS, τ_0 is the unirradiated initial CRSS and the other parameters have the same meaning as in Equations 8 and 9. Observe that the Taylor factor is not included in Eq. 10, since the geometric crystal orientation effects are accounted for by the polycrystal model. The critical stress τ is assigned to the 12 (110)[111] and the 12 (110)[112] slip systems of the BCC structure. The initial texture of the rolled ferritic steel is represented using 1000 crystallographic orientations. Each orientation is treated as an ellipsoidal inclusion embedded in and interacting with the effective medium that represents the aggregate. An incremental strain is enforced along the rolling direction, while leaving the lateral strains unconstrained. The stress and the strain is different from grain to grain, and the macroscopic (yield) stress is given by the average over all orientations.

Through Eq. 10 the model includes a dependence of the yield stress on the damage created due to radiation. Radiation damage is usually expressed as a statistical quantity describing the average number of displacements for each atom (dpa). The dpa influences the yield stress by determining the number density and the size of the defect clusters (obstacles) that impede the path of the dislocations and increase the critical stress required to move the dislocation.

It has commonly been assumed that the defect cluster density in irradiated metals increases linearly with increasing dose, up to the onset of cascade overlap which causes a saturation in the cluster density (Makin, Whapman et al. 1962; Koppenaal and Arsenault 1971; Trinkaus, Singh et al. 1996). However, in several pure FCC metals the defect accumulation as measured by electrical resistivity (Makin et al. 1962; Zinkle 1987) or transmission electron microscopy (Zinkle 1987; Muroga, Heinisch et al. 1992) often appears to exhibit an intermediate dose regime where the defect cluster density is proportional to the square root of dose. The defect accumulation behavior was found to be linear at very low doses (<0.0001 dpa, where the probability of uncorrelated point defect recombination is negligible), and proportional to the square root of dose at higher doses. According to simple kinetic models such as the unsaturable trap model (Thompson, Youngblood et al. 1973; Theis and Wollenberger 1980), the critical dose for transition from linear to square root behavior depends on specimen purity. In this model, the transition to square-root accumulation behavior can be delayed up to high doses if impurity trapping of migrating interstitial- type defects is dominant compared to interstitial– interstitial or interstitial–vacancy reactions.

The dependence on irradiation dose (expressed as dpa) of the defect cluster density (N) and the defect diameter (d) are taken from atomic level kinetic Monte Carlo (kMC) simulations and experimental observations (Deo et al. 2006; Deo, Baskes et al. 2007) The kMC model takes atomic level information of the migration energies and jump attempt frequencies of irradiation induced defects (interstitials, vacancies) and transmutation products (e.g., helium under high energy proton irradiation), and evolves the microstructure according to the rates of migration of these defects. The defects are allowed to cluster, and new irradiation damage is introduced during the simulation according to the irradiation dose rate. Our kMC simulations predict that the number density varies as the square root of the displacements per atom for the case of bcc iron irradiated up to 1 dpa by high energy proton irradiation.

The size dependence on irradiation dose is more complicated as the kMC simulations provide an entire distribution of defect cluster sizes. A single value of d as a function of dose is still a simplification of the kMC results. . The defect size usually increases with increasing dose (dpa) and can be fit by a power law; however the exponent of the power law expression can vary from 0 to 0.5 depending on initial simulation conditions (dose rate, temperature) and the defect cluster size considered. At low dpa, the exponent of the power law dependence is small for all defect sizes and increases at higher dpa.

The density of defects N is assumed to vary as the square root of the dpa while two cases of size dependence are considered, one in which the size is invariant with the dose (dpa) while the other in which the defect size varies as the square root of the dose. Additional systematic work is needed to confirm the presence and to understand the physical mechanisms responsible for this square root fluence-dependent defect cluster accumulation regime.

The link between the atomic level simulations and the VPSC calculations was established using the dispersed barrier hardening model. In this model, the vacancy /interstitial clusters produced in radiation cascades are assumed to act as barriers to the gliding dislocation in the slip plane and are therefore taken to be the main source of radiation hardening. A different model of radiation hardening postulates the formation of defect clouds along the length of the grown-in dislocation(see [4,5] for review). These clouds prevent the dislocation from acting as Frank Read dislocation sources and emitting more dislocations. Singh, Golubov et al. (1997) proposed the cascade induced source hardening model which accounted for interstitial cluster formation during radiation cascade formation. Such cluster formation has been observed in molecular dynamics simulations. In the CISH model, glissile loops produced directly in cascades are assumed to decorate grown-in dislocations so they cannot act as dislocation sources. The yield stress is related to the breakaway stress which is necessary to pull the dislocation away from the clusters/loops decorating it. Various aspects of the model (main assumptions and predictions) have been investigated by these researchers using analytical calculations, 3-D dislocation dynamics and molecular dynamics simulations It is possible to investigate such recent radiation hardening mechanisms by including them to develop the links between the atomic level understanding of defect sizes and concentrations and the VPSC model of polycrystalline hardening. Such mechanisms may also be investigated by atomic level simulations of single dislocation motion in the presence of defect impurities.

In a manner similar to the approach of Arsenelis and co-workers (Arsenlis, Wirth et al. 2004), the VPSC model can be used to combine microstructural input from both experimental observations and model predictions to evaluate the contributions from multiple defect cluster types. Although not all of the relevant parameters are currently known, such parameter-studies that can inform future atomic-scale studies of dislocation - obstacle interactions. The VPSC model could also incorporate experimentally observed defect cluster distributions, number densities to assess the effect of multiple defect types and distributions. A detailed multiscale study, wherein the dislocation-obstacle strength and the number density and size of defects are correlated to the increasing strain and each other, would then further explain the effect of irradiation on mechanical properties of ferritic steels.

The VPSC calculations provide a means to link atomistic first principles calculations to macroscopic observables. The formulation of the irradiation hardening law allows for the introduction of parameters such as the defect size and number density that can be calculated from evolution models and simulations such as the kinetic Monte Carlo method. The interaction of the dislocation with defect clusters can be investigated by using atomistic molecular dynamics calculations. In this document we have provided a framework for performing physically based modeling and simulations of hardening behavior observed during irradiation. Such modeling efforts can both reproduce experimental data and also guide future experiments of irradiation hardening. Performing modeling and simulation studies before initiating an expensive neutron or proton beam experiment would prove invaluable and cost-effective.

5. Conclusions

In this chapter, an overview of multiscale materials modeling tools used to simulate structural materials in irradiation conditions is presented. Next generation nuclear reactors will require a new generation of materials that can survive and function in extreme environments. Advanced modeling and simulation tools can study these materials at various length and time scale. Such varied methods are needed as radiation damage affects materials in excess of 10 orders of magnitude in length scale from the sub-atomic nuclear to structural component level, and span 22 orders of magnitude in time from the sub-picosecond of nuclear collisions to the decade-long component service lifetimes. The inherently wide range of time scales and the "rare-event" nature of the controlling mechanisms make modeling radiation effects in materials extremely challenging and experimental characterization is often unattainable. Thus, modeling and simulation of such materials holds great promise if coupled with suitably designed experiments in order to develop and sustain materials for advanced nuclear energy.

6. Acknowledgements

The author would like to thank Erin Hayward, Carlos Tome, Richardo Lebensohn, Stuart Maloy, Maria Okuniewski, Michael Baskes, Srinivasan Srivilliputhur and Michael James for useful discussions and acknowledge funding sources from the Department of Energy (DOE NEUP award DE-AC07-05ID14517 09-269) and a Nuclear Regulatory Commission Faculty Development Grant NRC-38-08-938

7. References

Arsenlis, A., B. D. Wirth, et al. (2004). "Dislocation density-based constitutive model for the mechanical behaviour of irradiated Cu." Philosophical Magazine 84(34): 3617-3635.

Bacon, D. J., A. F. Calder, et al. (1995). "COMPUTER-SIMULATION OF DEFECT PRODUCTION BY DISPLACEMENT CASCADES IN METALS." Nuclear Instruments & Methods in Physics Research Section B-Beam Interactions with Materials and Atoms 102(1-4): 37-46.

Bacon, D. J., F. Gao, et al. (2000). "The primary damage state in fcc, bcc and hcp metals as seen in molecular dynamics simulations." Journal of Nuclear Materials 276: 1-12.

Barashev, A. V., D. J. Bacon, et al. (1999). "Monte Carlo investigation of cascade damage effects in metals under low temperature irradiation." Materials Research Society Symposium - Symposium on Microstructural Processes in Irradiated Materials 540: 709-714.

Barashev, A. V., D. J. Bacon, et al. (2000). "Monte Carlo modelling of damage accumulation in metals under cascade irradiation." Journal of Nuclear Materials 276(1): 243-250.

Becquart, C. S., C. Domain, et al. (2005). "The influence of the internal displacement cascades structure on the growth of point defect clusters in radiation environment." Nuclear Instruments & Methods in Physics Research Section B-Beam Interactions With Materials and Atoms 228: 181-186.

Bortz, A. B., M. H. Kalos, et al. (1975). "A new algorithm for Monte Carlo simulation of Ising spin systems." Journal of Computational Physics 17(1): 10-18.

Bringa, E. M., B. D. Wirth, et al. (2003). "Metals far from equilibrium: From shocks to radiation damage." Nuclear Instruments and Methods in Physics Research, Section B: Beam Interactions with Materials and Atoms 202(SUPPL.): 56-63.

Calder, A. F. and D. J. Bacon (1993). "A molecular dynamics study of displacement cascades in alpha iron." Journal of Nuclear Materials 207: 25-45.

Calder, A. F. and D. J. Bacon (1994). "MD MODELING OF DISPLACEMENT CASCADES IN BCC IRON USING A MANY-BODY POTENTIAL." Radiation Effects and Defects in Solids 129(1-2): 65-68.

Caturla, M. J., N. Soneda, et al. (2000). "Comparative study of radiation damage accumulation in Cu and Fe." Journal of Nuclear Materials 276: 13-21.

Caturla, M. J., N. Soneda, et al. (2006). "Kinetic Monte Carlo simulations applied to irradiated materials: The effect of cascade damage in defect nucleation and growth." Journal of Nuclear Materials 351(1-3): 78-87.

Cirimello, P., D. G., et al. (2006). "Influence of metallurgical variables on delayed hydride carcking in Zr-Nb prssure tubes." Journal of Nuclear Materials 350: 135-146.

Cottrell, G. A. (2003). "Void migration, coalescence and swelling in fusion materials." Fusion Engineering and Design 66-8: 253-257.

de la Rubia, T. D., H. M. Zbib, et al. (2000). "Multiscale modelling of plastic flow localization in irradiated materials." NATURE 406(6798): 871-874.

Dederichs, P. and K. Schroeder (1978). "Anisotropic diusion in stress fields." Physical Review B 17(6): 2524-2536.

Deo, C., C. Tom, et al. (2008). "Modeling and simulation of irradiation hardening in structural ferritic steels for advanced nuclear reactors." Journal of Nuclear Materials 377(1): 136-140.

Deo, C. S., M. Baskes, et al. (2007). "Helium Bubble Nucleation in BCC Iron studied by kinetic Monte Carlo simulations." Journal of Nuclear Materials To be published.

Deo, C. S., M. A. Okuniewski, et al. (2007). "Helium bubble nucleation in bcc iron studied by kinetic Monte Carlo simulations." Journal of Nuclear Materials 361(2-3): 141-148.

Deo, C. S., S. G. Srinivasan, et al. (2007). "Kinetics of the Migration and Clustering of Extrinsic Gas in bcc Metals." Journal of ASTM International 4(9): 100698, 100691-100613.

Deo, C. S., S. G. Srivilliputhur, et al. (2006). Kinetics of the nucleation and growth of helium bubbles in bcc iron. Materials Research Society Symposium Proceedings.

Domain, C., C. S. Becquart, et al. (2004). "Simulation of radiation damage in Fe alloys: An object kinetic Monte Carlo approach." Journal of Nuclear Materials 335(1-3): 121-145.

Evans, J. H. (2004). "Breakaway bubble growth during the annealing of helium bubbles in metals." Journal of Nuclear Materials 334(1): 40-46.

Fichthorn, K. A. and W. H. Weinberg (1991). "Theoretical foundations of dynamical Monte Carlo simulations." Journal of Chemical Physics 95(2): 1090-1096.

Finnis, M. W. and J. E. Sinclair (1984). "A SIMPLE EMPIRICAL N-BODY POTENTIAL FOR TRANSITION-METALS." Philosophical Magazine a-Physics of Condensed Matter Structure Defects and Mechanical Properties 50(1): 45-55.

Friedel, J. (1955). "On the linear work hardening rate of face-centred cubic single crystals." Philosophical Magazine 46: 1169-1186.

Fu, C.-C. and F. Willaime (2005). "Ab initio study of helium in alpha -Fe: dissolution, migration, and clustering with vacancies." Physical Review B (Condensed Matter and Materials Physics) 72(6): 64117-64111.

Fu, C.-C., F. Willaime, et al. (2004). "Stability and mobility of mono- and di-interstitials in alpha -Fe." Physical Review Letters 92(17): 175503-175504.

Gao, F., D. J. Bacon, et al. (1999). "Kinetic Monte Carlo annealing simulation of damage produced by cascades in alpha-iron." Materials Research Society Symposium - Symposium on Microstructural Processes in Irradiated Materials 540: 703-708.

Ghoniem, N. M., E. P. Busso, et al. (2003). "Multiscale modelling of nanomechanics and micromechanics: an overview." Philosophical Magazine 83(31-34): 3475-3528.

Harrison, R. J., A. F. Voter, et al. (1989). Embedded atom potential for bcc iron. "Atomic Simulation of Materials - Beyond pair potentials": 219-222

Hasegawa, A., H. Shiraishi, et al. (1994). "Behavior of helium gas atoms and bubbles in low activation 9Cr martensitic steels." Journal of Nuclear Materials 212/215. 720-724.

Hayward, E. and C. Deo (2010). "A Molecular Dynamics Study of Irradiation Induced Cascades in Iron Containing Hydrogen." Cmc-Computers Materials & Continua 16(2): 101-116.

Heinisch, H. L. and B. N. Singh (2003). "Kinetic Monte Carlo simulations of void lattice formation during irradiation." Philosophical Magazine 83(31/34): 3661-3676.

Heinisch, H. L., B. N. Singh, et al. (2000). "Kinetic Monte Carlo studies of the effects of Burgers vector changes on the reaction kinetics of one-dimensionally gliding interstitial clusters." Journal of Nuclear Materials 276: 59-64.

Johnson, R. A. and D. J. Oh (1989). "Analytic embedded atom method model for bcc metals." Journal of Materials Research 4(5): 1195-1201.

Kinchin, G. H. and R. S. Pease (1955). "The displacement of atoms in solids by radiation." Reports on Progress in Physics 18: 1-51.
Kocks, U. F. (1977). "The theory of an obstacle-controlled yield strength-report after an international workshop." Material Science and Engineering 27(3): 291-298.
Koppenaal, T. J. and R. J. Arsenault (1971). "Neutron-irradiation-strengthening in f.c.c. single crystals." Metallurgical Reviews(157): 175-196.
Kroupa, F. and P. B. Hirsch (1964). "Elastic interaction between prismatic dislocation loops and straight dislocations." Discussions of the Faraday Society(38): 49-55.
Lebensohn, R. A. and C. N. Tome (1993). "A self-consistent anisotropic approach for the simulation of plastic deformation and texture development of polycrystals: application to zirconium alloys." Acta Metallurgica et Materialia 41(9): 2611-2624.
Makin, M. J., A. D. Whapman, et al. (1962). "Formation of dislocation loops in copper during neutron irradiation." Philosophical Magazine 7(74): 285-299.
Malerba, L. (2006). "Molecular dynamics simulation of displacement cascades in alpha-Fe: a critical review." Journal of Nuclear Materials 351(1-3): 28-38.
Marksworth, A. J. (1973). "Coarsening of gas filled pores in solids." Metallurgical Transactions 4(11): 2651-2656.
Matthews, J. R. and M. W. Finnis (1988). "Irradiation creep models - an overview." Journal of Nuclear Materials 159: 257-285.
Morishita, K., R. Sugano, et al. (2003a). "MD and KMC modeling of the growth and shrinkage mechanisms of helium-vacancy clusters in Fe." Journal of Nuclear Materials 323(2/3): 243-250.
Morishita, K., R. Sugano, et al. (2003b). "Thermal stability of helium-vacancy clusters in iron." Nuclear Instruments & Methods in Physics Research Section B-Beam Interactions With Materials and Atoms 202: 76-81.
Muroga, T., H. L. Heinisch, et al. (1992). "A comparison of microstructures in copper irradiated with fission, fusion and spallation neutrons." Journal of Nuclear Materials 191/194(pt B): 1150-1154.
Norgett, M. J., M. T. Robinson, et al. (1975). "Proposed method for calculating displacement dose rates." Nuclear Engineering and Design 33(1): 50-54.
Odette, G., B. D. Wirth, et al. (2001). "Multiscale-Multiphysics Modeling of Radiation-Damaged Materials: Embrittlement of Pressure Vessel Steels." MRS Bulletin 26: 176.
Olander, D. (1981). Fundamental Aspects of Nuclear Reactor Fuel Elements, National Technical Information Service.
Osetsky, Y. N., D. J. Bacon, et al. (1999). Atomistic simulation of mobile defect clusters in metals. Microstructural Processes in Irradiated Materials. S. J. Zinkle, G. E. Lucas, R. C. Ewing and J. S. Williams. 540: 649-654.
Phythian, W. J., R. E. Stoller, et al. (1995). "A comparison of displacement cascades in copper and iron by molecular dynamics and its application to microstructural evolution." Journal of Nuclear Materials 223(3): 245-261.
Rice, P. M. and S. J. Zinkle (1998). "Temperature dependence of the radiation damage microstructure in V-4Cr-4Ti neutron irradiated to low dose." Journal of Nuclear Materials 258/263(pt B): 1414-1419.

Seeger, A., J. Diehl, et al. (1957). "Work-hardening and Work-softening of face-centred cubic metal crystals." Philosophical Magazine 2: 323-350.

Sencer, B. H., G. M. Bond, et al. (2001). "Microstructural alteration of structural alloys by low temperature irradiation with high energy protons and spallation neutrons." American Society for Testing and Materials Special Technical Publication, 20th International Symposium on Effects of Radiation on Materials 1045: 588-611.

Sencer, B. H., F. A. Garner, et al. (2002). "Structural evolution in modified 9Cr-1Mo ferritic/martensitic steel irradiated with mixed high-energy proton and neutron spectra at low temperatures." Journal of Nuclear Materials 307: 266-271.

Simonelli, G., R. Pasianot, et al. (1994). "Point-Defect Computer Simulation Including Angular Forces in BCC Iron." Physical Review B 50(3): 727-738.

Singh, B. N., S. I. Golubov, et al. (1997). "Aspects of microstructure evolution under cascade damage conditions." Journal of Nuclear Materials 251: 107-122.

Singh B. N. and H. Trinkaus (1992). "An analysis of the bubble formation behaviour under different experimental conditions." Journal of Nuclear Materials 186(2): 153-165.

Stoller, R. E., G. R. Odette, et al. (1997). "Primary damage formation in bcc iron." Journal of Nuclear Materials 251: 49-60.

Terentyev, D., C. Lagerstedt, et al. (2006). "Effect of the interatomic potential on the features of displacement cascades in alpha-Fe: A molecular dynamics study." Journal of Nuclear Materials 351(1-3): 65-77.

Theis, U. and H. Wollenberger (1980). "Mobile interstitials produced by neutron irradiation in copper and aluminium." Journal of Nuclear Materials 88(1): 121-130.

Thompson, L., G. Youngblood, et al. (1973). "Defect retention in copper during electron irradiation at 80K." Radiation Effects 20(1/2): 111-134.

Tome, C. N., H. A. Cecatto, et al. (1982). "Point-Defect Diffusion in a strained crystal." Physical Review B 25(12): 7428-7440.

Trinkaus, H., B. Singh, et al. (1996). "Microstructural evolution adjacent to grain boundaries under cascade damage conditions and helium production." Journal of Nuclear Materials 237: 1089-1095.

Vitek, V. (1976). "Computer simulation of screw dislocation-motion in bcc metals under effect of external shear and uniaxial stresses." Proceedings of the Royal Society of London Series a-Mathematical Physical and Engineering Sciences 352(1668): 109-124.

Was, G. (2007). Fundamentals of Radiation Materials Science (Metals and Alloys), Springer-Verlag.

Wen, M., N. M. Ghoniem, et al. (2005). "Dislocation decoration and raft formation in irradiated materials." Philosophical Magazine 85(22): 2561-2580.

Wirth, B. D. and E. M. Bringa (2004). "A Kinetic Monte Carlo Model for Helium Diffusion and Clustering
in Fusion Environments." Physica Scripta T108: 80-84.

Wirth, B. D., O. G.R., et al. (2004). "Multiscale Modeling of Radiation Damage in Fe-based Alloys in the Fusion Environment." Journal of Nuclear Materials 329-333: 103.

Zinkle, S. J. (1987). "Microstructure and properties of copper alloys following 14-MeV neutron irradiation." Journal of Nuclear Materials 150(2): 140-158.

Zinkle, S. J., N. Hashimoto, et al. (2003). "Microstructures of irradiated and mechanically deformed metals and alloys: Fundamental aspects." Materials Research Society Symposium -Radiation Effects and Ion-Beam Processing of Materials; 2003; v.792; p.3-12 792: 3-12.

Zinkle, S. J. and Y. Matsukawa (2004). "Observation and analysis of defect cluster production and interactions with dislocations." Journal of Nuclear Materials 329-333(1-3 PART A): 88-96.

13

Application of Finite Symmetry Groups to Reactor Calculations

Yuri Orechwa[1,*] and Mihály Makai[2]
[1]*NRC, Washington DC*
[2]*BME Institute of Nuclear Techniques, Budapest*
[1]*USA*
[2]*Hungary*

1. Introduction

Group theory is a vast mathematical discipline that has found applications in most of physical science and particularly in physics and chemistry. We introduce a few of the basic concepts and tools that have been found to be useful in some nuclear engineering problems. In particular those problems that exhibit some symmetry in the form of material distribution and boundaries. We present the material on a very elementary level; an undergraduate student well versed in harmonic analysis of boundary value problems should be able to easily grasp and appreciate the central concepts.

The application of group theory to the solution of physical problems has had a curious history. In the first half of the 20th century it has been called by some the "Gruppen Pest" , while others embraced it and went on to win Noble prizes. This dichotomy in attitudes to a formal method for the solution of physical problems is possible in light of the fact that the results obtained with the application of group theory can also be obtained by standard methods. In the second half of the 20th century, however, it has been shown that the formal application of symmetry and invariance through group theory leads in complicated problems not only to deeper physical insight but also is a powerful tool in simplifying some solution methods.

In this chapter we present the essential group theoretic elements in the context of crystallographic point groups. Furthermore we present only a very small subset of group theory that generally forms the first third of the texts on group theory and its physical applications. In this way we hope, in short order, to answer some of the basic questions the reader might have with regard to the mechanical aspects of the application of group theory, in particular to the solution of boundary value problems in nuclear engineering, and the benefits that can accrue through its formal application. This we hope will stimulate the reader to look more deeply into the subject is some of the myriad of available texts.

The main illustration of the application of group theory to Nuclear Engineering is presented in Section 4 of this chapter through the development of an algorithm for the solution of the neutron diffusion equation. This problem has been central to Nuclear Engineering from the very beginning, and is thereby a useful platform for demonstrating the mechanics of bringing group theoretic information to bear. The benefits of group theory in Nuclear Engineering are

*The views expressed are those of the authors and do not reflect those of any government agency or any part thereof.

not restricted to solving the diffusion equation. We wish to also point the interested reader to other areas of Nuclear Engineering were group theory has proven useful.

An early application of group theory to Nuclear Engineering has been in the design of control systems for nuclear reactors (Nieva, 1997). Symmetry considerations allow the decoupling of the linear reactor model into decoupled models of lower order. Thereby, control systems can be developed for each submodel independently.

Similarly, group theoretic principles have been shown to allow the decomposition of solution algorithms of boundary value problems in Nuclear Engineering to be specified over decoupled symmetric domain. This decomposition makes the the problem amenable to implementation for parallel computation (Orechwa & Makai, 1997).

Group theory is applicable in the investigation of the homogenization problem. D. S. Selengut addressed the following problem (Selengut, 1960) in 1960. He formulated the following principle: If the response matrix of a homogeneous material distribution in a volume V can be substituted by the response matrix of a homogeneous material distribution in V, then there exists a homogeneous material with which one may replace V in the core. The validity of this principle is widely used in reactor physics, was investigated applying group theoretic principles (Makai, 1992),(Makai, 2010). It was shown that Selengut's principle is not exact; it is only a good approximation under specific circumstances. These are that the homogenization recipes preserve only specific reaction rates, but do not provide general equivalence.

Group theory has also been fruitfully applied to in-core signal processing (Makai & Orechwa, 2000). Core surveillance and monitoring are implemented in power reactors to detect any deviation from the nominal design state of the core. This state is defined by a field that is the solution of an equation that describes the physical system. Based on measurements of the field at limited positions the following issues can be addressed:

1. Determine whether the operating state is consistent with the design state.
2. Find out-of-calibration measurements.
3. Give an estimate of the values at non-metered locations.
4. Detect loss-of-margin as early as possible.
5. Obtain information as to the cause of a departure from the design state.

The solution to these problems requires a complex approach that incorporates numerical calculations incorporating group theoretic considerations and statistical analysis.

The benefits of group theory are not restricted to numerical problems. In 1985 Toshikazu Sunada (Sunada, 1985) made the following observation: If the operator of the equation over a volume V commutes with a symmetry group G, and the Green's function for the volume V is known and volume V can be tiled with copies tile t (subvolumes of V), then the Green's function of t can be obtained by a summation over the elements of the symmetry group G. Thus by means of group theory, one can separate the solution of a boundary value problem into a geometry dependent part, and a problem dependent part. The former one carries information on the structure of the volume in which the boundary value problem is studied, the latter on the physical processes taking place in the volume. That separation allows for extending the usage of the Green's function technique, as it is possible to derive Green's functions for a number of finite geometrical objects (square, rectangle, and regular triangle) as well as to relate Green's functions of finite objects, such as a disk, or disk sector, a regular hexagon and a trapezoid, etc. Such relations are needed in problems in heat conduction, diffusion, etc. as well.

An extensive discussion of the mathematics and application of group theory to engineering problems in general and nuclear engineering in particular is presented in (Makai, 2011).

2. Basic group theoretic tools

Although the basic mathematical definition of a group and much of the abstract algebraic machinery applies to finite, infinite, and continuous groups, our interest for applications in nuclear engineering is limited to finite point groups. Furthermore, it should be kept in mind that most of the necessary properties of the crystallographic point groups for applications, such as the group multiplication tables, the class structures, irreducible representations, and characters are tabulated in reference books or can be obtained with modern software such as MAPLE or MATHEMATICA for example.

2.1 Group definition

An abstract group G is a set of elements for which a law of composition or "product" is defined.

For illustrative purposes let us consider a simple set of three elements $\{E, A, B\}$. A law of composition for these three elements can be expressed in the form of a multiplication table, see Table 1. In position i, j of Table 1. we find the product of element i and element j with the numbering $1 \rightarrow E, 2 \rightarrow A, 3 \rightarrow B$. From the table we can read out that $B = AA$ because element $2, 2$ is B and the third line contains the products AE, AA, AB. The table is symmetric therefore $AB = BA$. Such a group is formed for example by the even permutations of three objects: $E = (a, b, c), A = (c, a, b), B = (b, c, a)$. The multiplication table reflects four necessary

	E	**A**	**B**
E	E	A	B
A	A	B	E
B	B	E	A

Table 1. Multiplication table for elements $\{E, A, B\}$

conditions that a set of elements must satisfy to form a group G. These four conditions are:

1. The product of any two elements of G is also an element of G. Such as for example $AB = E$.
2. Multiplication is associative. For example $(AB)E = A(BE)$.
3. G contains a unique element E called the identity element, such that for example $AE = EA = A$ and the same holds for every element of G.
4. For every element in G there exists another element in G, such that their product is the identity element. In our example $AB = E$ therefore B is called the inverse of A and is denoted $B = A^{-1}$.

The application of group theory to physical problems arises from the fact that many characteristics of physical problems, in particular symmetries and invariance, conform to the definition of groups, and thereby allows us to bring to bear on the solution of physical problems the machinery of abstract group theory.

For example, if we consider a characteristic of an equilateral triangle we observe the following with regard to the counter clockwise rotations by 120 degrees. Let us give the operations the following symbols: E-no rotations, C_3-rotation by 120°, $C_3C_3 = C_3^2$-rotation by 240°. The group operation is the sequential application of these operations, the leftmost operator should be applied first. The reader can easily check the multiplication table 2. applies to the

	E	C_3	C_3^2
E	E	C_3	C_3^2
C_3	C_3	C_3^2	E
C_3^2	C_3^2	E	C_3

Table 2. Multiplication table of $G = \{E, C_3, C_3^2\}$

	E	C_3	C_3^2	σ_v	σ'_v	σ''_v
E	E	C_3	C_3^2	σ_v	σ'_v	σ''_v
C_3	C_3	C_3^2	E	σ''_v	σ_v	σ'_v
C_3^2	C_3^2	E	C_3	σ'_v	σ''_v	σ_v
σ_v	σ_v	σ'_v	σ''_v	E	C_3	C_3^2
σ'_v	σ'_v	σ''_v	σ_v	C_3^2	E	C_3
σ''_v	σ''_v	σ_v	σ'_v	C_3	C_3^2	E

Table 3. Multiplication table of G_3

	E	A	B	a′	b′	c′
E	E	A	B	a′	b′	c′
A	A	B	E	b′	c′	a′
B	B	E	A	c′	a′	b′
a′	a′	c′	b′	E	B	A
b′	b′	a′	c′	A	E	B
c′	c′	b′	a′	B	A	E

Table 4. Multiplication table of a permutation group

group $G = \{E, C_3, C_3^2\}$. We see immediately that the multiplication table of the rotations of an equilateral triangle is identical to the multiplication table of the previous abstract group $G = \{E, C_3, C_3^2\}$. Thus there is a one-to-one correspondence (called isomorphism) between the abstract group of the previous example, and its rules.

2.2 Subgroups and classes

Groups can have more properties than just a multiplication table. The illustration in the previous subsection is not amenable to illustrating this; the groups are to small. However, if we again consider the equilateral triangle, we note that it has further symmetry operations. Namely, those associated with reflections through a vertical plane through each vertex. Let us give these reflection operations the symbols σ_v, σ'_v and σ''_v, reflection through planes through vertex a, b, c, respectively. We may describe the operation by the transformations of the vertices a, b, c. For example $\sigma_v : (a, b, c) \rightarrow (a, c, b)$. By adding these three reflections operations to the rotations, we form the larger group of symmetry operations of the equilateral triangle $G_3 = \{E, C_3, C_3^2, \sigma_v, \sigma'_v, \sigma''_v\}$. The multiplication table of the new group is given in Table 3. Another group with the same multiplication table as above can be constructed by considering the six permutations of the three letters a, b, c. Let $E = (a, b, c)$, $A = (c, a, b)$, $B = (b, c, a)$, $a' = (a, c, b)$, $b' = (c, b, a)$, $c' = (b, a, c)$ and $\{E, A, B\}$ are even permutations, $\{a', b', c'\}$ are odd. This leads to the multiplication table 4, which is isomorphic to Table 3. The two multiplication tables illustrate the concept of a subgroup that is defined as: A set S of elements in group G is considered as a subgroup of G if:

1. all elements in S are also elements in G

2. for any two elements in S their product is in S
3. all elements in S satisfy the four group postulates.

From the presented multiplication tables we see that $\{E, C_3, c_3^2\}$ and $\{E, A, B\}$ are subgroups; they are the only subgroups in their respective groups. These subgroups are the rotations in the former example, and the even permutations in the latter. While the remaining elements are associated with reflections and odd permutations, respectively. Furthermore, we see that two reflections are equivalent to a rotation, and two odd permutations are equivalent to an even permutation. Thus the operations $\{C_3, C_3^2\}$ and $\sigma_v, \sigma_v', \sigma''_v$ and similarly $\{A, B\}$ and $\{a', b', c'\}$ belong in some sense to different sets. This property is illustrated by taking the transform $T^{-1}QT$ of each element Q in G by all elements T in G. For group $\{E, C_3, C_3^2, \sigma_v, \sigma_v', \sigma''_v\}$ we obtain the following table of the transforms $T^{-1}QT$:

Q/T	E	C_3	C_3^2	σ_v	σ_v'	σ''_v
E	E	E	E	E	E	E
C_3	C_3	C_3	C_3	C_3^2	C_3^2	C_3^2
C_3^2	C_3^2	C_3^2	C_3^2	C_3	C_3	C_3
σ_v	σ_v	σ''_v	σ_v'	σ_v	σ''_v	σ_v'
σ_v'	σ_v'	σ_v	σ''_v	σ''_v	σ_v'	σ_v
σ''_v	σ''_v	σ_v'	σ_v	σ_v'	σ_v	σ''_v

We note that $\{E\}$, $\{C_3, C_3^2\}$, and $\{\sigma_v, \sigma_v', \sigma''_v\}$ transform into themselves and are thereby called classes. Classes play a leading role in the application of group theory to the solution of physical problems. In general physically significant properties can be associated with each class. In the solution of boundary value problems, different subspaces of the solution function space are assigned to each class.

2.3 Group representations

The application of the information in an abstract group to a physical problem, especially to the calculation of the solution of the boundary value problem that models the physical setting, requires a mathematical "connection" between the two. This connection originates with the transformations of coordinates that define the symmetry operations reflected in the actions of a point group.

As a simple illustration, let us again consider the abstract group $G = \{E, A, B, a', b', c'\}$ in the form of its realization in forms of rotations and reflections of an equilateral triangle, namely the point group $C_{3v} = \{E, C_3, C_3^2, \sigma_v, \sigma_v', \sigma''_v\}$. Let this group be consistent with a physical problem in terms of, for example, material distribution and the geometry of the boundary. Furthermore, let us consider a two-dimensional vector space with an orthonormal basis $\{\mathbf{e}_1, \mathbf{e}_2\}$ relative to which the physical model is defined. Each operation by an element g of the group C_{3v} can be represented by its action on an arbitrary vector $\mathbf{r}$ in a two-dimensional vector space. In the usual symbolic form we have

$$\mathbf{r}' = \mathrm{D}(g)\mathbf{r} \quad \text{for all } g \in C_{3v}$$

and where $\mathbf{r}' = r_1\mathbf{e}_1 + r_2\mathbf{e}_2$ is the transformed vector, and $\mathrm{D}(g)$ is the matrix operator associated with the action of group element $g \in C_{3v}$. It is well known from linear algebra that the matrix representation of operator $D(g)$ for each $g \in C_{3v}$ is obtained by its action on the basis vectors,

$$\mathbf{e}_i' = \sum_{j=1}^{2} D_{ij}(g)\mathbf{e}_j, \ \ i = 1,2,$$

and that the transpose of matrix $D_{ji}(g)$ gives the action of the group element g on the coordinates of the vector $\mathbf{r}$ as

$$r_i' = \sum_{j=1}^{2} D_{ij}^{-1}(g) r_j, \ \ i = 1,2. \tag{2.1}$$

For the point group C_{3v} we obtain the following six matrix representations. To spare room we replace the matrices by permutations:

$$\mathbf{E} = (1,2,3); \ \mathbf{D}(C_3) = (3,1,2); \ \mathbf{D}(C_3^2) = (2,3,1); \tag{2.2}$$

$$\mathbf{D}(\sigma_1) = (1,3,2); \ \mathbf{D}(\sigma_2) = (3,2,1); \ \mathbf{D}(\sigma_3) = (2,1,3). \tag{2.3}$$

These matrices satisfy the group multiplication table of C_{3v}, and therefore also the multiplication table of the abstract group G that is isomorphic to C_{3v}. We note that this is not the only matrix representation of C_{3v}. There are two one-dimensional representations, in particular that also satisfy the multiplication table of C_{3v}, and will be of interest later. These are

$$\mathrm{D}(E) = \mathrm{D}(C_3) = \mathrm{D}(C_3^2) = \mathrm{D}(\sigma_v) = \mathrm{D}(\sigma_v') = \mathrm{D}(\sigma''_v) = \mathbf{E}_2, \tag{2.4}$$

where $\mathbf{E}_2$ is the 2×2 identity matrix; and

$$\mathrm{D}(E) = \mathrm{D}(C_3) = \mathrm{D}(C_3^2) = \mathrm{E}_2 \ \ \mathrm{D}(\sigma_v) = \mathrm{D}(\sigma_v') = \mathrm{D}(\sigma''_v) = -\mathbf{E}_2. \tag{2.5}$$

The role played by these representations will become clear in later discussions of irreducible representations of groups, and their actions on function spaces.

2.4 Generation of group representations

To this point we have constructed the matrix representations the group elements of point groups such as C_{3v} in the usual physical space (two dimensional in our case). These representations were based on the transformations of the coordinates of an arbitrary vector in a physical space due to physical operations on the vector. Mathematical solutions to physical problems, however, are represented by functions in function spaces whose dimensions are generally much greater than three. Thus to bring the group matrix representations that act on coordinates to bear on the solution of physical problems in terms of functions, we need one more "connection" between symmetry operators on coordinates and symmetry operators on functions. This connection is defined as follows.

Let $f(\mathbf{r})$ be a function of a position vector $\mathbf{r} = (x,y)$ and $\mathrm{D}(g^{-1})$ be the matrix transformation associated with group element $g \in G$, such that $(x,y) \to (x',y')$ through

$$\mathbf{r}' = \mathrm{D}^{-1}(g)\mathbf{r}.$$

What we need is an algorithm that uses $\mathrm{D}^{-1}(g)$ to obtain a new function $h(\mathbf{r})$ from $f(\mathbf{r})$. To this end we define an operator $\mathbf{O}_g$ as

$$\mathbf{O}_g f(\mathbf{r}) = f(\mathbf{r}') = f(\mathrm{D}^{-1}(g)\mathbf{r}) = h(\mathbf{r}). \tag{2.6}$$

That is, operator $\mathbf{O}_g$ gives a new function $h(\mathbf{r})$ from $f(\mathbf{r})$ at $\mathbf{r}$, while f is unchanged at $\mathbf{r}'$. For example, let

$$f(x,y) = ax + by$$

and

$$\mathrm{D}(g) = \begin{pmatrix} 1/\sqrt{2} & -1/\sqrt{2} \\ 1/\sqrt{2} & 1/\sqrt{2} \end{pmatrix}$$

then

$$\mathbf{O}_g(ax + by) = ax' + by' = \frac{a+b}{\sqrt{2}}x + \frac{a-b}{\sqrt{2}}y = h(x,y)$$

For two group elements g_1 and g_2 in G, we obtain

$$\mathbf{O}_{g_1} f(\mathbf{r}) = f(\mathrm{D}^{-1}(g)\mathbf{r}) = h(\mathbf{r})$$

$$\mathbf{O}_{g_2}\mathbf{O}_{g_1} f(\mathbf{r}) = \mathbf{O}_{g_2}\left(\mathbf{O}_{g_1} f(\mathbf{r})\right) = \mathbf{O}_{g_2} h(\mathbf{r}) = h(\mathrm{D}^{-1}(g_2)\mathbf{r}) = f([\mathrm{D}^{-1}(g_1)\mathrm{D}^{-1}(g_2)]\mathbf{r}) \equiv f([\mathrm{D}(g_2)\mathrm{D}(g_1)]^{-1}\mathbf{r}).$$

Note: operator $\mathbf{O}_g$ acts on the coordinates of function f and not on the argument of f. Therefore

$$\mathbf{O}_{g_2}\mathbf{O}_{g_1} f(\mathbf{r}) = f(\left(D^{-1}(g_1)D^{-1}(g_2)\right)^{-1}\mathbf{r}) = f(D^{-1}(g1)\mathrm{D}^{-1}(g_2)\mathrm{D}^{-1}(g_1)\mathbf{r}) = f((\mathrm{D}(g_2)\mathrm{D}(g_1))^{-1}\mathbf{r}),$$

and thus we get

$$\mathbf{O}_{g2}\mathbf{O}_{g1} = \mathbf{O}_{g_2g_1}, \tag{2.7}$$

in words: the consecutive application of $\mathbf{O}_{g1}$ and $\mathbf{O}_{g2}$ is the same as the application of the transformation $\mathbf{O}_{g2g_1}$ belonging to group element g_1g_2, and the operators $\mathbf{O}_g, g \in G$ have the same multiplication table as G and any group isomorphic with G.

2.5 Invariant subspaces and regular representations

A common approach to the solution of physical problems is harmonic analysis, where a solution to the problem is sought in terms of functions that span the solution space. If the problem exhibits some symmetry, we would expect this symmetry to be reflected in the solution for this particular problem. Intuitively we would expect therefore the solution to belong to a subspace of the general solution space, and that the subspace be invariant under the symmetry operations exhibited by the problem.

As an illustration of this notion, we assume the problem has the symmetry of the cyclic permutation group $C_3 = \{E, C_3, C_3^2\}$ that was discussed previously. Let $f_E(r)$ be an arbitrary function that allows the operation of the operators in the group C_3 as discussed above. The action of each operator on f_E defines a new function that, is

$$\mathbf{O}_E f_E = f_E \quad \mathbf{O}_{C_3} f_E = f_{C_3} \quad \mathbf{O}_{C_3^2} f_E = f_{C_3^2}.$$

Based on this and the group multiplication table we get relations such as

$$\mathbf{O}_{C_3} f_{C_3} = \mathbf{O}_{C_3}\mathbf{O}_{C_3} f_E = \mathbf{O}_{C_3^2} f_E = f_{C_3^2},$$

etc. These observations can be summarized in a table: From that table we can construct matrix (permutation) representations of the operators $\mathbf{O}_E, \mathbf{O}_{C_3}, \mathbf{O}_{C_3^2}$ as for example

$$\mathbf{D}(C_3) = (2,3,1). \tag{2.8}$$

	f_E	f_{C_3}	$f_{C_3^2}$
f_E	f_E	f_{C_3}	$f_{C_3^2}$
f_{C_3}	f_{C_3}	$f_{C_3^2}$	f_E
$f_{C_3^2}$	$f_{C_3^2}$	f_E	f_{C_3}

This procedure gives the so-called regular representation for the group C_3 as

$$\mathbf{O}_E = (1,2,3); \; \mathbf{O}_{C_3} = (2,3,1); \; \mathbf{O}_{C_3^2} = (3,1,2). \tag{2.9}$$

The matrices, in general, satisfy the group multiplication table, and are characterized by only the one integer one in each column, the rest zeros, and the dimension of the matrix equals to the number of elements in the group. The functions $f_E, f_{C_3}, f_{C_3^2}$ that generate the regular representation, span the invariant subspace. They are not necessarily linearly independent basis functions.

2.6 Complete sets of linearly independent basis functions and irreducible representations

As was mentioned at the outset, symmetry as exemplified through group theory brings added information to the solution of physical problems, especially in the application of harmonic analysis. The heart of this information is encapsulated in the so called irreducible representations of the group elements. It should be stated at the outset that the irreducible representations used in most applications are readily available in tabulated form. Yet much of mathematical group theory is devoted to the derivation and properties of irreducible representations. We do not minimize in any way the importance of that material; it is necessary for a clear understanding of the applicability of the mathematical machinery and its physical interpretation. Our objective here is only to touch on a few of the central results used in the applications. Perhaps this may motivate the reader to look further into the subject.

The key property for the application of point groups to physical problems is that for a finite group all representations may be "built up" from a finite number of "distinct" irreducible representations. The number of distinct irreducible representations is equal to the number of classes in the group. Furthermore, the regular representation contains each irregular representation a number of times equal to the number of dimensions of that irreducible representation. Thus, if ℓ_α is the dimension of the α-th irreducible representation,

$$\sum_k \ell_\alpha^2 = |G|, \tag{2.10}$$

where $|G|$ is the order of the group G to be satisfied.

Let us illustrate this with the group C_3 that was discussed previously. To identify the classes in C_3, as before, we compute a table of $T^{-1}QT$, see Table 5. The elements that transform into

Q/T	E	C_3	C_3^2
E	E	E	E
C_3	C_3	C_3	C_3
C_3^2	C_3^2	C_3^2	C_3^2

Table 5. Classes of Group G_3

themselves form a class. There are three classes in C_3, denoted as E, C_3, and C_3^2 and therefore there are three irreducible representations in the regular representation. The condition

$$\ell_1^2 + \ell_2^2 + \ell_3^2 = 3$$

can only be satisfied by $\ell_1 = \ell_2 = \ell_3 = 1$. Therefore, there are three distinct one-dimensional representations. These are the building blocks for decomposing the regular representation to irreducible representations, and can be found in tables:

$$D^{(1)}(E) = 1 \;\; D^{(1)}(C_3) = 1 \;\; D^{(1)}(C_3^2) = 1 \tag{2.11}$$

$$D^{(2)}(E) = 1 \;\; D^{(2)}(C_3) = \omega \;\; D^{(2)}(C_3^2) = \omega^* \tag{2.12}$$

$$D^{(3)}(E) = 1 \;\; D^{(3)}(C_3) = \omega^* \;\; D^{(3)}(C_3^2) = \omega, \tag{2.13}$$

where $\omega = \exp(2\pi i/3)$. The element in each of the three irreducible representation conform to the multiplication of point group C_3.

These low dimension irreducible representations are used to build an irreducible representation from the regular representation of the operator O_{C_3} for example, as follows.

The regular representation has the form of a full matrix,

$$\begin{pmatrix} D_{11}(C_3) & D_{12}(C_3) & D_{13}(C_3) \\ D_{21}(C_3) & D_{22}(C_3) & D_{23}(C_3) \\ D_{31}(C_3) & D_{32}(C_3) & D_{33}(C_3) \end{pmatrix} = \begin{pmatrix} 0 & 1 & 0 \\ 0 & 0 & 1 \\ 1 & 0 & 0 \end{pmatrix}.$$

The irreducible representation has the form of a diagonal (block diagonal in the general case) matrix,

$$\begin{pmatrix} D^1(C_3) & 0) & 0 \\ 0 & D^2(C_3) & 0 \\ 0 & 0 & D^3(C_3) \end{pmatrix} = \begin{pmatrix} 1 & 0 & 0 \\ 0 & \omega & 0 \\ 0 & 0 & \omega^* \end{pmatrix}.$$

The mathematical relationship is discussed at length in all texts on the subject, and will not be repeated here. We assume the irreducible representations are known. Of interest is the information for the solution of physical problem, that is associated with irreducible representations.

Recall that starting with an arbitrary function $f(\mathbf{r})$ belonging to a function space $\mathbb{L}$ (a Hilbert space for example), we can generate a set of functions $f_1, \ldots, f_{|G|}$ that span an invariant subspace $\mathbb{L}_s \subset \mathbb{L}$. This process requires the matrices of coordinate transformations $g_1, \ldots, g_{|G|}$ that form the symmetry group G of interest. The diagonal structure of the irreducible representations of G tells us that there exists a set of basis functions $\{f_1, f_2, \ldots, f_n\}$ that split the subspace $\mathbb{L}_s$ further into subspaces invariant under the symmetry group G, and are associated with each irreducible representation $D^{(1)}(g), D^{(2)}(g), \ldots, D^{(n_c)}(g)$ where n_c is the number of classes in G. That is

$$\mathbb{L}_s = \mathbb{L}_1 \cup \mathbb{L}_2 \cup \ldots \mathbb{L}_{n_c} \tag{2.14}$$

and thus an arbitrary function $f(\mathbf{r}) \in \mathbb{L}_s$ is expressible as a sum of functions that act as basis function in the invariant subspaces associated with each irreducible representation $D^{(\alpha)}(g), \alpha = 1, \ldots, n_c$ as

$$f(\mathbf{r}) = \sum_{\alpha=1}^{n_c} f^{\alpha}(\mathbf{r}). \tag{2.15}$$

If the decomposition of the regular representation contains irreducible representations of dimension greater than one, we have for each basis function that "belongs to the α-th irreducible representation"

$$f^{\alpha}(\mathbf{r}) = \sum_{i=1}^{\ell_{\alpha}} f_{ii}\alpha(\mathbf{r}) \tag{2.16}$$

where ℓ_{α} is the dimension of the α-th irreducible representation.

The question now remains how do we obtain $f^{\alpha}(\mathbf{r})$, the basis function of each irreducible representation?

To this end we can apply a projection operator that resolves a given function $f(\mathbf{r})$ into basis functions associated with each irreducible representation. This projection operator is defined as

$$\mathbf{P}_i^{\alpha} = \frac{\ell_{\alpha}}{|G|} \sum_{g \in G} D_{ii}^{\alpha}(g)\mathbf{O}_g. \tag{2.17}$$

The information needed to construct this operator–the coordinate transformations, the irreducible representations–are known in the case of the point groups encountered in practice. So, for example, the i-th basis function of the α irreducible representation that is ℓ_{α} dimensional for a symmetry group with $|G|$ elements is constructed from an arbitrary function $f(\mathbf{r})$ in invariant space $\mathbb{L}_s$ as

$$f_i^{\alpha}(\mathbf{r}) = \frac{\ell_{\alpha}}{|G|} \sum_{g \in G} D_{ii}^{\alpha}(g)\mathbf{O}_g f(\mathbf{r}). \tag{2.18}$$

This decomposition creates a complete finite set of orthogonal basis functions.

In practice, a more simple projection operator is generally sufficient. This is due to the fact that the $D_{ii}^{\alpha}(g)$'s (the diagonal elements of a multidimensional irreducible representation) are quantities that are intrinsic properties of the irreducible representation $\mathrm{D}^{\alpha}(g)$. That is they are invariant under the change of coordinates.

Furthermore, the sum of the diagonal elements, or trace, of the irreducible representation $\mathrm{D}^{\alpha}(g)$ is also invariant under a change of coordinates. In group theory this trace is denoted by the symbol $\chi^{\alpha}(g)$ and

$$\chi^{\alpha}(g) = \sum_{i=1}^{\ell_{\alpha}} D_{ii}^{\alpha}(g), \tag{2.19}$$

and referred to as the character of element $g \in G$ in the α-th irreducible representation. There are tables of characters for all the point groups of physical interest.

The projection operator in terms of characters is given as

$$P^{\alpha} = \frac{\ell_{\alpha}}{|G|} \sum_{g \in G} \chi^{\alpha}(g)\mathbf{O}_g \tag{2.20}$$

so that the basis functions are

$$f^{\alpha}(\mathbf{r}) = \frac{\ell_{\alpha}}{|G|} \sum_{g \in G} \chi_g^{\alpha}\mathbf{O}_g f(\mathbf{r}), \tag{2.21}$$

and $f(\mathbf{r})$ is decomposed into a complete finite set of orthogonal functions, with one for each irreducible representation irrespective of its dimension.

3. Symmetries of a boundary value problem

Let us consider the following boundary value problem:

$$\mathbf{A}\phi(\mathbf{r}) = 0 \quad \mathbf{r} \in V \tag{3.1}$$
$$\mathbf{B}\phi(\mathbf{r}) = f(\mathbf{r}) \quad \mathbf{r} \in \partial V, \tag{3.2}$$

where **A** and **B** are linear operators. Group theory is not a panacea to the solution of boundary value problems; its application is limited. The main condition that must be met in nuclear engineering problems is that material distributions have symmetry. This is generally true in reactor cores, core cells and cell nodes.

In the following we give a heuristic outline of how the machinery presented above enters into the solution algorithm of a boundary value problem, and what benefits can be expected.

Symmetry is the key. If we have determined that the physical problem has symmetries these symmetries must form a group G. The symmetry operator $\mathbf{O}_g$ must commute for all $g \in G$ with the linear operators **A** and **B** for group theory to be applicable. That is

$$\mathbf{O}_g\mathbf{A} = \mathbf{A}\mathbf{O}_g \quad \text{and } \mathbf{O}_g\mathbf{B} = \mathbf{B}\mathbf{O}_g \tag{3.3}$$

must hold for all $g \in G$. If this condition is met, the boundary value problem can be written as

$$\mathbf{A}\mathbf{O}_g\psi(\mathbf{r}) = 0 \quad \mathbf{r} \in V \tag{3.4}$$
$$\mathbf{B}\mathbf{O}_g\psi(\mathbf{r}) = \mathbf{O}_g f(\mathbf{r}) \quad \mathbf{r} \in \partial V. \tag{3.5}$$

We can now use the projection operator (2.20) to form a set of boundary value problems

$$\mathbf{A}P^\alpha\psi(\mathbf{r}) = 0 \quad \mathbf{r} \in V \tag{3.6}$$
$$\mathbf{B}P^\alpha\psi(\mathbf{r}) = P^\alpha f(\mathbf{r}) \quad \mathbf{r} \in \partial V. \tag{3.7}$$

Since the projection operator creates linearly independent components, we have decomposed the boundary value problem into a number (equal to the number of irreducible components) of independent boundary value problems. These are

$$\mathbf{A}\psi^\alpha(\mathbf{r}) = 0 \quad \mathbf{r} \in V \tag{3.8}$$
$$\mathbf{B}\psi^\alpha(\mathbf{r}) = f^\alpha(\mathbf{r}) \quad \mathbf{r} \in \partial V, \tag{3.9}$$

whose solution $\psi^\alpha(\mathbf{r})$ belongs to the α-th irreducible representation. From this complete set of linearly independent orthogonal functions we reconstruct the solution to the original problem as

$$\psi(\mathbf{r}) = \sum_{\alpha=1}^{n_c} c_\alpha \psi^\alpha(\mathbf{r}), \tag{3.10}$$

where n_c is the number of classes in G.

Why is this better? Recall that we are applying harmonic analysis. The usual approach is to use some series that forms an incomplete set of expansion functions and results a coupled set of equations; one "large" matrix problem. With group theory, we find a relatively small set of complete basis functions that form the solution from symmetry considerations. These are found by solving a set of "small" boundary value problems. It is clear that the effectiveness of group theory is problem dependent. However, experience over the past half century has proven group theory's effectiveness in both nuclear engineering and other fields.

We present an especially simple example (Allgover et al., 1992) that demonstrates the advantages of symmetry considerations. The example is the solution of a linear system of equations with six unknowns:

$$\begin{pmatrix} 1\,5\,6\,2\,3\,4 \\ 5\,1\,4\,3\,2\,6 \\ 3\,4\,1\,5\,6\,2 \\ 2\,6\,5\,1\,4\,3 \\ 6\,2\,3\,4\,1\,5 \\ 4\,3\,2\,6\,5\,1 \end{pmatrix} \begin{pmatrix} x_1 \\ x_2 \\ x_3 \\ x_4 \\ x_5 \\ x_6 \end{pmatrix} = \begin{pmatrix} 9 \\ 14 \\ 21 \\ 15 \\ 14 \\ 11 \end{pmatrix}. \tag{3.11}$$

The example has been constructed so that the basis of the reduction is the observation that the matrix is invariant under the following permutations: $p_1 = (1,6)(2,5)(3,4)$ and $p_2 = (1,5,3)(2,6,4)$. As p_1 and p_2 generate a group D_6 of six element, the matrix commutes with the representation of group D_6 by matrices of order six. This suggests the application of group theory: decompose the matrix and the vector on the right hand side of the equation into irreducible components, and solve the resulting equations in the irreducible subspaces. The D_6 group is isomorphic to the symmetry group of the regular triangle discussed in Section 2.2.

The character table of the group D_6 can be found in tables (Atkins, 1970; Conway, 2003; Landau & Lifshitz, 1980), or, can be looked up in computer programs, or libraries (GAP, 2008).

Using the character table, and projector (2.17), one can carry out the following calculations. The observation that D_6 is isomorphic to the symmetry group of the equilateral triangle makes the problem easier. (Mackey, 1980) has made the observation: There is an analogy of the group characters and the Fourier transform. This allows the construction of irreducible vectors by the following ad hoc method. Form the following N-tuples ($N = |G|$):

$$\begin{aligned} \mathbf{e}_{2k-1} &= (\cos(2\pi/N*(2k-1)*1),\ldots,\cos(2\pi/N*(2k-1)*N), \\ \mathbf{e}_{2k} &= (\sin(2\pi/N*(2k)*1),\ldots,\sin(2\pi/N*(2k)*N), k=1,2,\ldots N. \end{aligned} \tag{3.12}$$

These vectors are orthonormal and can serve as an irreducible basis. After normalization, one gets a set of irreducible vectors in the N copies of the fundamental domain. Here one may exploit the isomorphism with the symmetry group of an equilateral triangle with the points positioned as shown in Fig. 1. Applying the above recipe to the points in the triangle, we get the following irreducible basis:

$$\mathbf{e}_1 = (1,1,1,1,1,1) \;\; \mathbf{e}_2 = (2,-1,-1,2,-1,-1) \;\; \mathbf{e}_3 = (0,1,-1,0,1,-1) \tag{3.13}$$

$$\mathbf{e}_4 = (2,1,-1,-2,-1,1) \;\; \mathbf{e}_5 = (0,1,1,0,-1,-1)) \;\; \mathbf{e}_6 = (1,-1,1,-1,1,-1). \tag{3.14}$$

We note that the points in the vectors $\mathbf{e}_i$ do not follow the order shown in Fig. 1. Thus we need to renumber the points, and normalize the vectors. For ease of interpolation we also renumber the vectors given above. It is clear that the vectors formed from cos and sin transform together. Thus they form a two-dimensional representation. We bring forward the one-dimensional representations. The projection to the irreducible basis is through a 6×6 matrix that contains

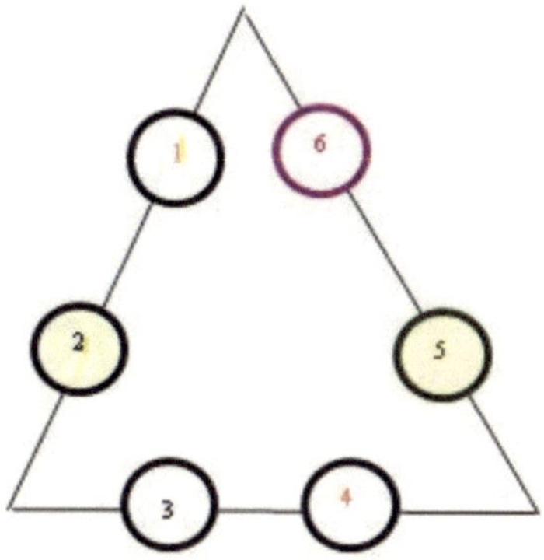

Fig. 1. Labeling Positions of Points on an Orbit

the orthonormal $\mathbf{e}'_i$ vectors:

$$\mathbf{O}^+ = \left(\mathbf{e}'^+_1, \mathbf{e}'^+_6, \mathbf{e}'^+_2, \mathbf{e}'^+_3, \mathbf{e}'^+_4, \mathbf{e}'^+_5 \right) \tag{3.15}$$

where the prime indicates rearranging in accordance with Fig. 1. Using the rearranging

$$\mathbf{A}x = b, \mathbf{OAO}^{-1}(\mathbf{O}x) = \mathbf{O}b,$$

we find[1]

$$\mathbf{OAO}^{-1} = \begin{pmatrix} 21 & 0 & 0 & 0 & 0 & 0 \\ 0 & -1 & 0 & 0 & 0 & 0 \\ 0 & 0 & -6 & 2a & 0 & 0 \\ 0 & 0 & -a & -1 & 0 & 0 \\ 0 & 0 & 0 & 0 & -6 & 2a \\ 0 & 0 & 0 & 0 & -a & -1 \end{pmatrix},$$

where $a = \sqrt{3}$. Compare the structure of the above matrix with that given in Section 3, where the similar form is achieved by geometrical similarity. In the present example there is no geometry, just a matrix invariant under a group of transformations.

In order to solve the resulting equations, we need the transformed right hand side of the equation:

$$\mathbf{O}b = \left(14\sqrt{6}, 2\sqrt{\frac{2}{3}}, 0, -8, 4, -\frac{2}{\sqrt{3}} \right)^+.$$

Finally, note that instead of solving one equation with six unknowns, we have four equations, two of them are solved by one division for each, and we have to solve two pairs of equations with two unknowns for each. At the end, we have to transform back from $\mathbf{O}x$ to x.

The Reader may ask: What is the benefit of the reduction? In a problem which is at the verge of solvability, that kind of reduction may become important.

[1] As matrix $\mathbf{O}$ is orthogonal, its inverse is just its transpose.

A more favorable situation is when there are geometric transformations leaving the equation and the volume under consideration, invariant. But before immersing into the symmetry hunting, we investigate the diffusion equation.

4. The multigroup diffusion equation

The diffusion equation is one of the most widely used reactor physical models. It describes the neutron balance in a volume V, the neutron energy may be continuous or discretized (multi group model). The multi group version is:

$$\frac{1}{v_k}\frac{\partial \Psi_k(\mathbf{r},t)}{\partial t} = \nabla(D_k(\mathbf{r})\nabla\Psi_k(\mathbf{r},t)) + \sum_{k'=1}^{G} T_{kk'}\Psi'_k(\mathbf{r},t), \tag{4.1}$$

where the processes leading to energy change are collected in $T_{kk'}$:

$$T_{kk'} = -\Sigma_{tk}\delta_{kk'} + \Sigma_{k'\to k} + \frac{\chi_k}{k_{eff}}\nu\Sigma_{fk'}, \tag{4.2}$$

where subscripts k, k' label the energy groups, v_k is the speed of neutrons in energy group k, $\Psi_k(\mathbf{r})$ is the space dependent neutron flux in group k, and $k_{eff} = 1$. In general, the cross-sections $D_k, \Sigma_{tk}, \Sigma_{k'\to k}, \Sigma_{fk'}$ are the space dependent diffusion constant, the total cross-section, the scattering cross-section, and the fission cross-section. χ_k is called the fission spectrum. Equation (4.1) is a set of partial differencial equations, to which the initial condition $\Psi_k(\mathbf{r},0), \mathbf{r} \in V$ and a suitable boundary condition, e.g. $\Psi_k(\mathbf{r},t), \mathbf{r} \in \partial V$ are given for every energy group k and every time t. The boundary conditions used in diffusion problems are of the type

$$(\nabla\mathbf{n})\Psi_k(\mathbf{r}) + b_k(\mathbf{r})\Psi_k(\mathbf{r}) = h_k(\mathbf{r}) \;\; k = 1,\ldots,G. \tag{4.3}$$

for $\mathbf{r} \in \partial V$. Here $b_k(\mathbf{r})$ depends on the boundary condition and may contain material properties, for example albedo.

The diffusion equation is a relationship between the cross-sections in V and the neutron flux $\Psi_k(\mathbf{r},t)$. The equation is linear in $\Psi_k(\mathbf{r},t)$. The main variants of equation (4.1) that are of interest in reactor physics are:

1. Static eigenvalue problem: When the flux does not depend on t, the left hand side is zero, and (4.1) has a nontrivial solution only if the cross-sections are interrelated. To this end, we free k_{eff} and the static diffusion equation is put in the form of an eigenvalue problem:

   $$\nabla(D_k(\mathbf{r})\nabla\Psi_k(\mathbf{r})) + \sum_{k=1}^{G} T_{kk'}(k_{eff})\Psi_{k'}(\mathbf{r}) = 0, \tag{4.4}$$

   where the eigenvalue k_{eff} introduced as a parameter in $T_{kk'}$ thus allowing for a non-trivial solution $\Psi_k(\mathbf{r})$. That usage is typical in core design calculations.
2. Time dependent solution allowing time dependence in some cross-sections. A typical application is transient analysis.
3. Equation (4.1) is homogeneous but it is possible to add an external source and to seek the response of V to the source.

The structure of the diffusion equation is simple. Mathematical operations, like summation and differentiation, and multiplication by material parameters (cross-sections) are applied to the neutron flux. In such equations the symmetries are mostly determined by the space dependence of the material properties. In the next subsection we investigate the possible symmetries of equation (4.4) and the exploitation of those symmetries.

When the solutions $\Psi_k(\mathbf{r}), k = 1, \ldots, G$ are known, not only the reaction rates, and net- and partial currents can be determined, but also matrices can be created to transform these quantities into each other. From diffusion theory it is known that the solution is determined by specifying the entering current along the boundary ∂V. Thus the boundary flux is also determined. But the given boundary flux also determines the solution everywhere in V. The solution is given formally by a Green's function as follows:

$$\Psi_k(\mathbf{r}) = \int_{\partial V} \sum_{k_0=1}^{G} \mathcal{G}_{k_0,k}(\mathbf{r}_0 \to \mathbf{r}) f_{k_0}(\mathbf{r}_0) d\mathbf{r}_0. \tag{4.5}$$

Here $\mathcal{G}_{k0,k}(\mathbf{r}_0 \to \mathbf{r})$ is the Green's function, it gives the neutron flux created at point $\mathbf{r}$ in energy group k by one neutron entering V at $\mathbf{r}_0$ in energy group k_0; and $f_{k_0}(\mathbf{r}_0)$ is the given flux in energy group k_0 at boundary point $\mathbf{r}_0$. Similarly the net current is obtained as

$$J_{nk}(\mathbf{r}) = -D_k \nabla \int_{\partial V} \sum_{k_0=1}^{G} \mathcal{G}_{k_0,k}(\mathbf{r}_0 \to \mathbf{r}) f_{k_0}(\mathbf{r}_0) d\mathbf{r}_0 \tag{4.6}$$

where the ∇ operator acts on variable $\mathbf{r}$.

4.1 Symmetries of the diffusion equation

First, the symmetry properties of the solution do not change in time because (4.1) is linear. This is not true for nonlinear equations. Secondly, the equations in (3.3) need to be satisfied. That is, the operations of the equation (4.4) and the boundary conditions must commute with the symmetry group elements. The symmetries of equation (4.4) are determined by the operators, the material parameters (cross-sections) and the geometry of V. The first term involves derivatives:

$$\nabla(D_k(\mathbf{r})\nabla\Psi_k(\mathbf{r},t)) = \nabla D_k(\mathbf{r})\nabla\Psi_k(\mathbf{r}) + D_k(\mathbf{r})\nabla^2\Psi_k(\mathbf{r}).$$

Here the first term contains a dot product which is invariant under rotations and reflections. The second term involves the laplace operator, which is also invariant under rotations and reflections. Thus, the major limiting symmetry factors are the material distributions, or the associated cross-sections as functions of space, and the shape of V. We assume the material distribution to be completely symmetric, thus for any cross-section $\Sigma(\mathbf{r})$ we assume the transformation property

$$\mathbf{O}_g\Sigma(\mathbf{r}) = \Sigma(\mathrm{D}(g)\mathbf{r}) = \Sigma(\mathbf{r}') = \Sigma(\mathbf{r}). \tag{4.7}$$

Here $\mathbf{O}_g$ is an operator applicable to the possible solutions. $\mathrm{D}(g)$ is a matrix representation of the symmetry group of the diffusion equation applicable to $\mathbf{r}$. The following operators are encountered in diffusion theory. The general form of a reaction rate at point $\mathbf{r} \in V$ can be expressed as

$$R(\mathbf{r}) = \sum_{k1} \Sigma_{k1}(\mathbf{r})\Psi_{k1}(\mathbf{r}). \tag{4.8}$$

Here subscript 1 refers to the symmetric component. Since

$$\mathbf{O}_g R(\mathbf{r}) = \mathbf{O}_g \sum_{k1} \Sigma_{k1}(\mathbf{r})\Psi_{k1}(\mathbf{r}) = \sum_{k1} \mathbf{O}_g \left(\Sigma_{k1}(\mathbf{r})\Psi_{k1}(\mathbf{r})\right) = \sum_{k1} \Sigma_{k1}(\mathbf{r})\mathbf{O}_g \Psi_{k1}(\mathbf{r})$$

because the material distribution is assumed symmetric hence $\mathbf{O_g}\boldsymbol{\Sigma}(\mathbf{r}) = \boldsymbol{\Sigma}(\mathbf{r})$ for every symmetry g, the transformation properties of a reaction rate are completely determined by the transformation properties of the flux $\Psi_{k1}(\mathbf{r})$. The normal component of the net current at $\mathbf{r} \in \partial V$ is

$$J_{nk}(\mathbf{r}) = -D_k(\mathbf{r})(\mathbf{n}\boldsymbol{\nabla})\Psi_k(\mathbf{r}), \tag{4.9}$$

where $\mathbf{n}$ is the normal vector at $\mathbf{r}$. We apply $\mathbf{O_g}$ to $J_{nk}(\mathbf{r})$ to obtain:

$$\mathbf{O}_g J_{nk}(\mathbf{r}) = -\mathbf{O}_g \left(D_k(\mathbf{r})(\mathbf{n}\boldsymbol{\nabla})\Psi_k(\mathbf{r})\right) = -D_k(\mathbf{r})(\mathbf{n}\boldsymbol{\nabla})\mathbf{O}_g\Psi_k(\mathbf{r}). \tag{4.10}$$

Thus, the transformation properties of the normal component of the net current agree with the transformation properties of the flux. In diffusion theory, the partial currents are defined as

$$I_k(\mathbf{r}) = \frac{1}{4}\left(\Psi_k(\mathbf{r}) - 2J_{nk}(\mathbf{r})\right); \; J_k(\mathbf{r}) = \frac{1}{4}\left(\Psi_k(\mathbf{r}) + 2J_{nk}(\mathbf{r})\right). \tag{4.11}$$

From (4.9) it follows that the transformation properties of the partial currents correspond to the transformation properties of the flux.

The boundary condition (4.3) commutes with rotations and reflections provided the material properties do. The same is true for the diffusion equation (4.1). Our first conclusion is that the material distribution may set a limit to the symmetry properties. As to the symmetries, the volume V under consideration may also be a limiting factor. Let $\mathbf{O}_g$ be an operator that commutes with the operations of the diffusion equation (4.1) and (4.3). Furthermore, the representation $\mathrm{D}(g)$ maps V into itself. The set of operators form a group; the group operation is the repeated application. That group is called the symmetry group of the diffusion equation.

Example 4.1 (Symmetries in a homogeneous square)**.** This symmetry group has eight elements, four rotations: E, C_4, C_4^2, C_4^3 and four reflections σ_x, σ_y, called of type σ_v and σ_{d1}, σ_{d2} called of type σ_d. Characters of a given class have identical values. This group is known as the symmetry group of the square and denoted as C_{4v}. The first column of a character table gives a mnemonic name to each representation, and a typical expression transforming according to the given representation. The first line is reserved for the most symmetric representation called unit representation. From the character table of the group C_{4v}, we learn that there are groups with the same character tables, there are five irreducible representations labeled A_1, A_2, B_1, B_2, E where As and Bs are one.dimensional and E is two-dimensional, it has two linearly independent components transforming as the x and y coordinates. □

Example 4.2 (Symmetries in a homogeneous equilateral triangle)**.** The group has six elements, three rotations: E, C_3, C_3^2, and three reflections through axis passing one edge: $\sigma_a, \sigma_b, \sigma_c$ called type σ_v. The symmetry group is isomorphic to the C_{3v} group and its character table is the same as that of the group D_3. The C_{3v} group is the symmetry group of the equilateral triangle, it has two one-dimensional and one two-dimensional representations. □

The key observation concerning the applications of symmetry considerations in boundary value problems is as follows. For a homogeneous problem (4.4) where there is no external source, the boundary condition is homogeneous, and every macroscopic cross-section $\Sigma(\mathbf{r}), \mathbf{r} \in V$ is such that

$$\mathbf{O}_g\Sigma(\mathbf{r}) = \Sigma(\mathbf{r})$$

for all $\mathbf{O}_g$ mapping V into itself. When the boundary conditions $h_k(\mathbf{r})$ in the expressions (4.3) transform according to an irreducible subspace $f^{\alpha}(\mathbf{r})$ then the neutron flux $\Phi(\mathbf{r})$, the partial currents $I(\mathbf{r}), J(\mathbf{r})$, the reaction rate

$$R(\mathbf{r}) = \sum_{k=1}^{G} \Sigma_g(\mathbf{r})\Psi_g(\mathbf{r})$$

all transform under the automorphism group of V as do the boundary conditions $h_k(\mathbf{r})$.

The symmetry group of the volume V makes it possible to reduce the domain on which we have to determine the solution of the diffusion theory problem. Once we know the transformation rule of the flux, for example, it suffices to calculate the flux in a part of V and exploit the transformation rules. That observation is formulated in the following concise way. Let $\mathbf{r} \in V$ a point in V and let $g \cdot \mathbf{r}$ be the image of $\mathbf{r}$ under $g \in G$. Then the set of points $g \cdot \mathbf{r}, g \in G$ is called the orbit of $\mathbf{r}$ under the group G. If there is a set $V_0 \in V$ such that the orbits of $\mathbf{r}_0 \in V_0$ give every point[2] of V we call V_0 the fundamental domain of V. It is thus sufficient to solve the problem on the fundamental domain V_0, and "continue" the solution to the whole volume V.

When the boundary condition is not homogeneous or there is an external source, we exploit the linearity of the diffusion equation. The general solution is the sum of two terms: one with external source but homogeneous boundary condition and one with no external source but with non-homogeneous boundary condition. In either case, it is the external term that determines the transformation properties of the respective solution component.

4.2 Selection of basis functions

The purely geometric symmetries of a suitable equation lead to a decomposition (2.16) of an arbitrary function in a function space, and thus the decomposition of the function space itself. The decomposed elements are linearly independent and can be arranged to form an orthonormal system. This can be exploited in the calculations.

In a homogeneous material one can readily construct trial functions that fulfill the diffusion equation at each point of V. For example consider

$$\nabla^2 \underline{\psi}(\mathbf{r}) + \mathbf{A}\underline{\psi}(\mathbf{r}) = 0 \tag{4.12}$$

where $\mathbf{A} = \left(\mathbf{\Sigma}_t - \mathbf{\Sigma}_s + \mathbf{\Sigma}_f\right)\mathbf{D}^{-1}$. The general solution to (4.12) takes the form of

$$\underline{\psi}(\mathbf{r}) = \sum_{k=1}^{G} \underline{t}_k \int_{|\mathbf{e}|=1} e^{i\lambda_k \mathbf{r}\cdot\mathbf{e}} W_k(\mathbf{e}) d\mathbf{e} \tag{4.13}$$

where the weight functions $W_k(\mathbf{e})$ are arbitrary suitable functions, $i^2 = -1$, and $\underline{t}_k$ signify the eigenvectors of matrix $\mathbf{A}$:

$$\mathbf{A}\underline{t}_k = \lambda_k^2 \underline{t}_k. \tag{4.14}$$

[2] Images of V_0 cover V.

When $\mathbf{e}(\theta) = (\cos\theta, \sin\theta)$, using (2.19), we build up a regular representation from (4.13) so that

$$\underline{\psi}_0(\mathbf{r}) = \sum_{k=1}^{G} t_k \int_0^{2\pi/|G|} e^{i\lambda_k \mathbf{r}\cdot\mathbf{e}(\theta)} W_k(\mathbf{e}(\theta))d\theta, \tag{4.15}$$

and the action of operators $\mathbf{O}_g$ on $\underline{\psi}_0(\mathbf{r})$ is defined as follows. $\mathbf{O}_g$ acts on variable $\mathbf{r}$, see (2.6), but in (4.13), $\mathbf{r}$ occurs only in the form of the dot product $\mathbf{re}(\theta)$, therefore action of $\mathbf{O}_g$ can be transferred to an action on θ. As a result, each $\mathbf{O}_g$ acts as

$$\mathbf{O}_g \underline{\psi}_0(\mathbf{r}) = \sum_{k=1}^{G} t_k \int_{I_g} e^{i\lambda_k \mathbf{r}\cdot\mathbf{e}(\theta)} W_k(\mathbf{e}(\theta))d\theta \tag{4.16}$$

where $\mathbf{O}_g$ maps the interval $0 \le \theta \le 2\pi/|G|$ into the interval I_g. In this manner we get the irreducible components of the solution as a linear combination of $|G|$ exponential function, it is only the coefficients in the linear combination that determine the irreducible components. The weight function $W_k(\theta)$ makes it possible to match the entering currents at given points of the boundary. Let $\theta = 0$ correspond to the middle of a side. Then choosing

$$W_k((e))(\theta) = W_k\delta(\theta), \tag{4.17}$$

we get by (4.15) the solution at face midpoints. The last step is the formation of the irreducible components. Observe that in projection (2.20) the solutions at different images of $\mathbf{r}$ are used in a linear combination, the coefficients of the linear combinations are the rows of the character table. But in the images (4.16), only the weight function changes. In each I_g interval the image of $W_k(\mathbf{e})$ is involved, which is a Dirac-delta function, only the place of the singularity changes as the group elements map the place of the singularity. A symmetry of the square maps a face center into another face center thus there will be four distinct positions and the space dependent part of the irreducible component of ψ_0 will contain four exponentials:

$$\pm e^{i\lambda_k x}, \pm e^{-i\lambda_k x}, \pm e^{i\lambda_k y}, \pm e^{-i\lambda_k y}. \tag{4.18}$$

From these expressions the following irreducible combinations can be formed:

$$A_1 : \cos\lambda_k x + \cos\lambda_k y; A_2 : \cos\lambda_k x - \cos\lambda_k y; E_1 : \sin\lambda_k x; E_2 : \sin\lambda_k y. \tag{4.19}$$

It is not surprising that when we represent a side by its midpoint the odd functions along the side are missing.

The above method may serve as a starting point for developing efficient numerical methods. The only approximation is in the continuity of the partial currents at the boundary of adjacent homogeneous nodes.

If elements of the function space are defined for all $\mathbf{r} \in V$, and if $f_1, f_2 \in \mathbb{L}_V$, then the following inner product is applicable:

$$(f_1, f_2) \equiv \int_V f_1(\mathbf{r}) f_2(\mathbf{r}) d^3\mathbf{r}. \tag{4.20}$$

Let $f_\ell^\alpha(\mathbf{r}), \ell = 1, \ldots, n_\alpha$ be a regular representation of group G. Then

$$(f_\ell^\alpha(\mathbf{r}), f_{\ell'}^\beta(\mathbf{r})) = \delta_{\alpha,\beta}\delta_{\ell,\ell'} \tag{4.21}$$

furthermore, for the reactions rates formed with the help of the cross-sections in (4.1), similar orthogonality relation holds. For the volume integrated reaction rates we have

$$(f_\ell^\alpha(\mathbf{r}), const) = \delta_{\alpha,1}\delta_{\ell,1}, \tag{4.22}$$

in other words: solely the most symmetric, one dimensional representation contributes to the volume integrated reaction rates. Note that as a result of the decomposition of the solution or its approximation into irreducible components not only that irreducible components of a given physical quantity (like flux, reaction rate, net current) but also the given irreducible component of every physical quantity fall into the same linearly independent irreducible subspace. As a consequence, the operators (matrices) mapping the flux into net currents (or vice versa) fall into the same irreducible subspace, therefore the mapping matrix automatically becomes diagonal.

Example 4.3 (Symmetry components of boundary fluxes). Consider the flux given along the boundary of a square. The flux is given by four functions corresponding to the flux along the four sides of the square. The flux along a given face is the sum of an even and an odd function with respect to the reflection through the midpoint of the face. The decomposition (2.21) gives the eight irreducible components shown in Figure 2. Note that the irreducible

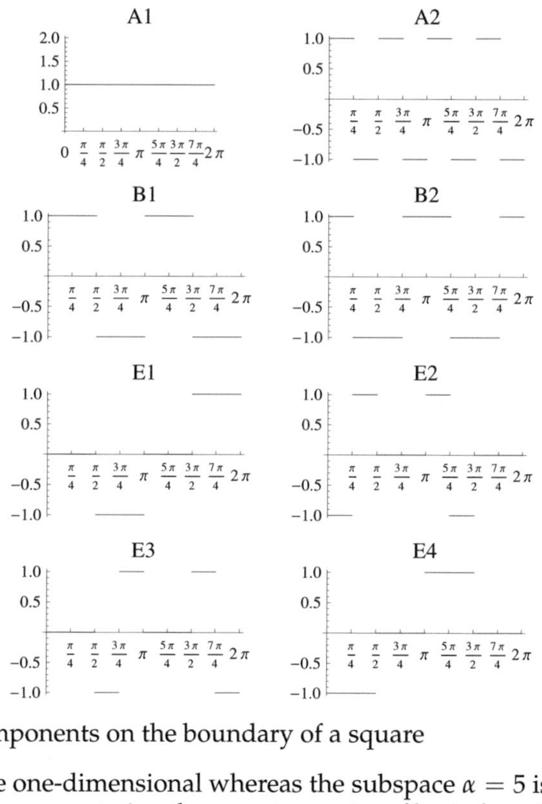

Fig. 2. Irreducible components on the boundary of a square

subspaces $\alpha_i, i < 5$ are one-dimensional whereas the subspace $\alpha = 5$ is two-dimensional, and in a two-dimensional representation there are two pairs of basis functions that are identical as to symmetry properties. Thus, we have altogether eight linearly independent basis functions.

The physical meaning of the irreducible components is that the flux distribution of a node is a combination of the flux distributions established by eight boundary condition types. The component A_1 represents a complete symmetry that is the same even distribution along each side. Component A_2 is also symmetric, but the boundary condition is an odd function on each side. Components B_1 and B_2 represent entering neutrons along one axes and exiting neutrons along the perpendicular axes, a realization of a second derivative with even functions over a face. B_2 is the same but with odd functions along a face. E_1 and E_4 represent streaming in the x and y directions with even distributions along a face, whereas E_2 and E_3 with odd distributions along a face.

The symmetry transformations of the square, map the functions given along the half faces into each other but they do not say anything about the function shape along a half face. Therefore, the functions in Fig. 2 serve only as patterns, the function shape is arbitrary along a half face. The corresponding mathematical term is the direct product; each function may be multiplied by a function $f(\xi), -h/2 \leq \xi \leq +h/2$. It is well known that a function along an interval can be approximated by a suitable polynomial (Weierstrass's theorem). We know from practice that in reactor calculations a second order polynomial suffices on a face for the precision needed in a power plant.

The invariant subspace means that the boundary flux, the net current, the partial currents must follow one of the patterns shown in Figure 2, the only difference may be in the shape function $f(\xi), -h/2 \leq \xi \leq +h/2$. This means the a constant flux may create a quadratic position dependent current, but the global structure of the flux, and current should belong to the same pattern of Figure 2.

Moreover, if we are interested in the solution inside the square, its pattern must also be the same although there the freedom allows a continuous function along 1/8-th of the square. These features are exploited in the calculation. □

4.3 Iteration

It is known that the diffusion (as well as the transport) equation has a well defined solution in V provided the entering current is given along the boundary ∂V. From the Green's function and from the operators in (4.11) we set up the following iteration scheme. To formalize this, we write the solution as

$$\Psi_k(\mathbf{r}) = \sum_{k'=1}^{G} \int_{\partial V} \mathcal{G}_{kk'}(\mathbf{r}' \to \mathbf{r}) I_{k'}(\mathbf{r}') d\mathbf{r}'. \tag{4.23}$$

Applying operator $\mathbf{F}$ that forms the exiting current from the flux, we obtain

$$J_k(\mathbf{r}) = \sum_{k'=1}^{G} \int_{\partial V} \mathbf{F}\mathcal{G}_{kk'}(\mathbf{r}' \to \mathbf{r}) I_{k'}(\mathbf{r}') d\mathbf{r}', \tag{4.24}$$

that can be put into the concise form

$$J_k = \sum_{k'=1}^{G} \mathbf{R}_{kk'} I_{k'}, \tag{4.25}$$

where we have suppressed that the partial currents depend on position along the boundary and the response matrix $\mathbf{R}$ includes an integration over variable $\mathbf{r}'$.

When volume V is large, we subdivide it into subvolumes (nodes) and determine the response matrices for each subvolume. At internal boundaries, the exiting current is the incident current of the adjacent subvolume. Thus in a composite volume the partial currents are connected by response matrices and adjacency. We collect the response matrices and adjacency into two big response matrices:

$$\underline{J} = \mathbf{R}\underline{I}; \ \underline{I} = \mathbf{H}\underline{J}, \tag{4.26}$$

and because the adjacency is an invertible relationship, we multiply the first expression by $\mathbf{H}$ and get

$$\underline{I} = \mathbf{HR}\underline{I}. \tag{4.27}$$

Since there is a free parameter k_{eff} in matrix $\mathbf{R}$, it makes the equation solvable. At external boundaries there is no adjacency, but the boundary condition there provides a rule to determine the entering current from the exiting current. With these supplements, the solution of equation (4.27) proceeds

$$\underline{I}^{(m+1)} = \mathbf{HR}^{(m)}\underline{I}^{(m)}. \tag{4.28}$$

The iteration starts with $m = 0$ with an initial guess for the k_{eff} and the entering currents $\underline{I}$. Let us assume that the needed matrices are available, their determinations are discussed in the subsequent Subsection. The iteration proceeds as follows. We sweep through the subvolumes in a given sequence and carry out the following actions (in node m):

- collect the actual incoming currents of subvolume m.
- determine the actual response matrix to calculate the new exiting currents and contributions to volume integrals[3].
- determine the new exiting currents ($\underline{J}$) from the entering currents and the response matrices using equation (4.26) and the contributions to the volume integrals.

After this, pass on to the next node. When the iteration reaches the last node, the sweep ends and the maximal difference is determined between the entering currents of the last two iterations. At the end of an iteration step, the parameter k_{eff} is re-evaluated from the condition that the largest eigenvalue of $\mathbf{HR}$ should equal one. If the difference of the last two estimates is greater than the given tolerance limit, a new iteration cycle is started, otherwise the iteration terminates. If we have a large number of nodes, the improvement after the calculations of a given node is small. This shows that the iteration process is rather slow, acceleration methods are required.

It has been proven (Mika, 1972) that the outlined iteration is convergent. The goal of the iteration is to determine the partial current vector. The length of vector $\mathbf{I}$ is $N_{node} \times n_F \times G$. From the point of view of mathematics, the iteration is a transformation of the following type:

$$\mathbf{A}(k_{eff})\mathbf{x}_m = a\mathbf{x}_{m+1}, \tag{4.29}$$

where m is the number of the iteration, matrix $\mathbf{A}(k_{eff})$ makes the new entering current vector $\mathbf{x}_{m+1}$ from the old entering current vector x_m. In the case of neutron diffusion or transport, operator $\mathbf{A}(k_{eff})$ maps positive vectors into positive vectors. In accordance with the Krein–Ruthman theorem, $\mathbf{A}(k_{eff})$ has a dominant eigenvalue and the associated eigenfunction[4]. When k_{eff} is a given value, the power method is a simple iteration technique to find a good estimate of $\mathbf{x} = \lim_{i\to\infty} \mathbf{x}_i$. Solution methods have been worked out for practical problems in nuclear reactor theory: for the solution of the diffusion and transport equations in

[3] We obtain reactions rates also from the Green's function.

[4] Actually a discretized eigenfunction, i.e. $\mathbf{x}$.

the core of a power reactor. The original numerical method is described elsewhere, see Refs. (Weiss, 1977), (Hegedus, 1991).

Note that the iteration (4.29) is just an example of the maps transforming an element of the solution space into another element. Thus in principle one can observe chaotic behavior, divergence, strange attractors[5] etc. Therefore it is especially important to design carefully the iteration scheme. The iteration includes derived quantities of two types: volume integrated and surface integrated. When you work with an analytical solution, the two are derived from the same analytical solution. But when you are using approximations (such as polynomial approximation), it has to be checked if the polynomials used inside the node and at the surface of the node are consistent. In an eigenvalue problem, parameter k_{eff} in equation (4.29)should be determined from the condition that the dominant eigenvalue a in (4.29) should equal one. First we deal with the general features of the iteration.

As has been mentioned, one iteration step (4.29) sweeps through all the subvolumes. The number of subvolumes (Gadó et al., 1994) varies between 590 and 7980, the number of unknowns is 9440 and 111680. At the boundary of two adjacent subvolumes, continuity of Φ and $D\partial_n\Phi$ (the normal current) is prescribed .

In node m in iteration i. In the derivation of the analytical solution we have assumed the node to be invariant under the group G_V. Actually, not the material properties are stored in a program because the material properties depend on:

- actual temperature of the node;
- the initial composition of the fuel (e.g. enrichment);
- the actual composition of the fuel as it may change with burn-up;
- the void content of the moderator;
- the power level;

In the calculations, app. 50–60% of the time is spent on finding the actual response matrix elements, because those depend on a number of local material parameters (e.g. density, temperature, void content). We mention this datum to underline how important it is to reduce the parametrization work in a production code.

4.4 Exploiting symmetries

In a given node, the response matrices are determined based on the analytical solution (4.19). We need an efficient recipe for decomposing the entering currents into irreps and reconstructing the exiting currents on the faces. Since the only approximation in the procedure requires the continuity of the partial currents, we need to specify the representation of the partial currents and how to represent them. The simplest is a representation by discrete points along the boundary, the minimal number is four, the maximal number depends on the computer capacity. An alternative choice is to represent the partial currents by moments over the faces. Usually average, first and second moment suffice to get the accuracy needed by practice. The representation fixes the number of points we need on a side and the number of points (n) on the node boundary.

To project the irreps, we may use (Mackey, 1980) the $\cos((k-1)2\pi/n), k = 1, \ldots, n/2$ and $\sin((k)2\pi/n), k = 1, \ldots, n/2$ vectors (after normalization). The following illustration shows

[5] Since k_{eff} depends on the entering currents, the problem is non-linear.

the case with $n = 4$, i.e. one value per face. In a square node we need the following matrix

$$\mathbf{\Omega}_4 = \begin{pmatrix} 1 & 1 & 1 & 1 \\ 1 & -1 & 1 & -1 \\ 1 & 0 & -1 & 0 \\ 0 & 1 & 0 & -1 \end{pmatrix}. \tag{4.30}$$

to project the irreducible components from the side-wise values. As (2.20) shows, irreducible components are linear combinations of the decomposable quantity [6]. The coefficients are given as rows in matrices $\mathbf{\Omega}_4$.

In a regular n-gonal node the response matrix has[7] $Ent\left[(n+2)/2\right]$ free parameters. The response matrix also has to be decomposed into irreps, this is done by a basis change. Let the response matrix give

$$\mathbf{J} = \mathbf{RI}$$

Multiply this expression by $\mathbf{\Omega}$ from the left:

$$\mathbf{\Omega J} = \left(\mathbf{\Omega R \Omega}^{-1}\right)\mathbf{\Omega I}, \tag{4.31}$$

and we see that for irreducible representations the response matrix is given by $\Omega\mathbf{R}\Omega^{+}$. In a square node:

$$\mathbf{R}_4 = \begin{pmatrix} r & t_1 & t_2 & t_1 \\ t_1 & r & t_1 & t_2 \\ t_2 & t_1 & r & t_1 \\ t_1 & t_2 & t_1 & r \end{pmatrix}, \tag{4.32}$$

and the irreducible representation of R_4 is diagonal:

$$\begin{pmatrix} A & 0 & 0 & 0 \\ 0 & B & 0 & 0 \\ 0 & 0 & C & 0 \\ 0 & 0 & 0 & C \end{pmatrix}, \tag{4.33}$$

where

$$A = r + 2t_1 + t_2, \quad B = r - 2t_1 + t_2, \quad C = r - t_2.$$

We summarize the following advantages of applying group theory:

- Irreducible components of various items play a central role in the method. The irreducible representations often have a physical meaning and make the calculations more effective (e.g. matrices transforming one irreducible component into another are diagonal).
- The irreducible representations of a given quantity are linearly independent and that is exploited in the analysis of convergence.
- The usage of linearly independent irreducible components is rather useful in the analysis of the iteration of a numerical process.
- In several problems of practical importance, the problem is almost symmetric, some perturbations occur. This makes the calculation more effective.

[6] After normalization of the row vectors, the $\mathbf{\Omega}$ matrices become orthogonal: $\mathbf{\Omega}^{+}\mathbf{\Omega}$ is the unit matrix.
[7] Here *Ent* is the integer division.

- It is more efficient to break up a problem into parts and solve each subproblem independently. Results have been reported for operational codes (Gadó et al., 1994).

The above considerations dealt with the local symmetries. However, if we decompose the partial currents into irreps, we get a decomposition of the global vector x in equation (4.28) as well. We exploit the linear independence of the irreducible components further on the global scale.

For most physical problems we have a priori knowledge about the solution to a given boundary value problem in the form of smoothness and boundedness. This is brought to bear through the choice of solution space. In the following, we introduce via group theoretical principles the additional information of the particular geometric symmetry of the node. This allows the decomposition of the solution space into irreducible subspaces, and leads, for a given geometry, not only to a rule for choosing the optimum combination of polynomial expansions on the surface and in the volume, but also elucidates the subtle effect that the geometry of the physical system can have on the algorithm for the solution of the associated mathematical boundary problem.

Consider the iteration (4.29) and decompose the iterated vector into irreducible component

$$\underline{x} = \sum_{\alpha} \underline{x}^{\alpha} \tag{4.34}$$

where because of the orthogonality of the irreducible components

$$\underline{x}^{\beta +} \underline{x}^{\alpha} = 0$$

when $\alpha \neq \beta$. The convergence of the iteration means that

$$\lim_{N \to \infty} \underline{x}_{N+k1} - \underline{x}_{N+k2} = 0 \tag{4.35}$$

for any k_1, k_2. But that entails that as the iteration proceeds, the difference between two iterated vector must tend to zero. In other words, the iteration must converge in every irreducible subspace. This observation may be violated when the iteration process has not been carefully designed.

Let us assume a method, see (Palmiotti, 1995), in which N basis functions are used to expand the solution along the boundary of a node and M basis functions to expand the solution inside the node. It is reasonable to use the approximation of same order along each face, hence, in a square node N is a multiple of four. For an Mth order approximation inside the node, the number of free coefficients is $(M+1)(M+2)/2$. It has been shown that an algorithm (Palmiotti, 1995) with a linear ($N = 1$) approximation along the four faces, with 8 free coefficients, of the boundary did not result in convergent algorithm unless $M = 4$ quartic polynomial, with 15 free coefficients, was used inside the node.

In such a code each node is considered to be homogeneous in composition. Central to the accuracy of the method are two approximations. In the first, we assume the solution on the boundary surface of the node to be expanded in a set of basis functions $(f_i(\xi); i = 1, \ldots, N)$. In the second, the solution inside the volume is expanded in another set of basis functions $(F_j(\mathbf{r}); j = 1, \ldots, M)$. Clearly the independent variable ξ is a limit of the independent variable $\mathbf{r}$.

Any iteration procedure, in principle, connects neighboring nodes through continuity and smoothness conditions. For an efficient numerical algorithm it is therefore desirable to have

i/Order	0	1	2	3	4
1	1	-	(x^2+y^2)	-	(x^2y^2), (x^4+y^4)
2	-	-	-	-	(x^3y-y^3x)
3	-	-	(x^2-y^2)	-	x^4-y^4
4	-	-	xy	-	(x^3y+y^3x)
5	-	x	-	x^3	-
6	-	-	-	xy^2	-
7	-	-	-	x^2y	-
8	-	y	-	y^3	-

Table 6. Irreducible components of at most fourth order polynomials under the symmetries of a square C_{4v}

the same number of degrees of freedom (i.e. coefficients in the expansion) on the surface of the node as within the node. With the help of Table 6, for the case of a square node, we compare the required number of coefficients for different orders of polynomial expansion. A linear approximation along the four faces of the square has at least one component in each irreducible subspace. At the same time the first polynomial contributing to the second irrep is fourth order. Convergence requires the convergence in each subspace thus the approximation inside the square must be at least of fourth order. There is no linear polynomial approximation that would use the same number of coefficients on the surface as inside the volume. The appropriate choice of order of expansion is thus not straightforward but it is important to the accuracy of the solution, because a mismatch of degrees of freedom inside and on the surface of the node is likely to lead to a loss of information in the computational step that passes from one node to the next. A lack of convergence has been observed, see (Palmiotti, 1995), in the case of calculations with a square node when using first order polynomials on the surface. A convergent solution is obtained only with fourth or higher order polynomial interpolation inside the node. Similar relationships apply to nodes of other geometry. For a hexagonal node that there is no polynomial where the number of coefficients on the surface matches the number of coefficients inside the node.

In a hexagonal node in (Palmiotti, 1995), the first convergent solution with a linear approximation on the surface requires at least a sixth order polynomial expansion within the node. Thus, in the case of a linear approximation on the surface, in the case of a square node a third order polynomial within the node does not lead to a convergent solution, although the number of coefficients is greater than those on the surface. In the case of the regular hexagonal node, a convergent solution is obtained only for the sixth order polynomial expansion in the node, while both a fourth and a fifth order polynomial have a greater number of coefficients inside the node than on the surface. It appears that some terms of the polynomial expansion contain less information than others, and are thus superfluous in the computational algorithm. If these terms can be "filtered out", a more efficient and convergent solution should result. The explanation becomes immediately clear from the decomposition of the trial functions inside the volume and on the boundary. In both the square and hexagon nodes, the first order approximation on the boundary is sufficient to furnish all irreducible subspaces whereas this is true for the interpolating polynomials inside V for surprisingly high order polynomials.

5. Reactor physics

In analogy with the application of group theory in particle physics, where group theory leads to insights into the relationships between elementary particles, we present an application of group theory to the solution to a specific reactor physics problem. The question is whether it is possible to replace a part of a heterogeneous core by a homogeneous material so that the solution outside the homogeneous region remains the same? This old problem is known as homogenization (Selengut, 1960).

In particular, for non-uniform lattices, asymptotic theory has shown that a lattice composed of identical cells has a solution that is composed of a periodic microflux and a slowly varying macroflux. What happens if the cell geometry is the same but the material composition varies?

In reactor calculations, we solve an equation derived from neutron balance. In that equation, we encounter reaction rates, currents or partial currents. It is reasonable to derive all the quantities we need from one given basic quantity, say from the neutron flux at given points of the boundary. The archetype of such relation is the exiting current determined from the entering current by a response matrix. We show that by using irreducible components of the partial currents, the response matrix becomes diagonal.

The Selengut principle is formulated: if the response matrix of a given heterogeneous material in V can be substituted by the response matrix of a homogeneous material in V, there exist an equivalent homogeneous material with which one may replace V. This principle simplifies calculations considerably, and, therefore, has been widely used in reactor physics. We investigate the Selengut principle more closely(Makai, 2010),(Makai, 1992).

The analysis is based on the analytical solution of the diffusion equation derived in the previous Section. The problem is considered in a few energy groups, the boundary flux $\underline{F}$ is a vector, as well as the volume averaged flux $\underline{\bar{\Phi}}$. Using that solution, we are able to derive matrices mapping into each other the volume integrated fluxes, the surface integrated partial and net currents. The derivation of the corresponding matrices is as follows. Our basis is the boundary flux, that we derive for each irrep i from (4.13). The expression (4.13) has three components. The first one is vector $\underline{t}_k$ which is independent of the position $\mathbf{r}$ and is multiplied by an exponential function with $\lambda_k \mathbf{r}$ in the exponent. The third component is the weight W_k which is independent of $\mathbf{r}$ but varies with subscript k. The product is summed for subscript k, that labels the eigenvalues of the cross-section matrix in (4.14). That expression can be put into the following concise form:

$$\underline{F}_i = \mathbf{T} < f_i > \underline{c}_i, \tag{5.1}$$

where $\underline{c}_i$ comprises the third component. Here $< f_i(\mathbf{r}) >$ is a diagonal matrix. Note that position dependent quantities like reaction rates, follow that structure. The normal component of the net current is $\underline{J}_{net}$ obtained from the flux by taking the derivative and is given in irrep i as

$$\underline{J}_{net,i} = -\mathbf{DT} < g_i > \underline{c}_i. \tag{5.2}$$

We eliminate $\underline{c}_i$ to get

$$\underline{J}_{net,i} = -\mathbf{DT} < g_i/f_i > \mathbf{T}^{-1}\underline{F}_i \equiv \mathbf{R}_i\underline{F}_i. \tag{5.3}$$

Here $\mathbf{n}$ is the outward normal to face F_i,

$$g_i = -\nabla \mathbf{n} f_i(r). \tag{5.4}$$

The volume integrated flux $\underline{\bar{\Phi}}$ is obtained after integration from (4.13) as

$$\underline{\bar{\Phi}} = \mathbf{T} < \bar{F}_{A1} > c_{A1}, \bar{F}_{A1} = \int f_{A1}(\mathbf{r}) d^3\mathbf{r}, \tag{5.5}$$

and the integration runs over volume V of the node. Note that only irrep $A1$ (i.e. complete symmetry) contributes to the average flux because of the orthogonality of the irreducible flux components. After eliminating c_{A1} from (5.1), we get the response matrix for determining the volume integrated flux $\underline{\bar{\Phi}}$ from the face integrated flux F_{A1}:

$$\underline{\bar{\Phi}} = \mathbf{T} < \overline{F}_{A1}/f_{A1} > \mathbf{T}^{-1} < \overline{F} > \equiv \mathbf{W} < \overline{F} > . \quad (5.6)$$

This assures that V is completely described by matrix $\mathbf{W}$ and the diagonal matrices $< F(r) >$ $, < f(r) >, < g(r) >$ for each irrep. For example, we are able to reconstruct the cross-section matrix $\mathbf{\Sigma}$ from them. Note that $\mathbf{WT} = \mathbf{T} < F/f >$, the eigenvectors of matrix $\mathbf{W}$ are the eigenvectors of $\mathbf{A}$. Now we need only a numerical procedure to find the eigenvalues λ_k from $< F/f >$.

The question is, under what conditions are the above calculations feasible. We count the number of response matrices. The matrix elements we need to characterize V may be all different and the number of matrices depends on the shape of V, since the number of irreducible components of the involved matrices depends on the geometry. In a square shaped homogeneous V, we have four $\mathbf{R}_i$ matrices and one $\mathbf{W}$. Altogether we have to determine $5 * G * G$ elements. In an inhomogeneous hexagonal volume, there are $6 * G * G$ matrix elements, whereas the homogeneous material is described by $G * (G+1)$ parameters as in a homogeneous material there are altogether $G * G$ cross-sections and G diffusion constants[8]. Therefore the Selengut principle is not exact it may only be a good approximation under specific circumstances. Homogenization recipes preserve only specific reaction rates, but they do not provide general equivalence.

6. Conclusions

The basic elements of the theory of finite symmetry groups has been introduced. In particular, the use of the machinery associated with the decomposition into irreducible representations, in analogy with harmonic analysis of functions in function space, in the analysis of Nuclear Engineering problems. The physical settings of many Nuclear Engineering problems exhibit symmetry, as for example in the solution of the multi-group neutron diffusion equation. This symmetry can be systematically exploited via group theory, and elicit information that leads to more efficient numerical algorithms and also to useful insights. This is a result due to the added information inherent in symmetry, and the ability of group theory to define the "rules" of the symmetry and allows one to exploit them.

7. References

Allgower, E. L., et al. (1992). Exploiting symmetry in boundary element methods, *SIAM J. Numer. Anal.*, 29, 534–552.

Atkins, P. W. et al. (1970). *Tables for group theory*, Oxford University Press, Oxford.

Brooks, R. (1988). Constructing isospectral manifolds, *Amer. Math. Monthly*, 95, 823–839.

Conway, J. H., et al. (2003). *The ATLAS of finite groups*, Oxford University Press, Oxford.

Deniz, V. C. (1986). *The Theory of Neutron Leakage in Reactor Lattices*, in CRC Handbook of Nuclear Reactor Calculations, vol. II, CRC Press, Boca Raton, (FL)

Falicov, L. M. (1996). *Group Theory and Its Physical Applications*, The University of Chicago Press, Chicago, (IL)

[8] Remember, here G is the number of energy groups.

Gadó, J. et al. (1994). KARATE-A Code for VVER-440 Core Calculation, *Trans. Am. Nucl. Soc.*, 71 , 485.
The GAP Group, (2008). *GAP – Groups, Algorithms, and Programming, Version 4.4.12*; 2008, (http://www.gap-system.org).
Hegedűs Cs. J. (1991). Generating conjugate directions for arbitrary matrices by matrix equations. I, *Comput. Math. Appl.*, 21, 71–85.
Korn, G. M. & Korn, T. M. (1975). *Mathematical handbook for scientist and engineers*, McGraw Hill.
Landau, L. D. & Lifshitz, E. M. (1980). *Theoretical Physics*, Volume VI, Pergamon, Oxford.
Mackey, G. W. (1980). Harmonic analysis as the exploitation of symmetry - A historical survey, *Bull. Am. Math. Soc.*, 3 543-698.
Makai M. & Orechwa Y. (2000). Field reconstruction from measured values in symmetric volumes, *Nuclear Engineer. Design*, 199 , 289–301.
Makai, M. (1992). Plane waves and response matrices, *Ann. Nucl. Energy*, 19, 715–736.
Makai, M. (2010). Thirty years of point group theory applied to reactor physics, *Il Nuovo Cimento*, 33C, No.1.
Makai, M. (2011). *Group Theory Applied to Boundary Value Problems with Applications to Reactor Physics*, Nova Science, New York
M. Makai (1996): Group Theory Applied to Boundary Value Problems, Report ANL-FRA-1996-5, Argonne National Laboratory
Borysiewicz, M. & Mika, J. (1972). Existence and uniqueness of the solution to the critical problem in the multigroup transport theory, *Tran. Theory Stat. Phys.*, 2, 243–270.
Nieva R. & Christensen , G. S. (1977). Symmetry Reduction of Reactor Systems, *Nucl. Sci. Eng.* 64, 79
Makai M. & Y. Orechwa (1977): *Problem Decomposition and Domain-Based Parallelism via Group Theoretic Principles*, Joint International Conference on Mathematical Methods and Supercomputing for Nuclear Applications, October 6-10, 1997, Saratoga Springs
Palmiotti, G. & al. (1995). *VARIANT*, Report ANL-95/40, Argonne National Laboratory, IL.
Seber, G. A. F & Wild, C. J. (1989). *Nonlinear regression*, John Wiley, New York.
Selengut, D. S. (1960). Diffusion Coefficients for Heterogeneous Systems, *Trans. Am. Nucl. Soc.* 3, 398
Sunada T. (1985). Riemannian coverings and isospectral manifolds, *Ann. Math.*, 121, 169–186.
Vladimirov V. S. (1971). *Equations of mathematical physics*, Marcel Dekker, New York.
Weiss Z. (1977). Some basic properties of the response matrix equations, *Nucl. Sci. Eng.*, 63, 457-472.

14

Neutron Shielding Properties of Some Vermiculite-Loaded New Samples

Turgay Korkut[1], Fuat Köksal[2] and Osman Gencel[3]
[1]*Faculty of Science and Art, Department of Physics, Ibrahim Cecen University, Ağrı*
[2]*Department of Civil Engineering, Faculty of Engineering and Architecture, Bozok University, Yozgat*
[3]*Department of Civil Engineering, Faculty of Engineering, Bartin University, Bartin*
Turkey

1. Introduction

Nuclear reactor technology is known as an emerging area of study from past to present. It is an implementation of the nuclear sciences including reactions about atomic nucleus and productions. During the construction of a nuclear reactor, the most important issue is nuclear safety. The term of security can be attributed radiation shielding processes. For nuclear reactors, there are several different materials used to radiation shielding. While determining the most appropriate material to shield, the type and energy of radiation is extremely important.

There are two types of nuclear reactions reveals very large energies. These are the disintegration of atomic nuclei (fission) and merging small atomic nuclei (fusion) reactions. Therefore, nuclear reactors can be divided into two groups according to the type of reaction occurred during as fission reactors and fusion reactors. Currently a nuclear reactor working with fusion reactions is not available. Today, there are the hundreds of nuclear reactors based on the fission reactions. For the realization of nuclear fission, a large fissile atomic nucleus such as ^{235}U can absorb a neutron particle. At the end of nuclear fission event, fission products (two or more light nucleus, kinetic energy, gamma radiation and free neutrons) arise. Fission reactions are controls by using neutron attenuators such as heavy water, cadmium, graphite, beryllium and several hydrocarbons. While designing a reactor shield materials against gamma and neutron radiations should be used.

Vermiculite is a monoclinic-prismatic crystal mineral including Al_2O_3, H_2O, MgO, FeO and SiO_2. It is used in heat applications, as soil conditioner, as loose-fill insulation, as absorber package material and lightweight aggregate for plaster etc. Its chemical formula is known as $(MgFe,Al)_3(Al,Si)_4O_{10}(OH)_2 \cdot 4H_2O$ and physical density of it about 2.5 g.cm^{-3}. Melting point of vermiculite is above 1350^{0}C. This mineral can be used as additive building material in terms of mineral properties.

Vermiculite is a component of the phyllosilicate or sheet silicate group of minerals. It has high-level exfoliation property. So if vermiculite is heated, it expands to many times its original volume. This feature is a striking ability for a mineral. Because it looks like *vermicularis*, its name is vermiculite. They molecular structure consists of two tetrahedral layers as silica and alumina and an octahedral layer including O, Mg, Fe and hydroxyl molecules. Water located between layers is an important member in the vermiculite. If mineral heats suddenly, inter-layer water transforms to steam and exfoliation feature occurs.

Vermiculite is clean to handle, odorless and mould durable. It has a wide range of uses as thermal insulation, fire durability, liquid absorption capability, low density and usefulness etc… The main uses of minerals are listed as follows;

i. Construction Industry (lightweight concretes, vermiculite-loaded plasters, loosefill insulation)
ii. Animal Feedstuff Industry
iii. Industrial Insulation for High Temperatures (up to 1100^{0}C)
iv. Automotive Industry
v. Packaging Materials
vi. Horticulture

In literature, there are several studies about vermiculite and its usage. A comparative study about the effects of grinding and ultrasonic treatment on vermiculite was done. The effect of mechanical treatment and cation type on the clay micro porosity of the Santa Olalla vermiculite untreated and mechanically treated (sonicated and ground) and saturated with different captions was investigated. An experimental study was performed about thermal conductivity of expanded vermiculite based samples. In this paper, measurements were carried out on samples in the temperature range of 300-1100 K. Because of the high heat insulation properties of vermiculite mineral, we frequently encounter studies on thermal properties of it. In another study, researchers produced new materials including vermiculite that can withstand up to 1150^{0}C. The cement-vermiculite composition was used to produce new materials to leach ^{54}Mn and ^{89}Sr radionuclide. Researchers have received the best results when using 95% Portland cement and 5% vermiculite composition. The effect on ultrasounds on natural macroscopic vermiculite flakes has been studied and effects of ultrasound treatment on the several parameters (particle sizes, crystal structure, surface area, etc...) were investigated. Finally, micron and submicron-sized vermiculites were prepared. High surface area silica was obtained by selectively leaching vermiculite. In this study, also the characteristics of the porous silica obtained from vermiculite are compared with those from other clay minerals. Studies are performed on the material composition and typical characteristics of micaceous minerals of the vermiculite series in the Tebinbulak deposit. Thermal treatments of nano-layered vermiculite samples were studied up to 900^{0}C. In another thermal effect study was achieved for 15-800^{0}C temperature range vermiculite originated by Tanzania region. Thermal properties of polypropylene-vermiculite composites were investigated using differential scanning calorimetry (DSC) and thermogravimetry (TG) techniques. At the end of this study, new composites with high thermal stability were produced. The effect of sodium ion exchange on the properties of vermiculite was studied by several methodology (scanning electron microscopy, X-ray fluorescence spectroscopy, inductively coupled plasma mass spectroscopy, X-ray diffraction and thermo mechanical analysis) and sodium exchange lowered the exfoliation onset temperature to below 300 °C. A comparative study about oil affinity of expanded and hydrophobized vermiculite was done. According to the results of this study, the expanded vermiculite had a greater affinity for oil than hydrophobized vermiculite. XRD characteristics of Poland vermiculites were studied and crystal structure of it was determined. A general study about typical properties and some parameters of different vermiculites was performed. In this study, especially heat conduction coefficients were commented and an evaluation was done about vermiculite based building materials. In another study, flyash-based fibre-reinforced hybrid phenolic composites filled with vermiculite were fabricated and characterized for their physical, thermal, mechanical and tribological performance.

Neutron shielding studies have a wide range of literature. For example, colemanite and epoxy resin mixtures have been prepared for neutron shielding applications by Okuno, 2005. Agosteo et al. have investigated double differential neutron distribution and neutron attenuation in concrete using 100-250 MeV proton accelerator. In another study, neutron transmission measurements were studied through pyrolytic graphite crystals by Adib et al. Neutron attenuation properties of zirconium borohydrite and zirconium hydride were determined by Hayashi et al., 2009. Sato et al. designed a new material evaluation method by using a pulsed neutron transmission with pixel type detectors.

In this paper; we investigated usability of vermiculite loaded samples for nuclear reactor shielding processes, because of excellent thermal insulation properties of it. This mineral was doped in cement and new samples including different vermiculite percents were produced. 4.5 MeV neutron dose transmission values were determined. Also 4.5 MeV neutron attenuation lengths were calculated for each sample.

2. Experiments

2.1 Sample preparation

Before the mixing procedure, a part of mixing water at the percentage of water absorption capacity of expanded vermiculite aggregate by weight was added to vermiculite to make it fully saturated with water. Fig.1 shows the expanded vermiculite particles saturated with water.

Fig. 1. Expanded vermiculite fine aggregate saturated with water.

Then, the rest of the mixing water cement and silica fume or steel fiber were mixed together for 1 minute in a mixer, and finally, expanded vermiculite aggregate saturated with water was added to cement slurry and mixed for 3 minutes again, to get a homogenous structure. Fig.2 shows the fresh state of the mixture of lightweight mortar.

Fig. 2. Fresh state of lightweight mortar prepared with expanded vermiculite aggregate.

The prepared fresh mortar were cast in standard cube (with an edge of 150 mm) molds, in two layers, each layer being compacted by self-weight on the shaker for 10 s. All the specimens were kept in moulds for 24 h at room temperature of about 20°C, and then demoulded, and after demoulding all specimens were cured in water at 23 ±2 °C for 27-days. After 28 days curing, three plate specimens with a dimension of 150x100x20mm for neutron dose transmission measurements were obtained by cutting the cube specimens using a stone saw. Plate specimens obtained by the way was illustrated in Fig.3. We obtained 12 different samples. Codes and contents of samples were shown in Table.1.

Fig. 3. View of the samples

Properties of samples including fiber steel (20^{0}C)		
Code of Sample	Vermiculite / Cement Ratio	Fiber Volume Fraction
4F0	4	0
4F15	4	1.5 %
6F0	6	0
6F15	6	1.5 %
8F0	8	0
8F15	8	1.5 %
Properties of samples including silica fume (20^{0}C)		
Code of Sample	Vermiculite / Cement Ratio	Silica Fume Contents
4S0	4	0
4S15	4	5 %
6S0	6	0
6S15	6	5 %
8S0	8	0
8S15	8	5 %

Table 1. Codes and Properties of Samples

2.2 Experimental design

For neutron transmission measurement, we used a ^{241}Am/Be neutron source and a Canberra portable neutron detector equipments. ^{241}Am/Be source emits 4.5 MeV neutron particles. Physical form of ^{241}Am/Be neutron source is compacted mixture of americium oxide with beryllium metal. Fast neutrons are produced by following nuclear reaction,

$$^{9}_{4}Be(\alpha, n)^{12}_{6}C$$

5.486 keV maximum energy alpha particles emitting from ^{241}Am. Neutron energy value produced by this nuclear reaction is 4.5 MeV. Radiation characteristics of ^{241}Am/Be neutron source are shown in Table.2 (Dose rate values have been obtained from The Health Physics and Radiological Health Handbook, Scintra _Inc., Revised Edition, 1992.).

The NP-100B detector provides us to detect slow and fast neutrons. Tissue equivalent dose rates of the neutron field can be measured by it. The detectors contain a proportional counter which produces pulses resulting from neutron interactions within it. The probes contain components to moderate and attenuate neutrons. So that the net incident flux at the proportional counter is a thermal and low epithermal flux representative of the tissue equivalent dose rate and the neutron field. Because of neutrons have no charge; they can only be detected indirectly through nuclear reactions that create charged particles. The NP100B detector uses ^{10}B as the conversion target. The charged particle – alpha or proton (respectively) created in the nuclear reaction ionizes the gas. Typical detector properties are shown in Table.3. Equivalent dose rate measurement results have read on RADACS program in system PC. Experimental design is shown in Fig.4.

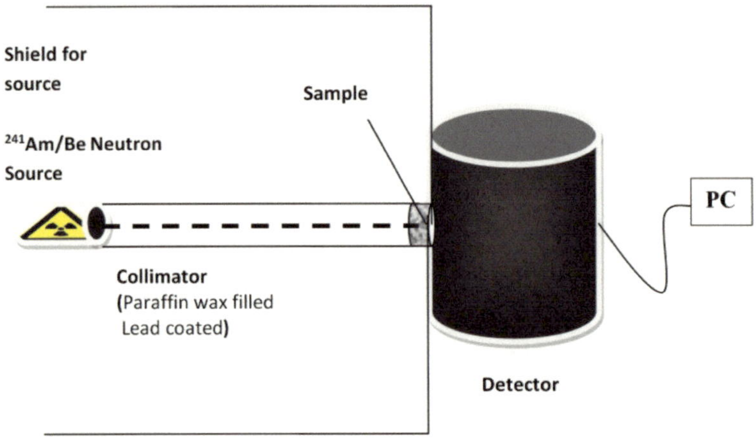

Fig. 4. Experimental Setup

Physical Half-Life: 432.2 years		Specific Activity: 127 GBq/g	
Principle Emissions	**E_{max} (keV)**	**E_{eff}**	**Dose Rate (μSv/h/GBq at 1m)**
Gamma/X-Rays	13.9 (42.7%) 59.5 (35.9%)	-	-
Alpha	5.443 (12.8%) 5.486 (85.2%)	-	85
Neutron	-	4.5 MeV	2

*http://www.stuarthunt.com/pdfs/Americium_241Beryllium.pdf

Table 2. Radiation characteristics of ^{241}Am-Be neutron source*

Specifications of Canberra NP100B Neutron Detector	
Detector Type	BF3 Proportional Counter
Detector Sensitivities	0–100 mSv/h (0–10 Rem/h)
Energy Range	0.025 eV - 15 MeV
Operating Temperature Range	-10 °C to +50 °C (+14 °F to +122 °F)
Size (mm.)	244 x 292 mm
(Dia. x inch)	(9.6 x 11.5 in.)
Weight kg (lb)	10 kg (22 lb)
Housing	Moisture Proof Aluminum
Operating Humidity	0–100% non-condensing
Detector Linearity	±5%
Accuracy	±10%
High Voltage Supply (internally generated)	1750–1950 V

*http://www.canberra.com/pdf/Products/RMS_pdf/NPSeries.pdf

Table 3. Typical Properties of Detector*

We determined dose transmission values of vermiculite loaded samples. Firstly, we counted equivalent dose rate by fast neutrons while there is no sample between source and detector system. And then we measured for each sample neutron equivalent dose rate while there is our sample between ^{241}Am-Be source box and detector probe. The ratio of two values is called dose transmission.

3. Results and discussion

Nowadays concrete is often used in radiation shielding process. In several studies, some additive materials were added in concrete to increase its radiation shielding capacity. In this study, as an additive material, we have used vermiculite mineral with a good heat insulation material. Produced samples have three different vermiculite and cement ratio values. 4.5 MeV neutron dose transmission values (Fig.5) and attenuation lengths of samples (Table.4) were obtained. Attenuation length is just equal to the average distance a particle travels before being scattered or absorbed. It is a useful parameter for shielding calculations. Also we calculated experimental 4.5 MeV neutron total macroscopic cross sections (μ) using by dose transmission values. The various types of interactions of neutrons with matter are combined into a total cross-section value:

$$\Sigma_{total} = \Sigma_{scatter} + \Sigma_{capture} + \Sigma_{fission} + \ldots.. \quad (1)$$

The attenuation relation in the case of neutrons is thus:

$$I_X = I_0 e^{-\Sigma_{total} x} \quad (2)$$

$$\Sigma_{total} = \frac{\ln\left(\frac{I_0}{I}\right)}{x} \quad (3)$$

where I_0 is known as beam intensity value, at a material thickness of $x = 0$. Equivalent dose rate has been used instead of beam intensity because of our equivalent dose rate measurements. Experimental 4.5 MeV neutron total macroscopic cross sections were shown in Table.5.

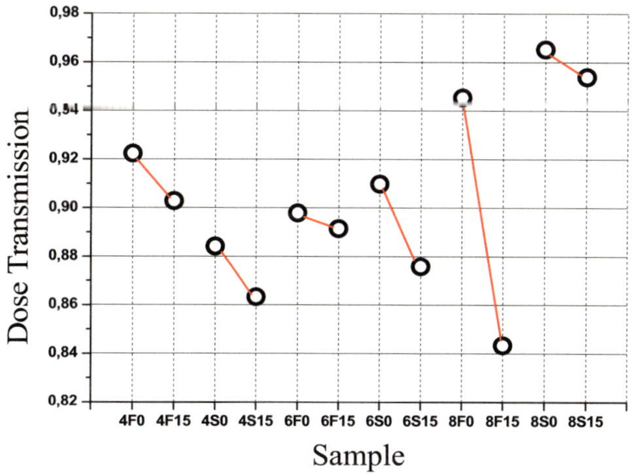

Fig. 5. 4.5 MeV neutron dose transmissions for each samples

Code of Sample	Attenuation Length (cm)
4F0	31.92848
4F15	25.25253
6F0	23.96932
6F15	22.47191
8F0	46.04052
8F15	15.15611
4S0	20.96876
4S15	17.55618
6S0	27.31494
6S15	19.46283
8S0	73.20644
8S15	54.79452

Table 4. 4.5 MeV neutrons attenuation lengths

Code of Sample	$\mu(cm^{-1})$
4F0	0.0313
4F15	0.0396
6F0	0.0417
6F15	0.0445
8F0	0.0217
8F15	0.0650
4S0	0.0477
4S15	0.0570
6S0	0.0366
6S15	0.0514
8S0	0.0137
8S15	0.0183

Table 5. 4.5 MeV neutron total macroscopic cross sections

As can be seen from Fig.5 and Table.4, dose transmission values and attenuation lengths decrease with increasing fiber steel and silica fume contents. This result indicates that neutron shielding capacity of samples is increased by silica and steel amount. According to the results, there is not a consistent relationship between vermiculite content and neutron shielding capacity of samples except of F15-samples. The sample named 8F15 is the best neutron attenuator in all specimens. The reason of this that, this sample has higher vermiculite and fiber steel content than others. The worst sample is 8S0 which has higher vermiculite but lower silica fume content. As a result, to increase neutron shielding capacity of sample, expanded vermiculite and fiber steel may be added in the mortar.

4. Conclussions

At the end of this experimental study, we reached the following outcomes;

1. Vermiculite mineral has high-level thermal insulation capacity. Concrete isn't decomposing with vermiculite addition. This mineral can be used as an additive for radiation shielding process.
2. According to the experimental results, neutron shielding property of concrete increase with increasing fiber steel and silica fume content.
3. To produce good materials which have high radiation shielding capacity and thermal insulation property, vermiculite and fiber steel may be doped in mortar. These materials can be used for neutronic and thermal applications.

5. References

B.A. Sakharov, E. Dubinska, P. Bylina, G Kapron, "Unusual X-Ray Characteristics of Vermiculite from Wiry, Lower Silesia, Poland", *Clay Clay Miner*, 49 (2001) (3), 197-203.

B.K. Satapathy, A. Patnaik, N. Dadkar, D.K. Kolluri, B.S. Tomar, "Influence of vermiculite on performance of flyash based fibre-reinforced hybrid composites as friction materials", *Mater Design*, 32 (2011) 4354–4361.

D. Mysore, T. Viraraghavan, Y.C. Jin, "Treatment of oily waters using vermiculite", *Water Res* 39 (2005) 2643–2653.

D.R. Ochbelagha, S. AzimKhani, H.G. Mosavinejad, "Effect of gamma and lead as an additive material on the resistance and strength of concrete", Nuclear Engineering and Design 241 (2011) 2359–2363.

E.M.M. Marwa, A.A. Meharg, C.M. Rice, "The effect of heating temperature on the properties of vermiculites from Tanzania with respect to potential agronomic applications", *Appl Clay Sci*, (2009) 376–382.

E.V.D. Gomes, L.L.Y. Visconte, E.B.A.V. Pacheco, "Thermal characterization of polypropylene/vermiculite composites", *J Therm Anal Calorim* 97 (2009) 571–575.

H. Sato, O. Takada, S. Satoh, T. Kamiyama, Y. Kiyangi. "Development of material evaluation method by using a pulsed neutron transmission with pixel type detectors". *Nucl Instrum Methods Phys Res A* 623(1) (2010) 597-599.

H.F. Muiambo, W.W. Focke, M. Atanasova, I. van der Westhuizen, L.R. Tiedt, "Thermal properties of sodium-exchanged palabora vermiculite", *Appl Clay Sci*, 50 (2010) 51–57.

http://www.ima-europe.eu/ (ESMA, The European Specialty Minerals Association)

I. Plecas, S. Dimovic, "Leaching kinetics of ^{54}Mn and ^{89}Sr radionuclide fixed in cement vermiculite composition", *J Radioanal Nucl Ch*, 264 (3) (2005) 687-689.

J. Temuujin, K. Okada, K. J.D. MacKenzie, "Preparation of porous silica from vermiculite by selective leaching", *Appl Clay Sci*, 22 (2003) 187- 195.

K.Okuno. "Neutron shielding material based on colemanite and epoxy resin" *Rad. Protec. Dos.*, 115(1-4) (2005) 58-61.

L.A. Perez-Maqueda, M.C. Jim´enez de Haro, J. Poyato, J.L. Perez-Rodriguez, "Comparative study of ground and sonicated vermiculite", *J Mater. Sci.*, 39 (2004) 5347-5351.

L.A. Perez-Maqueda, O.B.Caneo, J. Poyato, J.L. Perez-Rodrıguez, "Preparation and Characterization of Micron and Submicron-sized Vermiculite", *Phys Chem Minerals*, 28 (2001) 61-66.

M. Adib, N. Habib, , M. Fathaalla. "Neutron transmission through pyrolytic graphite crystals". *Ann. of Nucl. En.*, 33(7) (2007), 627-632.

M.C. Jimenez De Haro, J.M. Martinez Blanes, J. Poyato, L.A. Perez-Maqueda, A. Lerf, J.L. Perez-Rodriguez, "Effects of mechanical treatment and exchanged cation on the microporosity of vermiculite", *J Phys Chem Solids*, 65 (2004) 435-439.

M.E. Medhat, M. Fayez-Hassan, "Elemental analysis of cement used for radiation shielding by instrumental neutron activation analysis", *Nucl Eng Des*, 241 (2011) 2138 2142.

R. Şahin, R. Polat, O. Içelli, C. Çelik, "Determination of transmission factors of concretes with different water/cement ratio, curing condition, and dosage of cement and air entraining agent", *Ann of Nucl Energy*, 38 (2011) 1505–1511.

S. A. Suvorov and V. V. Skurikhin, "High-Temperature Heat Insulating Materials Based on Vermiculite", *Refract Ind Ceram+*, 43 (11) (2002) 383-389.

S. A. Suvorov and V. V. Skurikhin, "Vermiculite — A Promising Material for High Temperature Heat Insulators", *Refract Ind Ceram+*, 44 (3) (2003) 186-193.

S.Agosteo, M. Magistris, A. Mereghetti, M. Silari, Z. Zajacova. "Shielding data for 100–250 MeV proton accelerators: Double differential neutron distributions and attenuation in concrete. " *Nuclear Instruments and Methods in Physics Research B* 265 (2007) 581–598.

T. Korkut, A. Ün, F. Demir, A Karabulut, G. Budak, R. Şahin, M. Oltulu, "Neutron dose transmission measurements for several new concrete samples including colemanite". *Ann of Nucl Energy*, 37 (2010) 996–998.

V. É. Peletskii and B. A. Shur, "Experimental Study of the Thermal Conductivity of Heat Insulation Materials Based on Expanded Vermiculite", *Refract Ind Ceram+*, 48 (5) (2007) 356-358.

V. I. Andronova and P. A. Arifov, "Study of the Material Properties of Micaceous Vermiculite Minerals of the Tebinbulak Deposit", *Refract Ind Ceram+*, 48(5) (2007) 373 377.

Y. El Mouzdahir, A. Elmchaouri, R. Mahboub, A. Gil, S.A. Korili, "Synthesis of nano layered vermiculite of low density by thermal treatment", *Powder Technol*, 189 (2009) 2–5.

15

Development of ^{99}Mo Production Technology with Solution Irradiation Method

Yoshitomo Inaba
Japan Atomic Energy Agency
Japan

1. Introduction

Technetium-99m (^{99m}Tc, half-life: 6.01 hours) is the world's most widely used radiopharmaceutical for exams of cancer, bowel disease, brain faculty and so on, and it is used for more than twenty million exams per year in the world and more than one million exams per year in Japan. The demand for ^{99m}Tc is continuously growing up year by year. The features of ^{99m}Tc as the radiopharmaceutical are as follows:

1. It is easy to add ^{99m}Tc to diagnostic medicines.
2. It is easy to measure the γ-ray energy with 0.14 MeV generated by isomeric transition from outside the body.
3. β rays are not emitted.
4. The patients' exposure associated with the exams is kept to the minimum because of the short half-life.

The production of the short-lived ^{99m}Tc is conducted by extracting from molybdenum-99 (^{99}Mo, half-life: 65.94 hours), which is the parent nuclide of ^{99m}Tc. Therefore, the stable production and supply of ^{99}Mo is very important in every country. All of ^{99}Mo used in Japan is imported from foreign countries. However, a problem has emerged that the supply of ^{99}Mo is unstable due to troubles in the import and the aging production facility (Atomic Energy of Canada Limited [AECL], 2007, 2008). In order to solve the problem, the establishment of an efficient and low-cost ^{99}Mo production method and the domestic production of ^{99}Mo are needed in Japan.

As a major ^{99}Mo production method, the fission method ((n, f) method) exists, and as a minor ^{99}Mo production method, the neutron capture method ((n, γ) method) exists. In order to apply to the Japan Materials Testing Reactor (JMTR) of the Japan Atomic Energy Agency (JAEA), two types of ^{99}Mo production methods based on the (n, γ) method have been developed in JAEA (Inaba et al., 2011): one is a solid irradiation method, and the other is a solution irradiation method, which was proposed as a new ^{99}Mo production technique (Ishitsuka & Tatenuma, 2008).

The solution irradiation method aims to realize the efficient and low-cost production and the stable production and supply of ^{99}Mo, and the fundamental research and development for the practical application of the method has been started (Inaba et al., 2009).

In this paper, a comparison between ^{99}Mo production methods, an overview of the solution irradiation method containing the structure of ^{99}Mo production system with the method and the progress of the development made thus far, estimates of ^{99}Mo production with the method, and the results of a newly conducted test are described.

2. Comparison between ^{99}Mo production methods

A comparison between the three ^{99}Mo production methods (the fission method, the solid irradiation method and the solution irradiation method) is shown in Table 1, assuming the ^{99}Mo production in JMTR, which is a tank-type reactor.

2.1 Fission method ((n, f) method)

In the conventional fission method ((n, f) method), high-enriched uranium targets are irradiated with neutrons in a testing reactor, and ^{99}Mo is produced by the ^{235}U (n, f) ^{99}Mo reaction. Most of the world supply of ^{99}Mo is produced by the (n, f) method since ^{99}Mo with a high-level specific activity of 370 TBq/g-Mo is obtained. However, the method has problems about the nuclear nonproliferation and the generation of a significant amount of radioactive waste including Fission Products (FPs) and Pu. Caused by the radioactive waste, the separation process of ^{99}Mo is too complex, and ^{99}Mo production with the (n, f) method needs expensive facilities and extreme care to avoid contamination with FPs. The ^{99}Mo production cost by this method achieves 57 US$/37 GBq (Boyd, 1997), and it is too expensive.

2.2 Neutron capture method ((n, γ) method)

2.2.1 Solid irradiation method

In the conventional solid irradiation method, solid targets including natural molybdenum such as MoO_3 pellets are irradiated with neutrons in a testing reactor, and ^{99}Mo is produced by the ^{98}Mo (n, γ) ^{99}Mo reaction. The post-irradiation process is only dissolution of the irradiated solid targets with an alkaline solution, and only a small amount of radioactive waste is generated in the process compared with the (n, f) method. The ^{99}Mo production cost of this method or the (n, γ) method is only 0.83 US$/37 GBq (Boyd, 1997).

However, the (n, γ) method has the disadvantage of producing ^{99}Mo with a low-level specific activity of 37-74 GBq/g-Mo and therefore the method has not had practical application in earnest. In order to utilize ^{99}Mo with the low-level specific activity, a high-performance adsorbent for (n, γ) ^{99}Mo is needed. The Japan Atomic Energy Research Institute (the present organization: JAEA) and KAKEN Inc. had developed the high-performance molybdenum adsorbent of Poly-Zirconium Compound (PZC) in 1995 (Hasegawa et al., 1996) and improved PZC (Hasegawa et al., 1999), and then the practical application of the (n, γ) method is just in sight. The molybdenum adsorbent performance of PZC is over 100 times compared with the conventional molybdenum adsorbent of alumina.

^{99}Mo production in JMTR will start by using the solid irradiation method. JMTR aims to provide ^{99}Mo of 37 TBq/w (1,000 Ci/w), and it will cover about 20% of the ^{99}Mo imported into Japan (Inaba et al., 2011).

2.2.2 Solution irradiation method

In the new solution irradiation method, a solution target including natural molybdenum such as an aqueous solution of a molybdenum compound (aqueous molybdenum solution) is irradiated with neutrons in a testing reactor, and ^{99}Mo is produced by the ^{98}Mo (n, γ) ^{99}Mo reaction. This new method is the improved type of the solid irradiation method, and it is possible to enhance the ^{99}Mo production compared with the solid irradiation method. The solution irradiation method has the following advantages compared with the solid irradiation method:

1. It is easy to increase the irradiated volume by using a capsule with larger volume than that of a rabbit. The rabbit is a small sized (150 mm length) capsule (Inaba et al., 2011).
2. The separation and dissolution processes after the irradiation are not necessary because the irradiation target is an aqueous solution.
3. The amount of generated radioactive waste is smaller than that of the solid irradiation method.

Items	^{99}Mo production methods		
	Fission method ((n, f) method)	Neutron capture method ((n, γ) method)	
		Solid irradiation method	Solution irradiation method
<Irradiation target> •Chemical type •Form •Quality control	Enriched ^{235}U •U-Al alloy, UO_2 •Foil, pellet •Complex	Natural Mo •MoO_3, metal Mo •Powder, pellet, metal •Complex	Natural Mo •Molybdate •Aqueous solution •Simple
<Irradiation container>	Rabbit (30 cm^3)	Rabbit (30 cm^3)	Capsule (1,663 cm^3)
<Irradiation> •Pre-process of irradiation •Reaction of ^{99}Mo production •Irradiation time •Collection of target •Post-process of irradiation	•Adjustment of target and enclosing with container •^{235}U (n, f) •5-7 days •Batch collection •Isolation in hot lab. (Complex)	•Adjustment of target and enclosing with container •^{98}Mo (n, γ) ^{99}Mo •5-7 days •Batch collection •Dissolution in hot lab (Relatively simple)	•Adjustment of target •^{98}Mo (n, γ) ^{99}Mo •5-7 days •Continuous or batch collection •No special treatment
Generated ^{99}Mo •Specific activity •Activation by-product	•370 TBq/g-Mo •Quite many (^{131}I, ^{103}Ru, ^{89}Sr, ^{90}Sr, etc.)	•37-74 GBq/g-Mo •^{92m}Nb	•37-74 GBq/g-Mo •^{92m}Nb, ^{14}C, ^{42}K, etc. (depending on a target solution)
<Mo adsorbent>	Alumina (2 mg-Mo/g-Al_2O_3)	PZC (250 mg-Mo/g-PZC)	PZC (250 mg-Mo/g-PZC)
<Radioactive waste>	Rabbits with FPs and Pu (Generation every one irradiation)	Rabbits (Generation every one irradiation)	Capsule (Lifetime: about 15 operation cycles)
<Production in Japan>	Difficult	Possible	Possible

Table 1. Comparison between three ^{99}Mo production methods

In this new method, efficient and low-cost ^{99}Mo production compared with the conventional ^{99}Mo production can be realized by using the (n, γ) reaction and PZC. This new method aims to provide 100% of the ^{99}Mo imported into Japan.

3. Overview of solution irradiation method

3.1 Structure of ^{99}Mo production system with solution irradiation method

The schematic diagram of the ^{99}Mo production system with the solution irradiation method is shown in Fig. 1. The system consists of an irradiation system, a supply and circulation system, and a collection and subdivision system. In the irradiation system, an aqueous molybdenum solution in a capsule installed in a reactor core is irradiated with neutrons under static or circulation condition, and ^{99}Mo is generated. In the supply and circulation system, the solution is supplied to the capsule through pipes and is circulated by a circulator in irradiation operation. A gas disposal device and a heat exchanger are installed in order to take measures against the radiolysis gas and heat generated from the solution by irradiation. The system is designed so as to minimize unirradiated solution. In the collection and subdivision system, after the solution including the generated ^{99}Mo is collected from the capsule through pipes, it is treated so as to be products such as PZC-^{99}Mo columns or ^{99}Mo transport containers.

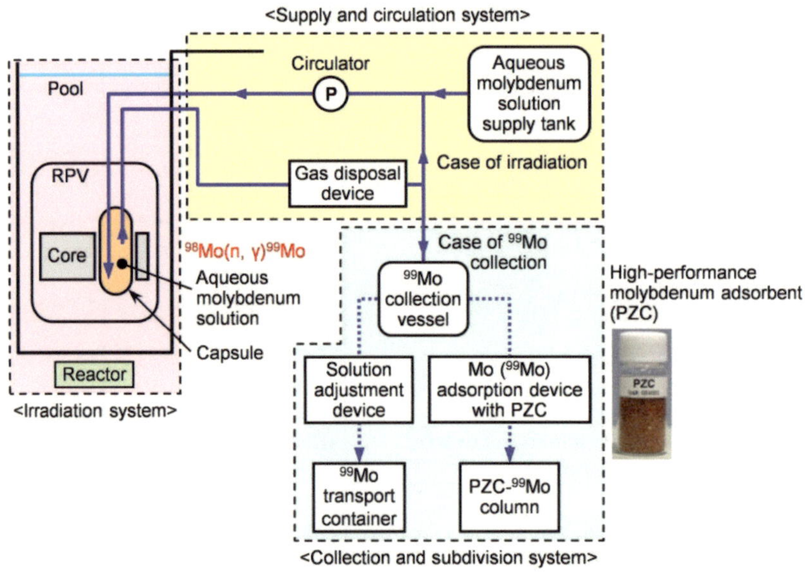

Fig. 1. Schematic diagram of ^{99}Mo production system with solution irradiation method

The detailed design of the ^{99}Mo production system is carried out based on the results of future investigations and tests.

3.2 Progress of the development made thus far

The most important element of the solution irradiation method is the aqueous molybdenum solution as the irradiation target. The solution with a high concentration near the saturation is used for efficient ^{99}Mo production, and the solution always is in contact with the structural materials of the capsule and the pipes in the ^{99}Mo production system under irradiation. Aqueous molybdate solutions are promising candidates for the irradiation target. The effect of the solutions on metals such as the structural materials has been researched, and molybdates are known as corrosion inhibitors (Kurosawa & Fukushima, 1987; Lu et al, 1989; McCune et al, 1982; Saremi et al, 2006). However, the behavior of aqueous molybdenum solutions including the aqueous molybdate solutions under such the conditions is not well understood. Therefore, the following subjects about the fundamental characteristics of the solutions should be investigated:

1. Selection of the aqueous molybdenum solutions as candidates for the irradiation target
2. Compatibility between the solutions and the structural materials
3. Chemical stability of the solutions
4. Effect of γ ray and neutron irradiation on the solutions such as the radiolysis, the γ heating and the activation by-products.

The some subjects described above had already investigated (Inaba et al., 2009), and the progress of the development made thus far is explained as below:

3.2.1 Selection of candidates for irradiation target

The selection of candidates for the irradiation target was carried out. The conditions required for the irradiation target solution are as follows:

1. The irradiation target solution has the high molybdenum content for the efficient production of ^{99}Mo.
2. Few activation by-products are generated by target solution irradiation for the prevention of radioactive contamination.
3. The solution has good compatibility with the structural materials of the capsule and the pipes for the prevention of corrosion.
4. The solution is chemically stable and has no generation of precipitation for the prevention of an obstruction to the solution's flow.

Based on the conditions (1) and (2), two aqueous molybdate solutions (aqueous ammonium molybdate and potassium molybdate solutions) were selected as the candidates for the irradiation target among the aqueous solutions of general molybdenum compounds.

The solubilities of ammonium molybdate ($(NH_4)_6Mo_7O_{24}\cdot 4H_2O$) and potassium molybdate (K_2MoO_4) for pure water are 44 g/100 g-H_2O and 182.4 g/100 g-H_2O respectively, and the molybdenum contents in the solubilities of $(NH_4)_6Mo_7O_{24}\cdot 4H_2O$ and K_2MoO_4 are 23.9 g and 73.5 g respectively.

The activation by-product of $(NH_4)_6Mo_7O_{24}\cdot 4H_2O$ is only ^{92m}Nb. The activation by-products of K_2MoO_4 are ^{42}K and ^{92m}Nb. The γ-ray energy emitted from ^{42}K is high. However, by using PZC, it is possible to remove ^{42}K and ^{92m}Nb from the two aqueous molybdate solutions irradiated with neutrons.

The conditions (3) and (4) were confirmed by tests with the two solutions.

3.2.2 Unirradiation and γ-ray irradiation tests

Unirradiation and γ-ray irradiation tests were carried out by using the selected two aqueous molybdate solutions (aqueous $(NH_4)_6Mo_7O_{24}\cdot 4H_2O$ and K_2MoO_4 solutions), and compatibility between the two solutions and the structural materials of stainless steel and aluminum, the chemical stability, the circulation characteristics, the radiolysis and the γ heating of the two solutions were investigated. In addition, the integrity of PZC was investigated under γ-ray irradiation. As a result, the following were found:

1. The compatibility between the two static aqueous molybdate solutions and stainless steel is very well under unirradiation and γ-ray irradiation.
2. The two solutions are chemically stable and have smooth circulation under unirradiation and γ-ray irradiation.
3. The ratios of hydrogen in the gases generated by the radiolysis of the two solutions are higher than that of pure water.
4. The effect of γ heating on the two solutions is the same level as that on pure water.
5. The integrity of PZC is maintained under γ-ray irradiation.

However, the pH of the aqueous $(NH_4)_6Mo_7O_{24}\cdot 4H_2O$ solution needs to be adjusted from weak alkaline to weak acid for the prevention of precipitation. This is a disadvantage as one of the candidates for the irradiation target.

At present, the aqueous K_2MoO_4 solution, which has no pH adjustment and has higher molybdenum content than that of the aqueous $(NH_4)_6Mo_7O_{24}\cdot 4H_2O$ solution, is investigated as the first candidate of the irradiation target.

4. Estimates of ^{99}Mo production rates by solution irradiation method

The estimates of ^{99}Mo production rates by the solution irradiation method are shown in cases using aqueous $(NH_4)_6Mo_7O_{24}\cdot 4H_2O$ and K_2MoO_4 solutions as irradiation targets, assuming the ^{99}Mo production in JMTR.

4.1 Conditions for estimates of ^{99}Mo production rates

The JMTR core arrangement is shown in Fig. 2. The capsule of the ^{99}Mo production system with the solution irradiation method is installed into the irradiation hole, M-9 with maximum and average thermal neutron fluxes of 3.5×10^{18} n/(m$^2\cdot$s) and 2.6×10^{18} n/(m$^2\cdot$s) (Department of JMTR Project, 1994). The capsule consists of inner and outer tubes, and an aqueous molybdate solution is irradiated with neutrons in the inner tube to prevent the solution from leaking into the reactor coolant. Table 2 shows the conditions of the capsule and the two irradiation targets of the aqueous $(NH_4)_6Mo_7O_{24}\cdot 4H_2O$ and K_2MoO_4 solutions.

^{99}Mo production rates are estimated based on the following conditions in addition to the conditions of Table 2:

- The generation and reduction of ^{99}Mo by the neutron capture reaction of ^{98}Mo (n, γ) ^{99}Mo and ^{99}Mo (n, γ) ^{100}Mo and the radioactive decay of ^{99}Mo are evaluated.
- ^{99}Mo doesn't exist in the initial stage of the calculation.
- The decay of neutron flux due to the inner and outer tubes of the capsule is considered.
- The circulation of the two irradiation targets is not considered.

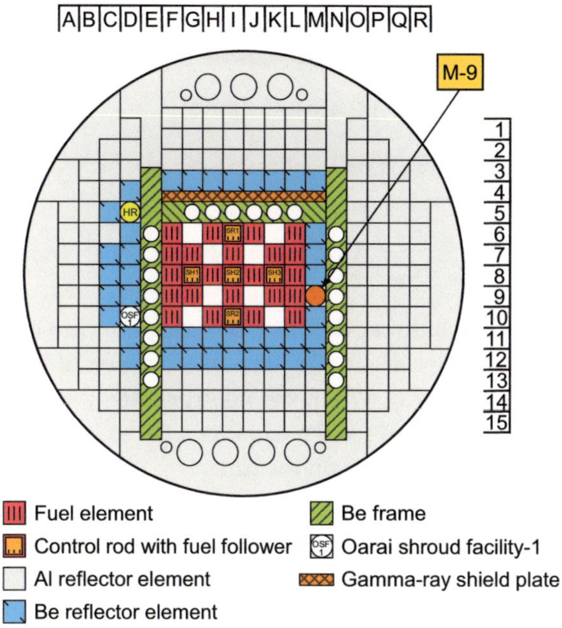

Fig. 2. JMTR core arrangement

Size and material of capsule	
Outer tube:	outer diameter of 65 mm, inner diameter of 61 mm
Inner tube:	outer diameter of 59 mm, inner diameter of 55 mm
Irradiation height:	700 mm
Irradiation volume:	1,663 cm^3
Material:	Stainless steel
Dissolved molybdenum in each irradiation target	
• Aqueous $(NH_4)_6Mo_7O_{24}\cdot 4H_2O$ solution (concentration: 90% of saturation): 372.8 g/1,663 cm^3	
• Aqueous K_2MoO_4 solution (concentration: 90% of saturation): 702.7 g/1,663 cm^3	

Table 2. Conditions of capsule and two irradiation targets of aqueous $(NH_4)_6Mo_7O_{24}\cdot 4H_2O$ and K_2MoO_4 solutions

4.2 Basic equations for estimates of ^{99}Mo production rates

The disintegration rates of ^{98}Mo (isotopic ratio: 24.138%) and ^{99}Mo are shown as following equations:

$$\frac{dN_{98}}{dt} = -\phi \sigma_{98} N_{98} \quad (1)$$

$$\frac{dN_{99}}{dt} = -(\lambda + \phi\sigma_{99})\, N_{99} + \phi\sigma_{98} N_{98} \tag{2}$$

The solutions of the equations (1) and (2) are as follows:

$$N_{98}(t) = N_{98}(0)\exp(-\phi\sigma_{98}t) \tag{3}$$

$$N_{99}(t) = \frac{\phi\sigma_{98}}{\lambda + \phi(\sigma_{99} - \sigma_{98})} N_{98}(0)\left[\exp(-\phi\sigma_{98}t) - \exp\{-(\lambda + \phi\sigma_{99})t\}\right] \tag{4}$$

where N_{98} and N_{99} are the atom number densities of ^{98}Mo and ^{99}Mo (n/cm^3), t is time (s), ϕ is neutron flux (n/(cm^2·s)), σ_{98} and σ_{99} are the capture cross section of ^{98}Mo and ^{99}Mo (cm^2), and λ is the decay constant of ^{99}Mo (1/s). When the neutron flux, the capture cross section, the decay constant and the time are given for the equations (3) and (4), ^{99}Mo generation rate per unit volume can be calculated depending on the time.

The specific activity of the generated ^{99}Mo is calculated from the following equation:

$$-\frac{dN_{99}}{dt} = \frac{W \times 4.17 \times 10^{23}}{AT} \tag{5}$$

where W is the mass of ^{99}Mo (g), A is the atomic mass number of ^{99}Mo, and T is the half-life of ^{99}Mo (s).

4.3 Estimated results of ^{99}Mo production rates

The relationship between the irradiation time and the calculated specific ^{99}Mo generation (generated ^{99}Mo activity per 1 g of molybdenum) is shown in Fig. 3. When the irradiation time is 6 days (144 h), the specific ^{99}Mo generation is 0.286 TBq/g-Mo as shown in Fig. 3.

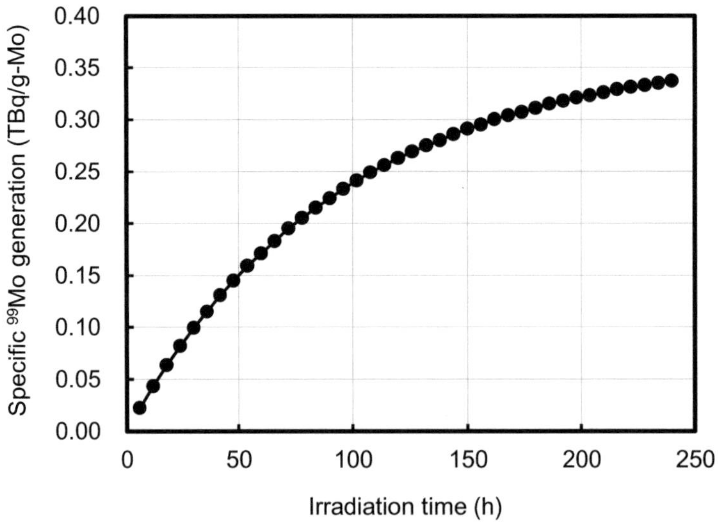

Fig. 3. Relationship between irradiation time and specific ^{99}Mo generation

Using the specific ^{99}Mo generation of 0.286 TBq/g-Mo, ^{99}Mo production rates are estimated. In the case using the aqueous $(NH_4)_6Mo_7O_{24}\cdot 4H_2O$ solution as the irradiation target,

(^{99}Mo production in the case using the aqueous $(NH_4)_6Mo_7O_{24}\cdot 4H_2O$ solution)

= 372.8 g × 0.286 TBq/g-Mo = 106.6 TBq = 2,881.9 Ci

In the case using the aqueous K_2MoO_4 solution as the irradiation target,

(^{99}Mo production in the case using the aqueous K_2MoO_4 solution)

= 702.7 g × 0.286 TBq/g-Mo = 201.0 TBq = 5,431.7 Ci

Here, the dilution effect by the unirradiated aqueous molybdate solution and the decay time of ^{99}Mo from the generation to the shipment are considered. It is assumed that the volume of the aqueous molybdate solution in the capsule and the pipes in the irradiation system and the supply and circulation system of the ^{99}Mo production system is about 2,500 cm^3 and that the time from the post-irradiation to the shipment is one day. After one day, ^{99}Mo decays to 0.78 times. Time from the irradiation to the shipment is one week. The ^{99}Mo production rates at the shipment are estimated as follows:

(^{99}Mo shipping activity in the case using the aqueous $(NH_4)_6Mo_7O_{24}\cdot 4H_2O$ solution)

= 2,881.9 Ci × 1,663/2,500 × 0.78 = 1,495.3 Ci/w

(^{99}Mo shipping activity in the case using the aqueous K_2MoO_4 solution)

= 5,431.7 Ci × 1,663/2,500 × 0.78 = 2818.3 Ci/w

The ^{99}Mo production rate in the case using the aqueous K_2MoO_4 solution is about twice compared with that in the case using the aqueous $(NH_4)_6Mo_7O_{24}\cdot 4H_2O$ solution. It is a distinct advantage of the aqueous K_2MoO_4 solution. However, in order to aim to provide 100% of the ^{99}Mo (5,000 Ci/w) imported into Japan and to increase the production rate, some ideas such as the concentration of ^{98}Mo are needed.

5. Compatibility test between flowing aqueous molybdate solution and structural material

In the ^{99}Mo production system with the solution irradiation method, a flowing target solution with a high concentration is in contact with the structural material of the capsule and the pipes, and then it is important to investigate compatibility between the flowing target solution and the structural material. In the previous tests (Inaba et al., 2009), the circulating solution test was carried out under γ-ray irradiation. However, the SUS304 specimen used in the test was only immersed in the bottom of an irradiation container with a volume of 2,000 cm^3, and the specimen had no influence of the circulating solution flow, and then the compatibility was not cleared. Therefore, the compatibility test between the flowing target solution and the structural material was carried out, and the corrosivity of the flowing target solution for the structural material as well as the chemical stability of the solution was investigated.

An aqueous K_2MoO_4 solution, which was the first candidate of the irradiation target, was used in the test. The purity of K_2MoO_4 used in the test was over 98%.

5.1 Test apparatus

Fig. 4 shows the schematic diagram of a test apparatus, which was used in order to investigate compatibility between a flowing aqueous K_2MoO_4 solution and a structural material and the chemical stability of the solution. The test apparatus consists of a immersion container for immersing specimens under flow, a glass storage tank with a volume of about 700 cm^3, a thermocouple inside the storage tank for solution temperature measurement, a feed pump to circulate the solution, a flowmeter, Teflon tubes with an inner diameter of 7.5 mm to connect each component, two syringes, which were used for depressurization, solution supply, air purge and solution sampling, a data logger to collect temperature data and to monitor the temperature and so on. Some components such as the immersion container and the storage tank were installed into a heating chamber to heat the solution.

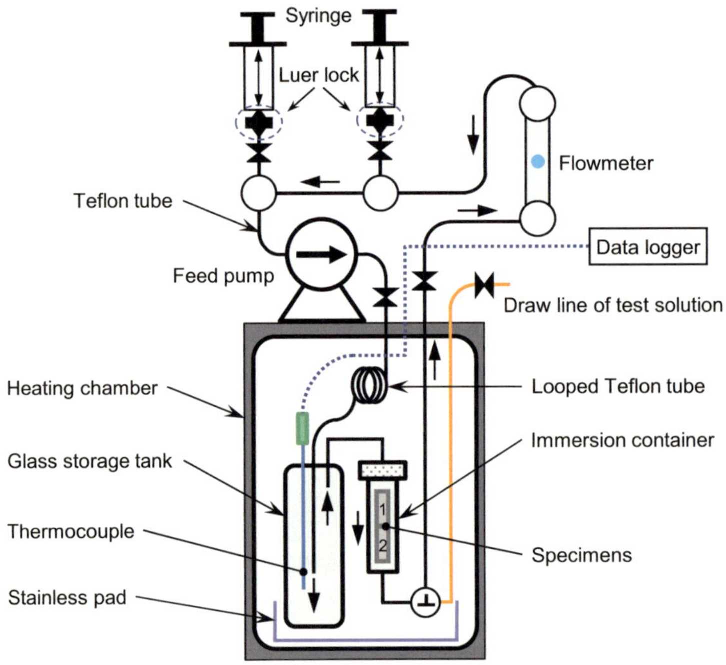

Fig. 4. Schematic diagram of test apparatus for compatibility test

The immersion container consists of a glass outer tube with an outer diameter of 22 mm and a height of 62.5 mm and a Teflon inner holder with an inner diameter of 13 mm and a height of 60 mm, and two specimens (specimen 1 and 2) were fixed in the center of the container by the holder as shown in Fig. 5 and they were arranged one above the other in the container. The storage tank was located upstream of the immersion container to keep the solution temperature constant and to prevent the solution from pulsating by the feed pump. In addition to the storage tank, a looped long Teflon tube connected between the pump and the storage tank was used to keep the solution temperature in the heating chamber. The total length of the circulation route of the solution was about 6.8 m, and the total quantity of circulating solution was about 300 cm^3 except the volume of the storage tank.

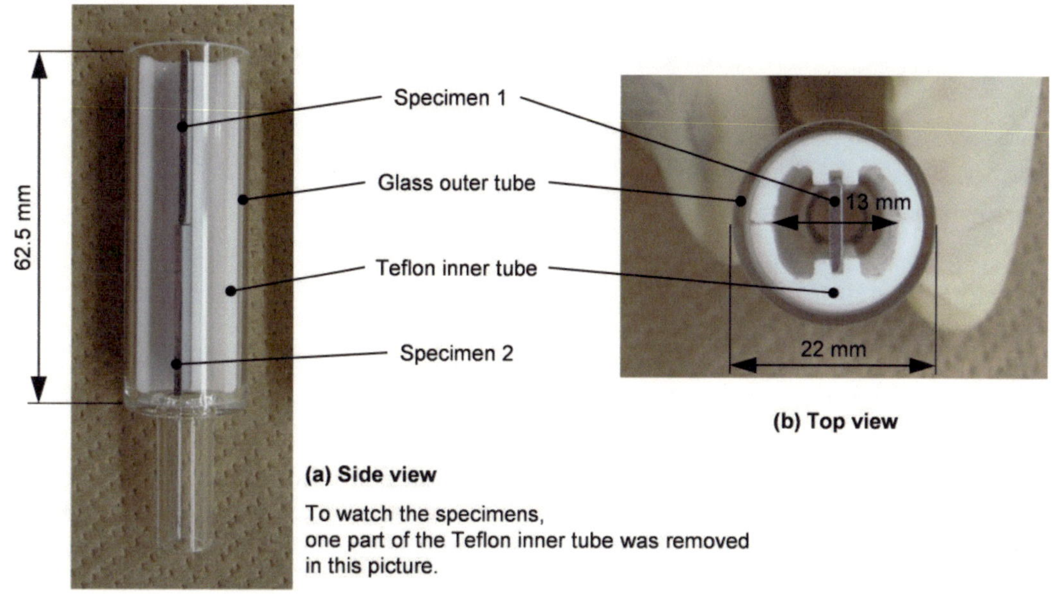

Fig. 5. Structure of immersion container

5.2 Test method and conditions

The compatibility test was carried out by using the test apparatus with a closed loop shown in Fig. 4. After the specimens were set in the immersion container, aqueous K_2MoO_4 solution was injected in the closed loop, and the solution was circulated at a constant flow rate. The flow rate was set at about 120 cm^3/min, considering the flow velocity assumed in an actual ^{99}Mo production system. The concentration of the solution was adjusted to about 90% of the saturation for the prevention of crystallization, and the temperature of the solution was maintained at about 80°C for the prevention of boiling. As the specimens immersed in the solution, stainless steel SUS304 was used based on the results of the previous immersion tests (Inaba et al., 2009). SUS304 has been used as the structural material of capsules and pipes in JMTR. The size of the specimens was $10^W \times 30^L \times 1.5^T$ mm. Table 3 shows the chemical composition of a SUS304 specimen. The total immersion time of the specimens was 112.7 days, and the immersion time under flow was 84.5 days out of a total of 112.7 days. The total immersion time was longer than the immersion time under flow because the feed pump was temporarily stopped by the planned blackouts and the pump troubles.

During the test, at regular intervals, the specimen 1 was taken from the immersion container, and the specimen's weight was measured after washing by pure water and drying, and its surface state was observed. In addition, the aqueous solution was sampled from the closed loop by using one of the syringes, and the pH and molybdenum concentration of the solution were measured, and the solution state was observed.

After the test, the specimen 1 and 2 were taken from the immersion container, and the specimens' weight was measured, and their surface states were observed. In addition, the aqueous solution was sampled from the closed loop, and the pH and molybdenum concentration of the solution were measured, and the solution state was observed.

C	Si	Mn	P	S	Ni	Cr	Fe
0.06	0.51	0.73	0.026	0.002	8.03	18.07	Balance

(Unit: wt%)

Table 3. Chemical composition of SUS304 specimen

5.3 Results and discussions

The average temperature and flow rate of the aqueous K_2MoO_4 solution used in the test were 81°C for a total immersion time of 112.7 days and 123 cm^3/min for a total immersion time under flow of 85.5 days respectively.

5.3.1 Corrosivity of flowing aqueous K_2MoO_4 solution for SUS304

The surface states of the two SUS304 specimens before and after the immersion in the flowing aqueous K_2MoO_4 solution for a total of 84.5 days are shown in the Fig. 6, and the relationships between the immersion time and corrosion rates of the specimens are shown in Fig. 7. The corrosion rates were estimated by the following equation:

$$Corrosion\ rate = \frac{Weight\ change}{Surface\ area \times Immersion\ time \times Density} \tag{6}$$

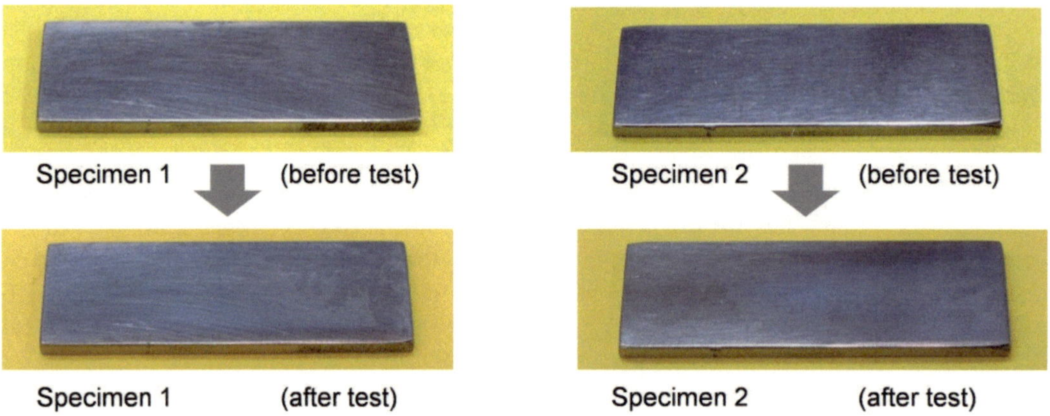

Fig. 6. Surface states of SUS304 specimens before and after compatibility test

The equation (6) shows the wastage thickness per unit time. In the visual observation and comparison of the two specimens' surfaces before and after the compatibility test, whereas streamlined patterns, partly slight tarnish and the partly slight loss of metallic luster were found on the surfaces, obvious corrosion such as corrosion products was not found. The corrosion rate of the specimen 1 increased temporarily to 0.10 mm/y in the initial stage of the test (an immersion time of 21 days) and decreased finally to 0.02 mm/y. On the other hand, the corrosion rate of the specimen 2 was 0 mm/y at the beginning and end of the test. There was no change in the state of the specimen 1 surface in the initial stage of the test, and the temporary increase of the specimen 1 corrosion rate might be affected by taking out from the immersion container.

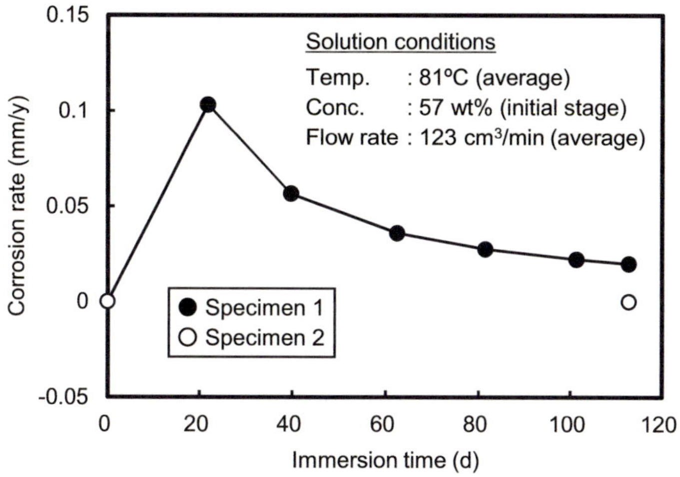

Fig. 7. Relationships between immersion time and corrosion rates of SUS304 specimens immersed in flowing aqueous K_2MoO_4 solution for 84.5 days

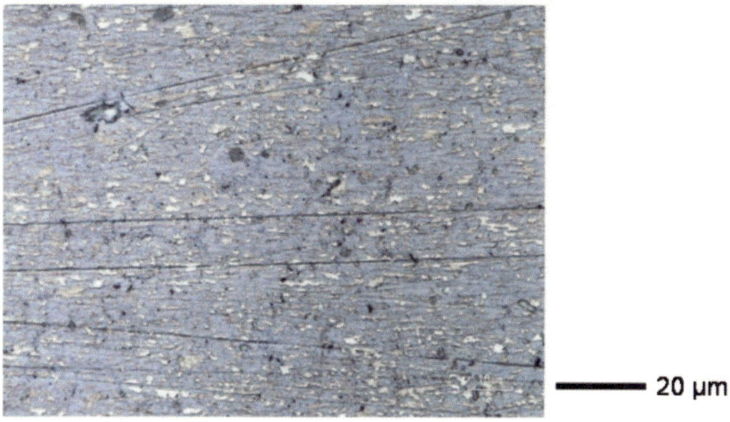

Fig. 8. Inverted materials microscope photograph of specimen 2 surface immersed in flowing aqueous K_2MoO_4 solution for 84.5 days

For the confirmation of the detailed surface states, the specimen 2 as the representative of the two specimens were observed and analyzed with an inverted materials microscope and a field emission Electron Probe Micro Analyzer (EPMA). Fig. 8 shows the inverted materials microscope photograph of the specimen 2 surface. The black lines and dots in Fig. 8 are preexistent scratches and hollows. Tarnish is recognized on the surface. Fig. 9 shows the Scanning Electron Microscope (SEM) photograph of the specimen 2 cross-section surface taken with the EPMA, and Fig. 10 shows the color map of the specimen 2 cross-section surface analyzed with the EPMA. The cross-section surface was prepared by cutting the center of the specimen 2, mounting in a resin and polishing. A thin coating layer, which is thought to be the cause of the tarnish, is found on the surface as shown in Fig. 9. To see Fig. 10, K and Mo, which are the main components of K_2MoO_4, are not detected and a relatively-high level of Si is detected on the surface. After the test, the corrosion of the glass outer tube in the immersion container was found, and then it is considered that the main component of the coating layer is Si eluted from the tube. This Si coating layer might inhibit the corrosion of the specimens. In any case, the progress of the corrosion was not observed in the SUS304 specimens, and SUS304 has good compatibility with a flowing aqueous K_2MoO_4 solution.

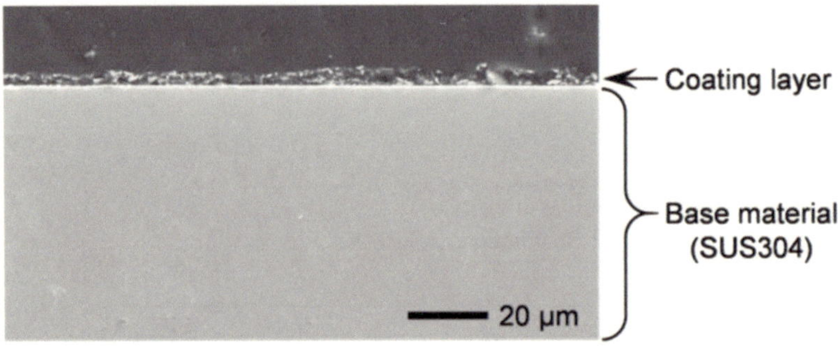

Fig. 9. SEM photograph of specimen 2 cross-section surface immersed in flowing aqueous K_2MoO_4 solution for 84.5 days

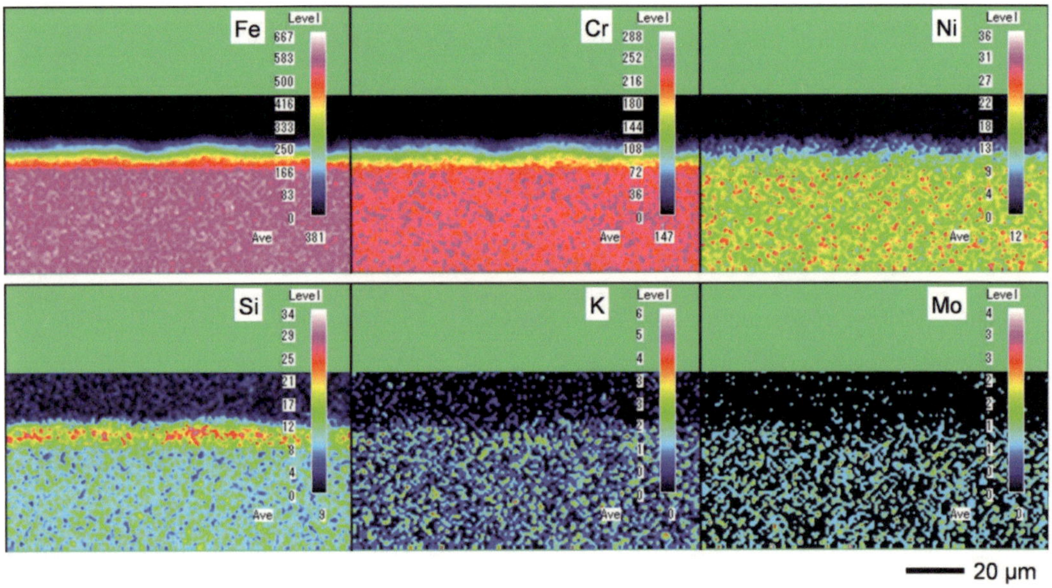

Fig. 10. EPMA color map of specimen 2 cross-section surface immersed in flowing aqueous K_2MoO_4 solution for 84.5 days

5.3.2 Chemical stability of flowing aqueous K_2MoO_4 solution

During the test term, the aqueous K_2MoO_4 solution was chemically stable, and the precipitation or the deposit was not generated in the solution. Then the molybdenum concentration of the solution was almost constant before and after the test, and the concentrations before and after the test were 396.2 mg/mℓ and 384.0 mg/mℓ respectively. The concentrations were measured with an Inductively Coupled Plasma Atomic Emission Spectrometer (ICP-AES). The pH of the solution was also almost constant at pH9.5-9.7.

6. Conclusion

In the ^{99}Mo production system with the solution irradiation method, a static or flowing aqueous molybdenum solution in a capsule is irradiated with neutrons in a testing reactor, and ^{99}Mo is produced by the ^{98}Mo (n, γ) ^{99}Mo reaction. The system aims to provide 100% of the ^{99}Mo imported into Japan. As a part of the technology development, aqueous $(NH_4)_6Mo_7O_{24}\cdot 4H_2O$ and K_2MoO_4 solutions were selected as candidates for the irradiation target of the system, and compatibility between the static two solutions and the structural materials of the capsule and pipes in the system, the chemical stability, the radiolysis and the γ heating of the solutions were investigated. As a result, it was found that the solutions are promising as the target. In addition, compatibility between a flowing aqueous K_2MoO_4 solution, which was the first candidate for the irradiation target in terms of a ^{99}Mo production rate, and the structural material and the chemical stability of the flowing solution were investigated. As a result, it was found that stainless steel SUS304 has good compatibility with a flowing aqueous K_2MoO_4 solution and that the solution is chemically stable. The fundamental characteristics of the selected aqueous molybdate solutions became clear, and SUS304 can be used as the structural material of the capsule and the pipes.

In the future, a neutron irradiation test will be carried out as an overall test of ^{99}Mo production system with the solution irradiation method, and ^{99}Mo production, the separation of activation by-products, the quantity of radiolysis gas, nuclear heating and so on will be investigated.

Aiming at the domestic production of ^{99}Mo in Japan, the development of ^{99}Mo production with the solution irradiation method is kept going.

7. Acknowledgment

The author would like to thank Dr. Tsuchiya, K. and Mr. Ishida, T. of JAEA and Mr. Ishikawa, K. of KAKEN. Inc. for their valuable comments.

8. References

AECL (December 2007). AECL Provides Status Report on NRU Reactor, In: *AECL Web Page,* 27.06.2011, Available from http://www.aecl.ca/NewsRoom/News/Press-2007/071204.htm

AECL (May 2008). AECL to Discontinue Development of the MAPLE Reactors, In: *AECL Web Page,* 27.06.2011, Available from http://www.aecl.ca/NewsRoom/News/Press-2008/080516.htm

Inaba, Y.; Ishikawa, K.; Tatenuma, K. & Ishitsuka, E. (2009). Development of ^{99}Mo Production Technique by Solution Irradiation Method, *Transactions of the Atomic Energy Society of Japan,* Vol. 8, No. 2, (June 2009), pp. 142-153 (in Japanese)

Inaba, Y.; Iimura, K.; Hosokawa, J.; Izumo, H.; Hori, N. & Ishitsuka, E. (2011). Status of Development on ^{99}Mo Production Technologies in JMTR, *IEEE Transactions on Nuclear Science,* Vol. 58, No. 3-3, (June 2011), pp. 1151-1158, ISSN 0018-9499

Ishitsuka, E. & Tatenuma, K. (2008). Manufacturing Method of Radioactive Molybdenum, Manufacturing Apparatus and Radioactive Molybdenum Manufactured Thereby, *Japanese Patent,* 2008-102078

Boyd, R. E. (1997). The Gel Generator: a Viable Alternative Source of ^{99m}Tc for Nuclear Medicine, *Applied Radiation and Isotopes,* Vol. 48, No. 8, (August 1997), pp. 1027-1033

Department of JMTR Project (1994). JMTR Irradiation Handbook, JAERI-M 94-023, (March 1994), Japan Atomic Energy Agency (in Japanese)

Hasegawa, Y.; Nishino, M.; Takeuchi, T.; Ishikawa, K.; Tatenuma, K.; Tanase, M. & Kurosawa, K. (1996). Synthesis of New Adsorbents for Mo as an RI Generator, *Nippon Kagaku Kaishi,* Vol. 1996, No. 10, pp. 888-894 (in Japanese)

Hasegawa, Y.; Nishino, M.; Ishikawa, K.; Tatenuma, K.; Tanase, M. & Kurosawa, K. (1999). Synthesis and Characteristics of High-performance Mo Adsorbent for ^{99m}Tc Generators, *Nippon Kagaku Kaishi,* Vol. 1999, No. 12, pp. 805-811 (in Japanese)

Kurosawa, K. & Fukushima, T. (1987). Treatment of Mild Steels by Chemical Conversion in Aqueous Molybdate Solutions, *Nippon Kagaku Kaishi,* Vol. 1987, No. 10, pp. 1822-1827 (in Japanese)

Lu, Y. C.; Clayton, C. R. & Brooks, A. R. (1989). A Bipolar Model of the Passivity of Stainless Steels - II. The Influence of Aqueous Molybdate, *Corrosion Science,* Vol. 29, No. 7, pp. 863-880

McCune, R. C.; Shilts, R. L. & Ferguson, S. M. (1982). A study of Film Formation on Aluminum in Aqueous Solutions Using Rutherford Backscattering Spectroscopy, *Corrosion Science,* Vol. 22, No. 11, pp. 1049-1065

Saremi, M.; Dehghanian, C. & Mohammadi Sabet, M. (2006). The Effect of Molybdate Concentration and Hydrodynamic Effect on Mild Steel Corrosion Inhibition in Simulated Cooling Water, *Corrosion Science,* Vol. 48, pp. 1404-1412